Symmetries of Algebraic Varieties in Two and Three Dimensions

Symmetries of Algebraic Varieties in Two and Three Dimensions explores the computation of the symmetries of algebraic varieties in two and three dimensions. Most of the attention goes to curves (both planar and spatial) and surfaces, either implicitly defined or defined by means of a rational parametrization. But other algebraic objects, like planar vector fields and finite sets of points, are also covered.

The book is primarily addressed to researchers and post-graduates, but the content is also accessible to advanced undergraduates with a modest background in computer algebra and algebraic geometry. A primer on these topics is also provided.

Key Features:

- Treats problems in constructive (algebraic) geometry from a computer algebra perspective.
- Accessible and self-contained, assuming only a minimal background in advanced mathematics.
- Each chapter contains a list of open problems related to the chapter, providing topics for further research and inviting the reader to work on specific questions.

Juan Gerardo Alcázar is a Full Professor of Applied Mathematics in the Department of Physics and Mathematics of the Universidad de Alcalá (Alcalá de Henares, Madrid, Spain), where he got his Ph.D. in Mathematics in 2007. He has authored and co-authored more than fifty research papers on Computer Algebra and Constructive Algebraic Geometry, most of them related to the interplay between Symbolic Computation and the effective solution of geometric problems on algebraic varieties, mainly curves and surfaces. His most cherished research topics are the effective computation of the topology of algebraic curves and surfaces, offset varieties, detection of special types of curves and surfaces, and efficient checking of symmetries, affine and projective equivalences.

Symmetries of Algebraic Varieties in Two and Three Dimensions

A Computational Approach

Juan Gerardo Alcázar

CRC Press
Taylor & Francis Group
Boca Raton London New York

CRC Press is an imprint of the
Taylor & Francis Group, an **informa** business
A CHAPMAN & HALL BOOK

Designed cover image: Shutterstock

First edition published 2027
by CRC Press
2385 NW Executive Center Drive, Suite 320, Boca Raton, FL 33431

and by CRC Press
4 Park Square, Milton Park, Abingdon, Oxon, OX14 4RN

CRC Press is an imprint of Taylor & Francis Group, LLC

Library of Congress Cataloging-in-Publication Data
Names: Alcázar, Juan Gerardo author
Title: Symmetries of algebraic varieties in two and three dimensions : a computational approach / Juan Gerardo Alcázar.
Description: First edition. | Boca Raton, FL ; Abingdon, Oxon : CRC Press, 2026. | Includes bibliographical references and index.
Identifiers: LCCN 2026012291 (print) | LCCN 2026012292 (ebook) | ISBN 9781041088202 hardback | ISBN 9781041091509 paperback | ISBN 9781003648710 ebook
Subjects: LCSH: Symmetry (Mathematics) | Algebraic varieties | Curves, Algebraic | Surfaces, Algebraic
Classification: LCC QA564 .A3537 2026 (print) | LCC QA564 (ebook)
LC record available at https://lccn.loc.gov/2026012291
LC ebook record available at https://lccn.loc.gov/2026012292

ISBN: 978-1-041-08820-2 (hbk)
ISBN: 978-1-041-09150-9 (pbk)
ISBN: 978-1-003-64871-0 (ebk)

DOI: 10.1201/9781003648710

Typeset in CMR12 font
by KnowledgeWorks Global Ltd.

Publisher's note: This book has been prepared from camera-ready copy provided by the authors.

Contents

Preface

Symmetry is a property that we associate with the idea of beauty. We find it in nature, in manufactured objects, in art, and of course in Mathematics. Although the notion of symmetry appears in various branches of Mathematics and even in Physics with different meanings, here we will be concerned with the classical *Euclidean* symmetries (translations, rotations, reflections, central symmetries, and so forth) and their occurrence in algebraic varieties of the plane and the space, meaning geometric objects which are zerosets of bivariate or trivariate polynomials. Our main interest will be in algebraic planar curves, algebraic space curves, and algebraic surfaces, although other algebraic objects, like vector fields or finite sets of points, are also treated.

Although the study of the symmetries (and related questions) of planar and three-dimensional objects has been for decades a question of interest in applied fields like Computer Aided Geometric Design, Pattern Recognition and Computer Vision, in the last twenty years and for the case of algebraic varieties, there have been a number of contributions to this problem from the field of Computer Algebra and Symbolic Computation. This book aims at collecting the ideas, results and algorithms on the matter appearing in this field. Thus, we focus on the properties of the symmetries of these varieties and on their effective computation. The kind of representation we have for the considered object will also be relevant, so we will be discussing the two main representations we can encounter, implicit and rational.

This book should be accessible to a great variety of readers: we might say that it follows the trend of other books, papers and so forth related to Computer Algebra, where there is an emphasis on constructivity (meaning algorithms) and also a kind of nineteenth century mathematical flavor. Although this book is perhaps more addressed toward the graduate level, especially for researchers in Mathematics with an interest in computational aspects, it is also suitable for undergraduates with some background in Linear Algebra, Complex Analysis, Differential Geometry, basic Algebraic Geometry and some notions of Computer Algebra, interested in geometry and computation. In any case, several techniques and notions typically used in Algebraic Geometry, Differential Geometry and Computer Algebra are reviewed in the text, and proper references on these fields are provided. In the case of undergraduates, this book can help them to see how certain Computer Algebra techniques are used to effectively solve problems in Geometry, and how solving effective problems, i.e., building algorithmic solutions to concrete questions, requires collecting ideas, results and so forth from very different areas in Mathematics.

Additionally, every chapter ends with a brief collection of open problems related to the topic treated in the chapter, therefore inviting the reader to go further.

The book starts providing the background in Euclidean planar and space symmetries. Then we move to algebraic planar objects, namely plane algebraic curves, implicit and rational, and planar vector fields. From there we jump to algebraic varieties in space, first space algebraic curves, implicit and rational, and then algebraic surfaces, also either implicitly defined, or defined by means of a rational parametrization. We finish with finite sets of points, which may seem funny, taking into account that point clouds are simpler objects compared to algebraic curves or algebraic surfaces: this last thought is definitely true, and it would have made perfect sense to have started with point clouds and have moved afterwards to non-finite sets of points. However, it is for curves and surfaces that the algebraic structure is essential, so we opted for postponing finite sets, where this structure is not really important, for the end.

Although part of the content of the book is classical, most of it comes from material developed by different authors, including myself, in the last twenty years. Among those authors I count several colleagues whose collaboration over the years, and sometimes strong friendship, I want to acknowledge: so let special thanks be given to Carlos Hermoso, Georg Muntingh, Uğur Gözütok, Hüsnü A. Çoban, Emily Quintero, Gema Díaz-Toca, Jorge Caravantes, Miroslav Lávička and Jan Vršek. I also want to thank Ron Goldman not only for his collaboration but also because, long ago, it was one of his papers that brought my attention to the topic of this book. So to all of them, thanks for all the fun and the nice discussions we shared.

Juan Gerardo Alcázar, Madrid, January 12th, 2026

CHAPTER 1

Euclidean symmetries

The notion of *symmetry* is very general and appears in many different contexts in Mathematics and also in Physics. Here we will begin with a very abstract definition of symmetry, but soon we will make precise the kind of symmetries we are interested in, which are the Euclidean symmetries, i.e., the isometries of the Euclidean n-space. We will see the properties of these transformations, and we will classify and examine in detail the Euclidean symmetries of the plane and the space, i.e., the cases $n = 2$ and $n = 3$.

1.1 VECTOR ISOMETRIES

Let $\mathcal{E}$ be a space, i.e., a very general set, whose elements we do not need to describe right now, and let $\mathcal{V} \subset \mathcal{E}$ be a subset of $\mathcal{E}$. Furthermore, let $\mathcal{G}$ be a transformation group in $\mathcal{E}$, i.e., a set of transformations $g : \mathcal{E} \to \mathcal{E}$ furnished with a group structure. Then we have the following definition.

Definition 1. *We say that g is a* symmetry *of $\mathcal{V}$ if $g(\mathcal{V}) = \mathcal{V}$. We say that g is a* non-trivial *symmetry if $g \neq \mathrm{id}_{\mathcal{E}}$. If $\mathcal{V}$ admits a non-trivial symmetry, we say that $\mathcal{V}$ is* symmetric.

It is clear that the notion of a symmetry refers to a certain group $\mathcal{G}$, so the precise meaning changes according to the group. Assuming that $\mathcal{G}$ is fixed and since $\mathcal{G}$ has a group structure, the set $\mathrm{Sym}(\mathcal{V})$ of all the symmetries of $\mathcal{V}$ also inherits a group structure and is a subgroup of $\mathcal{G}$; we call this group the *symmetry group* of $\mathcal{V}$. Therefore, the symmetries of $\mathcal{V}$ are all the transformations

DOI: 10.1201/9781003648710-1

belonging to a certain group of interest $\mathcal{G}$ that leave $\mathcal{V}$ invariant. In particular, the *potential* symmetries that a subset $\mathcal{V}$ can have are nothing else than the elements of $\mathcal{G}$.

In our case $\mathcal{E} = \mathbb{A}_n$, i.e., the Euclidean affine n-space. In most cases, we will consider the *real* affine n-space, so that the underlying vector space will be $\mathbb{V}_n = \mathbb{R}^n$. The essential question is what group $\mathcal{G}$ we are interested in when $\mathcal{E} = \mathbb{A}_n$.

We will certainly answer this question, but before that we must direct our attention toward $\mathbb{V}_n = \mathbb{R}^n$. In other words, coming back to Definition 1, we must consider $\mathcal{E} = \mathbb{V}_n = \mathbb{R}^n$, fix a certain group of interest there, and study the symmetries of $\mathbb{V}_n$ with respect to this group. To do this, we will denote the elements of $\mathbb{V}_n$ by $\bar{u}, \bar{v}$, etc., where, say, $\bar{u} = (u_1, \ldots, u_n)$ with $u_i \in \mathbb{R}$ for $i = 1, \ldots, n$. Furthermore, we will consider $\mathbb{V}_n$ furnished with a Euclidean structure provided by the *standard* dot product

$$\langle \bar{u}, \bar{v} \rangle = \bar{u} \cdot \bar{v} = u_1 v_1 + \cdots + u_n v_n$$

for $\bar{u}, \bar{v} \in \mathbb{V}_n$, u_i, v_j the coordinates of $\bar{u}, \bar{v}$ in an orthonormal basis, and the norm associated with this product,

$$||\bar{u}|| = \sqrt{u_1^2 + \cdots + u_n^2}.$$

The following definition makes precise the group of transformations of $\mathbb{V}_n$ we will be interested in.

Definition 2. *We say that $\mathcal{T} : \mathbb{V}_n \to \mathbb{V}_n$ is an* orthogonal *transformation, also called a* vector isometry, *if $\mathcal{T}$ is a linear mapping preserving the dot product, i.e., if for $\bar{u}, \bar{v} \in \mathbb{V}_n$,*

$$\langle \mathcal{T}(\bar{u}), \mathcal{T}(\bar{v}) \rangle = \langle \bar{u}, \bar{v} \rangle. \tag{1.1}$$

There is an equivalent definition of orthogonal transformations in terms of norms, as shown by the next proposition.

Proposition 3. *A linear mapping $\mathcal{T} : \mathbb{V}_n \to \mathbb{V}_n$ is orthogonal if and only if $\mathcal{T}$ preserves norms.*

Proof. The implication ($\Rightarrow$) follows by making $\bar{u} = \bar{v}$ in Eq. (1.1). In order to prove ($\Leftarrow$), assume that $\mathcal{T}$ is a linear mapping preserving norms, and consider the dot product

$$\langle \mathcal{T}(\bar{u} + \bar{v}), \mathcal{T}(\bar{u} + \bar{v}) \rangle, \tag{1.2}$$

which is equal to $\|\mathcal{T}(\bar{u}+\bar{v})\|^2$. Since by hypothesis $\mathcal{T}$ preserves the norm,

$$\langle \mathcal{T}(\bar{u} + \bar{v}), \mathcal{T}(\bar{u} + \bar{v}) \rangle = \|\mathcal{T}(\bar{u} + \bar{v})\|^2 = \|\bar{u} + \bar{v}\|^2 = \langle \bar{u} + \bar{v}, \bar{u} + \bar{v} \rangle. \tag{1.3}$$

Since $\mathcal{T}$ is linear, we have

$$\langle \mathcal{T}(\bar{u} + \bar{v}), \mathcal{T}(\bar{u} + \bar{v}) \rangle = \langle \mathcal{T}(\bar{u}) + \mathcal{T}(\bar{v}), \mathcal{T}(\bar{u}) + \mathcal{T}(\bar{v}) \rangle, \tag{1.4}$$

and since the dot product is distributive and commutative,

$$\langle \mathcal{T}(\bar{u}) + \mathcal{T}(\bar{v}), \mathcal{T}(\bar{u}) + \mathcal{T}(\bar{v}) \rangle = \|\mathcal{T}(\bar{u})\|^2 + \|\mathcal{T}(\bar{v})\|^2 + 2\langle \mathcal{T}(\bar{u}), \mathcal{T}(\bar{v}) \rangle. \tag{1.5}$$

Taking Eqs. (1.3), (1.4), and (1.5) into account and the fact that $\mathcal{T}$ preserves norms, we get, after expanding $\langle \bar{u} + \bar{v}, \bar{u} + \bar{v} \rangle$, that

$$\langle \mathcal{T}(\bar{u}), \mathcal{T}(\bar{v}) \rangle = \langle \bar{u}, \bar{v} \rangle,$$

which implies that $\mathcal{T}$ is orthogonal. □

Corollary 4. *An orthogonal linear mapping $\mathcal{T} : \mathbb{V}_n \to \mathbb{V}_n$ preserves angles.*

Furthermore, orthogonal transformations are vector isomorphisms (i.e., bijections), as shown by the next proposition.

Proposition 5. *Let $\mathcal{T} : \mathbb{V}_n \to \mathbb{V}_n$ be an orthogonal transformation in $\mathbb{V}_n$. Then $\mathcal{T}$ is a vector isomorphism.*

Proof. Since the original and final vector spaces are the same, it suffices to see that $\mathcal{T}$ is injective. Thus, let $\mathrm{Ker}(\mathcal{T})$ denote the kernel of $\mathcal{T}$, and let $\bar{v} \in \mathrm{Ker}(\mathcal{T})$, so $\mathcal{T}(\bar{v}) = \bar{0}$. Then $\|\mathcal{T}(\bar{v})\| = 0$, and since $\mathcal{T}$ preserves norms by Proposition 3, $\|\bar{v}\| = 0$. But this implies that $\bar{v} = \bar{0}$. Thus, $\mathrm{Ker}(\mathcal{T}) = \{\bar{0}\}$, so $\mathcal{T}$ is injective. Since $\mathcal{T} : \mathbb{V}_n \to \mathbb{V}_n$, $\mathcal{T}$ is bijective. □

Because of Proposition 5, orthogonal transformations map bases of $\mathbb{V}_n$ onto bases of $\mathbb{V}_n$. Furthermore, since orthogonal transformations preserve norms and angles, we get the following.

Corollary 6. *Orthogonal transformations in $\mathbb{V}_n$ transform orthonormal bases onto orthonormal bases.*

It is easy to see that orthogonal transformations in $\mathbb{V}_n$ form a group that we will denote by $\mathbb{O}_n$. The elements of this group will be called *Euclidean vector symmetries*. To explore this group further, recall that since orthogonal transformations are linear mappings, they have an associated matrix A when a basis of $\mathbb{V}_n$, in our case an orthonormal basis, is fixed. Let us see what the properties of this matrix can be. We begin with the following definition. Here $\mathcal{M}_{n\times n}(\mathbb{R})$ denotes the set of real matrices of dimension n; the identity matrix is denoted $\mathrm{Id}_{n\times n}$, or simply I.

Definition 7. *Let $A \in \mathcal{M}_{n\times n}(\mathbb{R})$. We say that A is an* orthogonal *matrix if $A^T A = I$.*

The following result is a direct consequence of the previous definition.

Proposition 8. *Let $A \in \mathcal{M}_{n\times n}(\mathbb{R})$ be an orthogonal matrix.*

(1) A is regular, and the determinant of A is $|A| = \pm 1$.

(2) A is orthogonal if and only if $A^T = A^{-1}$.

(3) A is orthogonal if and only if A^T is orthogonal.

Furthermore, one can see that the set of orthogonal matrices of dimension n has a group structure with respect to the product of matrices. This group will be denoted by $O(n)$ and called the *orthogonal group*. The following proposition brings some light on the shape of orthogonal matrices.

Proposition 9. *Let $A \in \mathcal{M}_{n\times n}(\mathbb{R})$. Then A is orthogonal if and only if the columns of A form an orthonormal basis of $\mathbb{V}_n$.*

Proof. (⇒) Since by statement (1) in Proposition 8 the matrix A is regular, its columns form a basis B of $\mathbb{V}_n$. Let us see that B is orthonormal.

In order to see this, let $\bar{c}_i, \bar{c}_j$, with $i \neq j$, be the elements of $\mathbb{R}^n$ corresponding to the i-th and j-th columns of A, and let us consider the dot product $\bar{c}_i \cdot \bar{c}_j$. This value coincides with the result of multiplying $c_i \cdot c_j^T$, where now c_i, c_j are row matrices whose elements are the elements of the i-th and j-th columns of A. Next let $M = A^T A$. The element m_{ij} of M is the result of multiplying the i-th row of A^T times the j-th column of A. However, the i-th row of A^T is precisely the i-th column of A, so m_{ij} coincides with $c_i \cdot c_j^T$. Since A is orthogonal $M = I$, and therefore

$$\begin{cases} m_{ij} = 1 & \text{if } i = j, \\ m_{ij} = 0 & \text{if } i \neq j. \end{cases}$$

Thus, $\bar{c}_i \cdot \bar{c}_j = 1$ when $i = j$, which implies that the elements of B are unitary, and $\bar{c}_i \cdot \bar{c}_j = 0$ when $i \neq j$, which implies that the elements of B are orthogonal. Then we conclude that B is orthonormal.

(⇐) Let $\bar{c}_i$ denote the element of $\mathbb{V}_n$ corresponding to the i-th column of A, and let

$$M = A^T A, \ M = (m_{ij})_{1 \leq i,j \leq n} \, .$$

We observe that m_{ij} is the product of the row matrix corresponding to $\bar{c}_i$, and the column matrix corresponding to $\bar{c}_j$, so $m_{ij} = \bar{c}_i \cdot \bar{c}_j$. Since by hypothesis the $\bar{c}_i$s form an orthonormal basis of $\mathbb{V}_n$, the dot product $\bar{c}_i \cdot \bar{c}_j$ is equal to 1 whenever $i = j$ and is equal to 0 whenever $i \neq j$. Thus, M is the identity matrix, so by definition A is orthogonal. □

Therefore, we can easily generate orthogonal matrices: we just pick an orthonormal basis of $\mathbb{V}^n$, and dispose the elements of this basis as columns of a matrix. We can wonder now about the properties of a linear mapping whose associated matrix, in an orthonormal basis, is orthogonal. So consider now the linear mapping $f(\bar{u}) = A\bar{u}$ in a certain orthonormal basis B of $\mathbb{V}_n$, with A orthogonal. By identifying the coordinates of the vectors of $\mathbb{V}_n$ in the basis B with column matrices, since B is orthonormal we get that for $\bar{u}, \bar{v} \in \mathbb{V}_n$,

$$\bar{u} \cdot \bar{v} = \bar{u}^T \bar{v},$$

and thus, taking into account that A is orthogonal and therefore $A^T A = I$,

$$f(\bar{u}) \cdot f(\bar{v}) = f(\bar{u})^T f(\bar{v}) = (A\bar{u})^T A\bar{v} = \bar{u}^T A^T A\bar{v} = \bar{u}^T \bar{v} = \bar{u} \cdot \bar{v},$$

so f is an orthogonal transformation in $\mathbb{V}_n$, i.e., a vector isometry. So we have proved the following result.

Proposition 10. *Let $f : \mathbb{V}_n \to \mathbb{V}_n$ be a linear mapping. If there exists an orthonormal basis B of $\mathbb{V}_n$ such that the matrix associated with f in the basis B is orthogonal, then f is a vector isometry.*

The converse of Proposition 10 is true, so there is a complete identification between the vector isometries of $\mathbb{V}_n$ and orthogonal matrices.

Theorem 11. *Let $f : \mathbb{V}_n \to \mathbb{V}_n$ be a linear mapping. Then f is a vector isometry of $\mathbb{V}_n$ if and only if there exists an orthonormal basis B of $\mathbb{V}_n$ such that the matrix associated with f in the basis B is orthogonal.*

Proof. We just need to prove ($\Rightarrow$). Let $B = \{\bar{u}_1, \ldots, \bar{u}_n\}$ be an orthonormal basis of $\mathbb{V}_n$, and let A be the matrix associated with f in the basis B. The columns of A are the coordinates of $f(\bar{u}_1), \ldots, f(\bar{u}_n)$ in the basis B. However, since f is a vector isometry, by Corollary 6 we conclude that $\{f(\bar{u}_1), \ldots, f(\bar{u}_n)\}$ is also an orthonormal basis of $\mathbb{V}_n$. Then the implication follows from Proposition 9. □

Remark 1. *In fact, if $f : \mathbb{V}_n \to \mathbb{V}_n$ is an orthogonal transformation, the matrix associated with f in* any *orthonormal basis is also orthogonal. This follows from the fact that the change of basis matrix between two orthonormal bases is orthogonal.*

Therefore, because of Theorem 11 and Remark 1, the groups $\mathbb{O}_n$, of orthogonal transformations in $\mathbb{V}_n$, and $O(n)$, of orthogonal matrices of dimension n, are isomorphic. Thus, one can write

$$\mathbb{O}_n = \mathbb{O}_n^+ \cup \mathbb{O}_n^-,$$

where $\mathbb{O}_n^+$ (resp. $\mathbb{O}_n^-$) is the subset of elements in $\mathbb{O}_n$ whose associated matrix has determinant $+1$ (resp. -1). We say that the elements of $\mathbb{O}_n^+$ are *orientation-preserving*, and the elements of $\mathbb{O}_n^-$ are *orientation-reversing*. This corresponds to whether or not the orientation of a basis of $\mathbb{V}_n$, i.e., the sign of the determinant whose columns are the vectors in the basis, is changed when an orthogonal transformation f is applied: if it is not, then $f \in \mathbb{O}_n^+$, and otherwise $f \in \mathbb{O}_n^-$.

Before finishing this section, let us introduce some properties of eigenvalues and eigenspaces of vector isometries that we will need later.

Proposition 12. *Let $f : \mathbb{V}_n \to \mathbb{V}_n$ be a vector isometry, and let A be the (orthogonal) matrix associated with f in an orthonormal basis B.*

(1) If $\lambda \in \mathbb{R}$ is an eigenvalue of f, then $\lambda = \pm 1$.

(2) If $\lambda = 1$, $\lambda = -1$ are both eigenvalues of f, and V_1, V_{-1} denote the eigenspaces corresponding to $\lambda = 1$, $\lambda = -1$, then V_1, V_{-1} are orthogonal.

Proof. We begin with (1). Let $\lambda \in \mathbb{R}$ be an eigenvalue of f, and let V_λ be the corresponding eigenspace. Let $\bar{u} \in V_\lambda$, $\bar{u} \neq \bar{0}$. Since f is a vector isometry, f preserves norms, so

$$\|f(\bar{u})\|^2 = \|\bar{u}\|^2. \tag{1.6}$$

Furthermore, since $\bar{u} \in V_\lambda$, $f(\bar{u}) = \lambda\bar{u}$, and therefore

$$\|f(\bar{u})\|^2 = \|\lambda\bar{u}\|^2 = \lambda^2\|\bar{u}\|^2. \tag{1.7}$$

From Eqs. (1.6) and (1.7), we get that $(\lambda^2 - 1)\|\bar{u}\|^2 = 0$. Since $\bar{u} \neq \bar{0}$, $\|\bar{u}\| \neq 0$, so $\lambda^2 - 1 = 0$, i.e., $\lambda = \pm 1$.

Now let us see (2). Let $\bar{u}_1 \in V_1$, $\bar{u}_{-1} \in V_{-1}$, both of them nonzero. Since f is a vector isometry, f preserves dot products, so

$$f(\bar{u}_1) \cdot f(\bar{u}_{-1}) = \bar{u}_1 \cdot \bar{u}_{-1}. \tag{1.8}$$

Additionally, since $\bar{u}_1 \in V_1$, we get that $f(\bar{u}_1) = \bar{u}_1$, and since $\bar{u}_{-1} \in V_{-1}$, we get that $f(\bar{u}_{-1}) = -\bar{u}_{-1}$. Because of this and taking into account that the dot product is linear, we deduce that

$$f(\bar{u}_1) \cdot f(\bar{u}_{-1}) = \bar{u}_1 \cdot (-\bar{u}_{-1}) = -\bar{u}_1 \cdot \bar{u}_{-1}. \tag{1.9}$$

From Eqs. (1.8) and (1.9), we conclude that $\bar{u}_1 \cdot \bar{u}_{-1} = 0$, so V_1 and V_{-1} are orthogonal. □

Remark 2. *Notice that the elements of V_1 are* fixed *or* invariant, *i.e., left unchanged, by f, and therefore V_1 itself is also invariant. Although the elements of V_{-1} are not fixed by f, because their images under f are the opposite, the eigenspace V_{-1}, as a whole,* is *certainly fixed, i.e., again, invariant or left unchanged. Furthermore, let $V_1^\perp, V_{-1}^\perp$ be the* orthogonal subspaces *to V_1, V_{-1}, i.e., the vector subspaces consisting of all the vectors in $\mathbb{V}_n$ orthogonal to V_1, V_{-1}, respectively. Since f preserves the dot product, f preserves orthogonality, which implies that not only V_1, V_{-1} are invariant under f, but $V_1^\perp, V_{-1}^\perp$ are invariant too.*

One can go further and prove that the complex eigenvalues of an orthogonal matrix must have modulus equal to 1, so only n-th roots of unity are permitted. In fact, this is true not only for orthogonal matrices but also for their complex counterpart, *unitary* matrices, where the inverse of the matrix is not the transpose, but the conjugate of the transpose. The interested reader can check [119, §5.5] for all these questions.

1.2 CLASSIFICATION OF PLANAR AND SPACE EUCLIDEAN VECTOR ISOMETRIES

We learnt before that the elements of $\mathbb{O}_n$ are the Euclidean symmetries of $\mathbb{V}_n$. Now we want to classify the elements of $\mathbb{O}_n$, in order to find out what kind of symmetries we can have in $\mathbb{V}_n$. Because of Theorem 11, this is equivalent to classifying the matrices in the orthogonal group $O(n)$. Here we will do this for $n = 2$ and $n = 3$.

1.2.1 Planar Euclidean vector isometries

Let us begin with the plane, $n = 2$. In order to do this, we must classify orthogonal matrices $A \in \mathcal{M}_{2\times 2}(\mathbb{R})$; f will denote the transformation in $\mathbb{V}_2$ defined by A. Because of Proposition 9, the columns of A must form an orthonormal basis, so we have just two possibilities,

$$A = \begin{bmatrix} a & -b \\ b & a \end{bmatrix}, \ A = \begin{bmatrix} a & b \\ b & -a \end{bmatrix}, \tag{1.10}$$

where $a^2 + b^2 = 1$, and therefore $a = \cos(\theta)$, $b = \sin(\theta)$ for some $\theta \in [0, 2\pi)$. Notice that the matrix at the left of Eq. (1.10) satisfies that $|A| = 1$, while the matrix at the right of Eq. (1.10) satisfies that $|A| = -1$. Thus, any element of $\mathbb{O}_2^+$ corresponds to a matrix like Eq. (1.10), left, and any element of $\mathbb{O}_2^-$ corresponds to a matrix like Eq. (1.10), right.

Orientation-preserving planar vector isometries

These vector isometries correspond to Eq. (1.10), left. Notice that the characteristic polynomial of the matrix in Eq. (1.10), left, is

$$p_A(\lambda) = (a - \lambda)^2 + b^2 = \lambda^2 - 2a\lambda + 1,$$

which only has real eigenvalues when $a = \pm 1$, i.e., when $A = \pm I$. Thus, for $a \neq \pm 1$ no element $\bar{u} \in \mathbb{V}_2$ is fixed. In order to recognize what kind of transformation we have, we can write $\bar{u}$ in the canonical basis as

$$\bar{u} = (r\cos(\alpha), r\sin(\alpha)),$$

so the extreme point of $\bar{u}$ has polar coordinates (r, α). Replacing $a = \cos(\theta)$, $b = \sin(\theta)$ in Eq. (1.10), left, we get that

$$\begin{aligned} A\bar{u} &= \begin{bmatrix} \cos(\theta) & -\sin(\theta) \\ \sin(\theta) & \cos(\theta) \end{bmatrix} \begin{bmatrix} r\cos(\alpha) \\ r\sin(\alpha) \end{bmatrix} \\ &= \begin{bmatrix} r(\cos(\alpha)\cos(\theta) - \sin(\alpha)\sin(\theta)) \\ r(\cos(\alpha)\sin(\theta) + \sin(\alpha)\cos(\theta)) \end{bmatrix} = \begin{bmatrix} r\cos(\alpha + \theta) \\ r\sin(\alpha + \theta) \end{bmatrix}. \end{aligned}$$

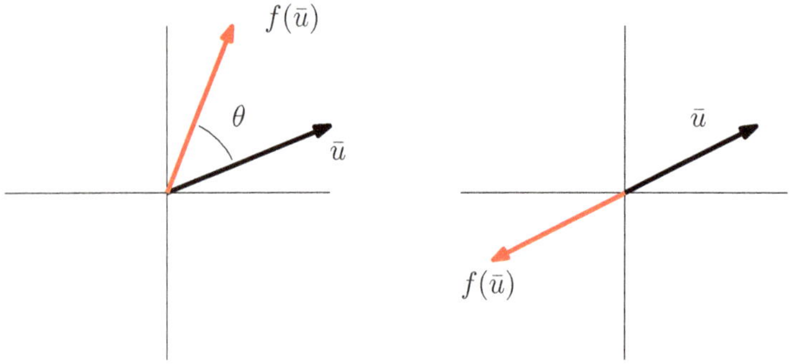

FIGURE 1.1 Rotation of angle θ (left) and central symmetry (right)

Therefore, $\bar{u}$ has undergone a *rotation (around the origin) of angle* θ; we say that θ is a *rotation angle*; for $\theta \neq 0, \pi$, i.e., for $a \neq \pm 1$, this certainly implies that no direction is fixed. If $a = 1$, then $A = I$, so f is the identity. And if $a = -1$, then $A = -I$, in which case $f(\bar{u}) = -\bar{u}$ for $\bar{u} \in \mathbb{V}_2$. In this last case, f represents a rotation of angle π, which sometimes receives the name of *central symmetry* with respect to the origin. Fig. 1.1 illustrates all these transformations.

Orientation-reversing planar vector isometries

These vector isometries correspond to Eq. (1.10), right. Taking into account that $a^2 + b^2 = 1$, the characteristic polynomial of the matrix in Eq. (1.10), right, is

$$p_A(\lambda) = 1 - \lambda^2,$$

so A has two different, simple eigenvalues, $\lambda = 1$ and $\lambda = -1$. The eigenspace V_1 associated with $\lambda = 1$, which has dimension 1, is invariant, and all the elements of V_1 are fixed by f. Thus, we have a vector line $\mathcal{L} = V_1$ which is left unchanged by f. Furthermore, the eigenspace V_{-1} associated with $\lambda = -1$, which also has dimension 1 and by Proposition 12 corresponds to a line orthogonal to $\mathcal{L}$, is also invariant, although not pointwise invariant. Since $V_1 \oplus V_{-1} = \mathbb{V}_2$, where $\oplus$ denotes the direct sum of subspaces, given $\bar{u} \in \mathbb{V}_2$ we can write

$$\bar{u} = \bar{u}_1 + \bar{u}_{-1}, \ \bar{u}_1 \in V_1, \ u_{-1} \in V_{-1}.$$

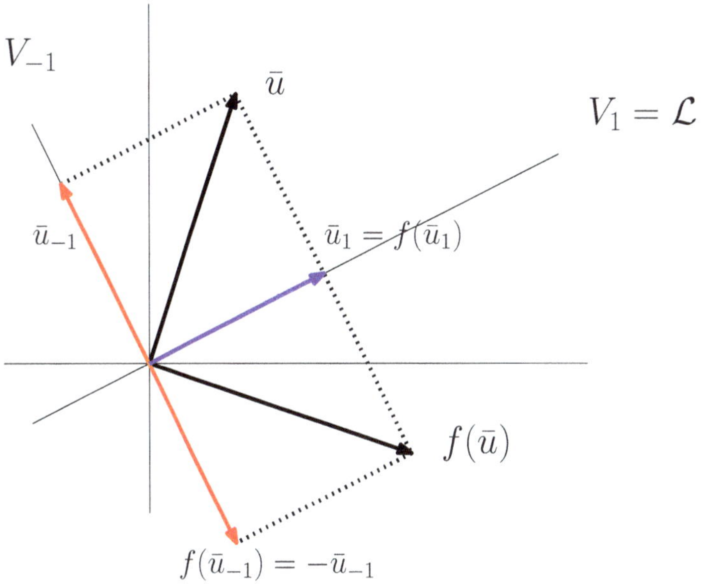

FIGURE 1.2 Reflection on a vector line

Therefore, since f is linear,

$$f(u) = f(\bar{u}_1) + f(\bar{u}_{-1}) = \bar{u}_1 - \bar{u}_{-1}.$$

Fig. 1.2 illustrates this process. From Fig. 1.2, we observe that in this case f represents a *reflection* on the vector line $\mathcal{L} = V_1$, sometimes also called *mirror symmetry* or (planar) *axial symmetry*, so the reflection axis corresponds to the eigenspace associated with $\lambda = 1$. In fact, in a basis of $\mathbb{V}_2$ consisting of eigenvectors, the matrix associated with f is

$$A' = \begin{bmatrix} 1 & 0 \\ 0 & -1 \end{bmatrix}.$$

We summarize the classification of planar Euclidean vector isometries in Table 1.1.

1.2.2 Space Euclidean vector isometries

Now we move to $n = 3$, so we must classify orthogonal matrices $A \in \mathcal{M}_{3\times 3}(\mathbb{R})$. This is more complicated, compared to the case

TABLE 1.1 Classification of planar vector Euclidean isometries

$\|A\|$	**Eigenvalues**	**Vector Symmetry**
1	$\lambda = 1$ (double)	Trivial symmetry (identity)
1	Complex conjugate	Rotation of angle $\theta \neq \pi$
1	$\lambda = -1$ (double)	Rotation of angle π (central sym.)
-1	$\lambda = \pm 1$ (both simple)	Reflection on a vector line (V_1)

$n = 2$. In order to carry out this classification , we observe that the characteristic polynomial $p_A(\lambda)$ of these matrices is a polynomial of degree 3, so it must have at least one real root, which according to Proposition 12 must be either $\lambda = 1$ or $\lambda = -1$.

We begin with the matrices that have $\lambda = 1$ as an eigenvalue. Notice that the eigenspace V_1 associated with $\lambda = 1$ is fixed, i.e., the elements in V_1 are invariant. However, now the dimension of V_1 can be 1, 2 or 3, which results in different transformations. As in the planar case, we denote by f the transformation in $\mathbb{V}_3$ corresponding to A. Furthermore, we will denote the eigenvalues of A as $\lambda_1, \lambda_2, \lambda_3$, some of which can coincide.

$\lambda_1 = 1$ (simple), λ_2, λ_3 complex conjugate

Since $\lambda = 1$ is simple, $\dim(V_1) = 1$ and V_1 corresponds to an invariant vector line. By Remark 2 the orthogonal subspace $V_1^\perp$, which has dimension 2 because $\mathbb{V}_3 = V_1 \oplus V_1^\perp$, is also invariant, although not pointwise. Considering the restriction $f^\star = f|_{V_1^\perp}$, we know that $f^\star(V_1^\perp) = V_1^\perp$, so $f^\star$ is a planar vector isometry without fixed directions. Thus, according to the results in the previous subsection $f^\star$ must be a rotation, so in an orthonormal basis $B = \{\bar{u}, \bar{v}_1, \bar{v}_2\}$ where $\bar{u} \in V_1$, $\bar{v}_1, \bar{v}_2 \in V_1^\perp$, the matrix associated with f is

$$A' = \begin{bmatrix} 1 & 0 & 0 \\ 0 & \cos(\theta) & -\sin(\theta) \\ 0 & \sin(\theta) & \cos(\theta) \end{bmatrix} \tag{1.11}$$

with $\theta \in (0, 2\pi)$, $\theta \neq \pi$; notice that for $\theta = 0$ or $\theta = \pi$, the matrix A' has no complex eigenvalues. Since the vector line V_1 stays invariant and the restriction of f to $V_1^\perp$ is a rotation, in this

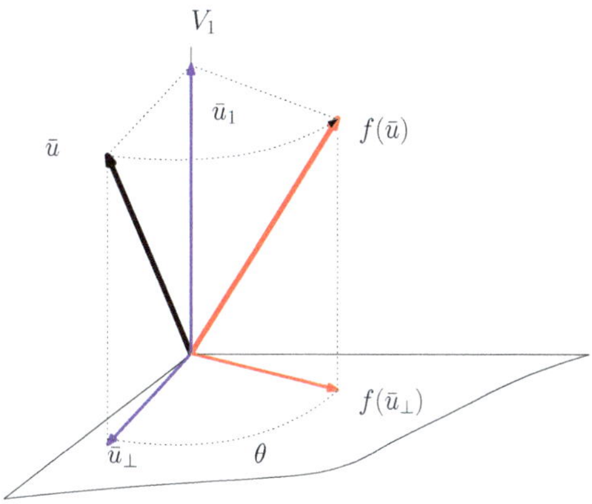

FIGURE 1.3 3D Rotation about a vector line

case f corresponds to a *rotation* about the vector line defined by V_1, of rotation angle θ (see Fig. 1.3).

$\lambda_1 = 1$ (simple), $\lambda_2 = -1$ (double)

Since the algebraic multiplicity of $\lambda_2 = -1$ is two, one can wonder whether or not the matrix A is diagonalizable. Let us see that the answer is affirmative. Indeed, assume that the dimension of the eigenspace V_{-1} associated to the eigenvalue $\lambda_1 = -1$ is one. Since the eigenspaces V_1, V_{-1} associated with $\lambda_1 = 1$ and $\lambda_2 = -1$ are orthogonal, one can form an orthonormal basis B of $\mathbb{V}_3$ by taking unitary vectors $\bar{u} \in V_1$, $\bar{v} \in V_{-1}$, and $\bar{w} = \bar{u} \times \bar{v}$. By Remark 1, the matrix associated with f in the basis B is also orthogonal. Since the columns of this matrix are the images $f(\bar{u})$, $f(\bar{v})$, $f(\bar{w})$ written in the basis B, by Proposition 9 we deduce that $\{f(\bar{u}), f(\bar{v}), f(\bar{w})\}$ is an orthonormal basis. Finally, since $f(\bar{u}) = \bar{u}$, which has coordinates $(1, 0, 0)$ in the basis $B = \{\bar{u}, \bar{v}, \bar{w}\}$, and $f(\bar{v}) = -\bar{v}$, which has coordinates $(0, -1, 0)$ in B, and since $f(\bar{w})$ is orthogonal to $f(\bar{u})$ and $f(\bar{v})$, we deduce that the coordinates of $f(\bar{w})$ in B must be $(0, 0, \kappa)$, implying that $\bar{w}$ is also an eigenvector. Since $\lambda_1 = 1$ is simple by hypothesis, the only possibility is that $\bar{w}$ is also associated with $\lambda_2 = -1$. Finally, since V_{-1} contains two independent vectors, $\bar{v}$ and $\bar{w}$, its dimension must be two, and A is diagonalizable.

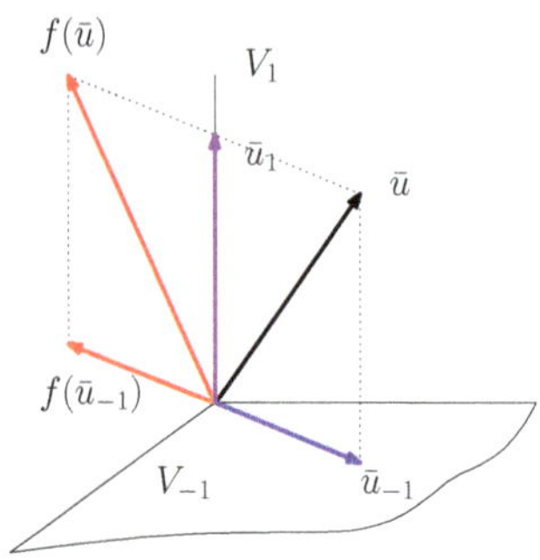

FIGURE 1.4 Axial symmetry about a vector line

Hence, in a basis consisting of eigenvectors, the matrix associated with f is

$$A' = \begin{bmatrix} 1 & 0 & 0 \\ 0 & -1 & 0 \\ 0 & 0 & -1 \end{bmatrix}. \tag{1.12}$$

Now we can argue as in the case of orientation-reversing planar vector isometries to conclude that A defines an *axial symmetry* on the vector line defined by V_1, i.e., a rotation of angle π about the vector line defined by V_1 (see Fig. 1.4). This transformation is also sometimes referred to as a *reflection on the line* defined by V_1, or a *half-turn*.

$\lambda_1 = 1$ (double), $\lambda_2 = -1$ (simple)

As in the previous case, here we can also prove that A is diagonalizable. Thus, in a basis consisting of eigenvectors, the matrix associated with f is

$$A' = \begin{bmatrix} -1 & 0 & 0 \\ 0 & 1 & 0 \\ 0 & 0 & 1 \end{bmatrix}. \tag{1.13}$$

Arguing as in previous cases, we conclude that A defines a *reflection* on the vector plane defined by V_1 (see Fig. 1.5), or just a *planar symmetry* with respect to V_1.

$\lambda_1 = 1$ (triple)

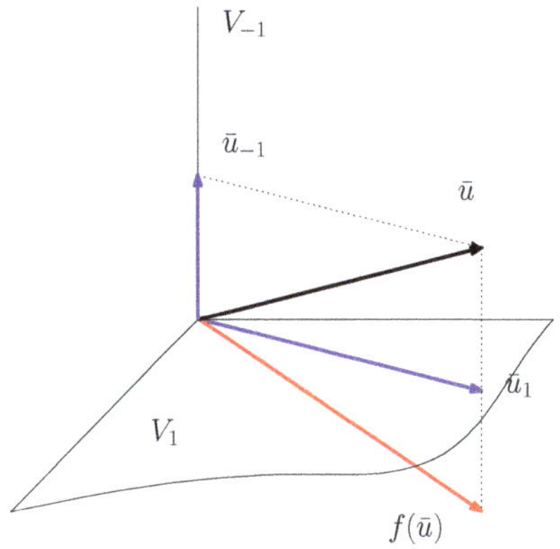

FIGURE 1.5 Reflection on a vector plane

In this case one can prove that A is also diagonalizable, and therefore similar to the identity matrix. But this implies that A is the identity itself, so this case corresponds to the *trivial* symmetry, when $A = I$.

Now let us consider the matrices where $\lambda = 1$ is not present as an eigenvalue, but $\lambda = -1$ is. Since $\lambda = 1$ is not an eigenvalue, there are no pointwise invariant sets.

$\lambda_1 = -1$ (simple), λ_2, λ_3 complex conjugate

In the basis $B = \{\bar{u}, \bar{v}_1, \bar{v}_2\}$ where $\bar{u} \in V_{-1}$ and $\{\bar{v}_1, \bar{v}_2\}$ is a basis of $V_{-1}^{\perp}$, arguing as in the case when A has $\lambda = 1$ as a simple eigenvalue and two complex conjugate eigenvalues, the matrix associated with f is

$$A' = \begin{bmatrix} -1 & 0 & 0 \\ 0 & \cos(\theta) & -\sin(\theta) \\ 0 & \sin(\theta) & \cos(\theta) \end{bmatrix},$$

This matrix is the result of multiplying the matrix in Eq. (1.13), corresponding to a reflection on a vector plane, in this case $V_{-1}^{\perp}$, and the matrix in Eq. (1.11), corresponding to a rotation about a vector line, in this case V_{-1}. Thus, here we have the composition of a rotation about a vector line (the axis) and a reflection on a vector plane orthogonal to the axis; this transformation is called a *rotoreflection*.

$\lambda_1 = -1$ (triple)

Here, as in previous cases, we can prove that A is diagonalizable, and therefore similar to $-I$, which in turn implies that $A = -I$. This transformation maps every vector $\bar{u}$ onto $-\bar{u}$, so geometrically corresponds to a *central symmetry* with respect to the origin.

We summarize the classification of space Euclidean vector isometries in Table 1.2.

1.3 AFFINE EUCLIDEAN SYMMETRIES

We move now to $\mathbb{A}_n$, the affine Euclidean n-space, built over $\mathbb{V}_n$ endowed with the usual dot product. The elements $\boldsymbol{x}$ of $\mathbb{A}_n$ are now points, and not vectors, as is the case of $\mathbb{V}_n$, and have coordinates $\boldsymbol{x} = (x_1, x_2, \dots, x_n)$ in a fixed reference system. We recall that in an affine space a reference system $\mathcal{R}$ consists of two things: an origin O, and a basis $B = \{\bar{e}_1, \dots, \bar{e}_n\}$ of the underlying vector space, in our case $\mathbb{V}_n$, so we write

$$\mathcal{R} = \{O; \bar{e}_1, \dots, \bar{e}_n\}.$$

If B is an orthonormal basis, we say that $\mathcal{R}$ is an orthonormal reference system, and we will assume to be working in such a reference system. Then, the coordinates $x_1, \dots, x_n$ in $\mathcal{R}$ of a point $\boldsymbol{x} \in \mathbb{A}_n$ are exactly the coordinates of the vector connecting O and $\boldsymbol{x}$ in the basis B. The directions of the $\bar{e}_i$ define the axes associated with the reference system, i.e., the lines through O parallel to the $\bar{e}_i$.

The natural mappings in $\mathbb{A}_n$ are the *affine mappings* , *affine transformations* or *affinities*, which correspond to mappings $f : \mathbb{A}_n \to \mathbb{A}_n$ of the type

$$f(\boldsymbol{x}) = A\boldsymbol{x} + \boldsymbol{b}, \tag{1.14}$$

where A is an $n \times n$ matrix, and $\boldsymbol{b} \in \mathbb{A}_n$. The notion of *distance* in

TABLE 1.2 Classification of space vector Euclidean isometries

$\|\mathbf{A}\|$	**Eigenvalues**	**Vector Symmetry**
1	$\lambda_1 = 1$ (simple), λ_2, λ_3 complex conjugate	Rotation about a vector line (V_1)
1	$\lambda_1 = 1$ (simple), $\lambda_2 = -1$ (double)	Axial symmetry about a vector line (V_1)
−1	$\lambda_1 = 1$ (double), $\lambda_2 = -1$ (simple)	Reflection on a vector plane (V_1)
1	$\lambda = 1$ (triple)	Trivial symmetry (Identity)
−1	$\lambda_1 = -1$ (simple), λ_2, λ_3 complex conjugate	Rotoreflection ($V_{-1}, V_{-1}^{\perp}$)
−1	$\lambda_1 = -1$ (triple)	Central symmetry

$\mathbb{A}_n$ is inherited from the norm of $\mathbb{V}_n$, and in our case is the usual *Euclidean* distance, i.e.,

$$d(\boldsymbol{x}, \boldsymbol{y}) = \|\boldsymbol{x} - \boldsymbol{y}\|,$$

where $\boldsymbol{x} - \boldsymbol{y}$ corresponds to the vector in $\mathbb{V}_n$ connecting the points $\boldsymbol{x}, \boldsymbol{y}$. In this new space $\mathcal{E} = \mathbb{A}_n$ (see the beginning of the chapter), we want to fix a group of interest that determines the kind of symmetries we want to study. In order to do this, we need the following definition.

Definition 13. *Let $f : \mathbb{A}_n \to \mathbb{A}_n$ be a mapping. We say that f is a* Euclidean symmetry *(also called* isometry, *or* rigid motion*) if f preserves distances, i.e., if $\forall \boldsymbol{x}, \boldsymbol{y} \in \mathbb{A}_n$, we have $d(f(\boldsymbol{x}), f(\boldsymbol{y})) = d(\boldsymbol{x}, \boldsymbol{y})$.*

Notice that in Definition 13 we are not requiring that f is an affine mapping. However, it must be: the proof can be found in Theorem 1.7 of [93]. As a consequence, we can indistinctly speak about (global) isometries of $\mathbb{A}_n$, or affine isometries of $\mathbb{A}_n$. Sometimes the term "orthogonal transformation" is also used to refer to affine isometries. Euclidean symmetries form a group $\mathbb{E}(n)$, the *Euclidean group* , which is the group of interest in this book. In order to describe this group, the following result is crucial.

Proposition 14. *Let $f(\boldsymbol{x}) = A\boldsymbol{x} + \boldsymbol{b}$ be an affine mapping in $\mathbb{A}_n$. Then f is a Euclidean symmetry if and only if the matrix A defines a vector isometry of $\mathbb{V}_n$.*

Proof. Let $\boldsymbol{x}, \boldsymbol{y} \in \mathbb{A}_n$. Then

$$d(f(\boldsymbol{x}), f(\boldsymbol{y})) = \|f(\boldsymbol{x}) - f(\boldsymbol{y})\| = \|A(\boldsymbol{x} - \boldsymbol{y})\|.$$

Let us see ($\Rightarrow$). If f is a Euclidean symmetry, then $d(f(\boldsymbol{x}), f(\boldsymbol{y})) = d(\boldsymbol{x}, \boldsymbol{y})$. Since $d(\boldsymbol{x}, \boldsymbol{y}) = \|\boldsymbol{x}-\boldsymbol{y}\|$, we get that $\|A(\boldsymbol{x}-\boldsymbol{y})\| = \|\boldsymbol{x}-\boldsymbol{y}\|$, and therefore that A preserves norms. Thus, by Proposition 3, A defines a vector isometry. Now let us see ($\Leftarrow$). If A defines a vector isometry then, by Proposition 3, A preserves norms, and therefore

$$\|A(\boldsymbol{x} - \boldsymbol{y})\| = \|\boldsymbol{x} - \boldsymbol{y}\|.$$

Since $d(\boldsymbol{x}, \boldsymbol{y}) = \|\boldsymbol{x} - \boldsymbol{y}\|$, we get $d(f(\boldsymbol{x}), f(\boldsymbol{y})) = d(\boldsymbol{x}, \boldsymbol{y})$, so f is a Euclidean symmetry. □

From Proposition 14 we know that the matrix A in $f(\boldsymbol{x}) = A\boldsymbol{x} + \boldsymbol{b}$ must be orthogonal, so $|A| = 1$ or $|A| = -1$. Transferring the terminology from vector isometries, if $|A| = 1$ we say that f is *orientation-preserving* (or *direct*), and if $|A| = -1$ we say that f is *orientation-reversing* (or *reverse*). Furthermore, orientation-preserving Euclidean symmetries form a subgroup of $\mathbb{E}(n)$; this not true for orientation-reversing symmetries, though.

The goal now is to explore what kind of transformations we have in the group of Euclidean symmetries of the plane and the space. Because of Proposition 14 and since we have already classified vector isometries, part of the job is already done, but not all of it. To complete this task we need to pay attention to *fixed points*. We say that $\boldsymbol{x}_0 \in \mathbb{A}_n$ is a fixed point of a Euclidean symmetry f if $f(\boldsymbol{x}_0) = A\boldsymbol{x}_0 + \boldsymbol{b} = \boldsymbol{x}_0$, i.e., if $(I - A)\boldsymbol{x}_0 = \boldsymbol{b}$. Thus, the fixed points of $f(\boldsymbol{x}) = A\boldsymbol{x} + \boldsymbol{b}$ are the solutions of the linear system

$$(I - A)\boldsymbol{x} = \boldsymbol{b}. \tag{1.15}$$

A first observation is that whenever $\lambda = 1$ is not an eigenvalue of A, there is a unique fixed point. Indeed, $\lambda = 1$ is an eigenvalue of A if and only if $|A - I| = |I - A| = 0$, i.e., if and only if the coefficient matrix $I - A$ of the system in Eq. (1.15) does not have full rank. Thus, if $|A - I| \neq 0$, i.e., if $\lambda = 1$ is not an eigenvalue of A, the system in Eq. (1.15), which is a square system, has full rank and therefore, by Rouché-Fröbenius Theorem, has a unique solution.

If $\lambda = 1$ is, however, an eigenvalue of A, then the system in Eq. (1.15) may have no solution, in which case there are no fixed points, or may have infinitely many solutions, in which case we may have, for $n = 2$ or $n = 3$, a line or a plane of fixed points. We are in the first or the second situation depending on whether the ranks of the coefficient matrix and the augmented matrix $\operatorname{rank}(I - A)$, $\operatorname{rank}(I - A|\boldsymbol{b})$ of Eq. (1.15) are equal or not. It is the dimension of the set of fixed points we have in each case that allows us to completely classify the Euclidean symmetries of the plane and the space.

Furthermore, the following result will be useful.

Lemma 15. *Assume that the solution set $\mathcal{S}$ of Eq. (1.15) is infinite, and let $\mathbf{p}_1, \mathbf{p}_2 \in \mathcal{S}$. Then the vector $\mathbf{p}_2 - \mathbf{p}_1$ is a fixed vector of the matrix A, i.e., is an eigenvector associated with $\lambda = 1$.*

Proof. Since $\mathbf{p}_1, \mathbf{p}_2 \in \mathcal{S}$, $(I - A)\mathbf{p}_1 = \boldsymbol{b}$ and $(I - A)\mathbf{p}_2 = \boldsymbol{b}$. Subtracting these two equalities, we get that $(I - A)(\mathbf{p}_2 - \mathbf{p}_1) = \mathbf{0}$, i.e., $A(\mathbf{p}_2 - \mathbf{p}_1) = \mathbf{p}_2 - \mathbf{p}_1$. □

1.4 CLASSIFICATION OF PLANAR AND SPACE EUCLIDEAN SYMMETRIES

The two ingredients to classify planar and space Euclidean symmetries $f(\boldsymbol{x}) = A\boldsymbol{x} + \boldsymbol{b}$ are, on one hand, the fact that A is orthogonal and the classification of orthogonal matrices that we already carried out in Section 1.2, and on the other hand, the analysis of fixed points, which in turn is related to whether or not $\text{rank}(I - A)$ and $\text{rank}(I - A|\boldsymbol{b})$ are equal.

1.4.1 Classification of planar Euclidean symmetries

The matrix A in $f(\boldsymbol{x}) = A\boldsymbol{x} + \boldsymbol{b}$ must be of one of the two types in Eq. (1.10): the matrix at the left for orientation-preserving symmetries, the matrix at the right for orientation-reversing symmetries.

Orientation-preserving planar Euclidean symmetries

$A \neq I$

If $A \neq I$, from the analysis carried out at the orientation-preserving case in Section 1.2.1 we know that A has either complex conjugate eigenvalues, or the double eigenvalue $\lambda = -1$. In any case, $\lambda = 1$ is not eigenvalue of A, so $\text{rank}(I - A) = \text{rank}(I - A|\boldsymbol{b}) = 2$ and f has a unique fixed point $\mathbf{c}$, which is the solution to Eq. (1.15). By moving the origin of the reference system to $\mathbf{c}$, we realize that f represents a *rotation about the point* $\mathbf{c}$, which is the fixed point of the transformation (see Fig. 1.6). The point $\mathbf{c}$ is called the *center of rotation.* If $A = -I$ we have a rotation of angle $\theta = \pi$,

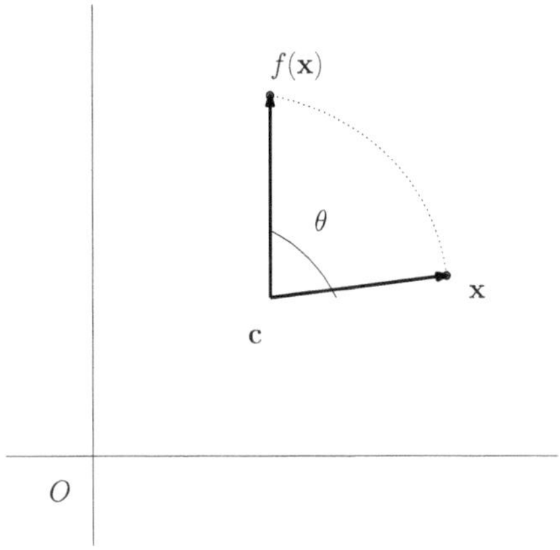

FIGURE 1.6 Rotation about a point **c**

and therefore a *central symmetry with respect to the point* **c**, which is called the *symmetry center.*

$\boxed{A = I}$

In this case, it is obvious that $\lambda = 1$ is an eigenvalue of A, so we must analyze the existence of fixed points, which depends on whether or not $\text{rank}(I - A) = \text{rank}(I - A|\boldsymbol{b})$.

(i) $\text{rank}(I - A) = \text{rank}(I - A|\boldsymbol{b})$: since $A = I$, $\text{rank}(I - A) = 0$; thus, $\text{rank}(I - A|\boldsymbol{b}) = 0$ if and only if $\boldsymbol{b} = \mathbf{0}$. Since $A = I$, $\boldsymbol{b} = \mathbf{0}$ we get that $f(\boldsymbol{x}) = \boldsymbol{x}$, so f is the *trivial symmetry*, i.e., the identity.

(ii) $\text{rank}(I - A) \neq \text{rank}(I - A|\boldsymbol{b})$: from (i) this happens when $\boldsymbol{b} \neq \mathbf{0}$, in which case $f(\boldsymbol{x}) = \boldsymbol{x} + \boldsymbol{b}$. Here the image of each point $\boldsymbol{x} \in \mathbb{A}_2$ is displaced according to the vector $\boldsymbol{b}$, so we have a *translation* by the vector $\boldsymbol{b}$ (see Fig. 1.7).

Orientation-reversing planar Euclidean symmetries

Here the matrix A is the matrix at the right in Eq. (1.10). From the analysis carried out in Section 1.2.1, we know that the eigenvalues of A are $\lambda = \pm 1$, so $\lambda = 1$ is an eigenvalue and therefore it

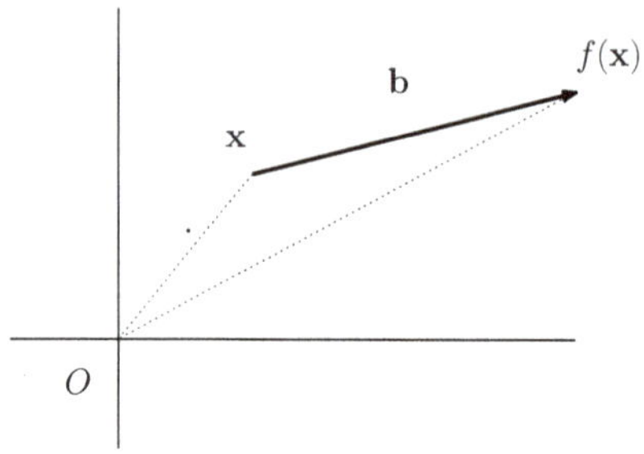

FIGURE 1.7 Translation by a vector $\boldsymbol{b}$

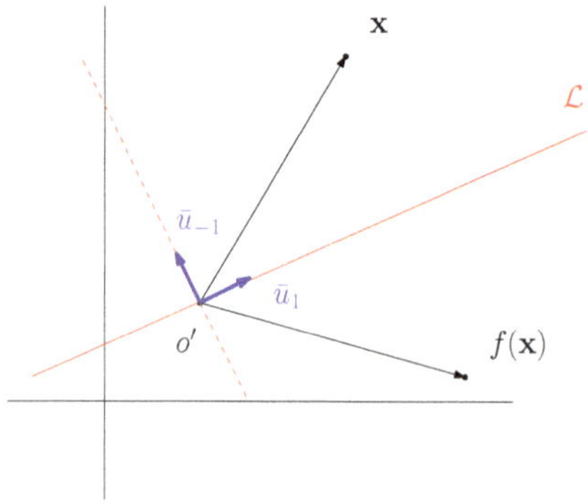

FIGURE 1.8 Affine reflection on a line $\mathcal{L}$

is not guaranteed that $f(\boldsymbol{x}) = A\boldsymbol{x} + \boldsymbol{b}$ has fixed points. Thus, we need to distinguish two cases.

(i) $\text{rank}(I - A) = \text{rank}(I - A|\boldsymbol{b})$: since $\lambda = 1$ is simple, the dimension of the eigenspace associated with $\lambda = 1$ is one, and therefore $\text{rank}(I - A) = 1$. If $\text{rank}(I - A) = \text{rank}(I - A|\boldsymbol{b})$, the system in Eq. (1.15) has a solution consisting of an affine line, i.e., there is a fixed line $\mathcal{L}$. By moving the origin to any point O' on $\mathcal{L}$, we notice that we have a *reflection* (or *mirror symmetry*) *on the line* $\mathcal{L}$, i.e., on the line of fixed points. Because of Lemma 15, the direction of $\mathcal{L}$ is that of the eigenspace V_1 of A associated with $\lambda = 1$, i.e., $\mathcal{L}$ is parallel to the symmetry axis of the vector isometry defined by A (see Fig. 1.8). The line $\mathcal{L}$ is called the *symmetry axis*.

(ii) $\text{rank}(I-A) \neq \text{rank}(I-A|\boldsymbol{b})$: from (i) we know that $\text{rank}(I-A) = 1$, so $\text{rank}(I-A|\boldsymbol{b}) = 2$. There are no fixed points in this case. To discern what kind of transformation we have here, we consider a reference system with axes parallel to the eigenvectors $\bar{u}_1$, associated with $\lambda = 1$, and $\bar{u}_{-1}$, associated with $\lambda = -1$, which must be orthogonal. In that new system of reference the affine mapping now looks like

$$f(\boldsymbol{x}) = \begin{bmatrix} b'_1 \\ b'_2 \end{bmatrix} + \begin{bmatrix} 1 & 0 \\ 0 & -1 \end{bmatrix} \boldsymbol{x}.$$

Thus, we can write

$$f(\boldsymbol{x}) = \begin{bmatrix} b'_1 \\ 0 \end{bmatrix} + \underbrace{\begin{bmatrix} 0 \\ b'_2 \end{bmatrix} + \begin{bmatrix} 1 & 0 \\ 0 & -1 \end{bmatrix} \boldsymbol{x}}_{f^\star(\boldsymbol{x})},$$

which corresponds to the composition of a translation and another affine mapping $f^\star(\boldsymbol{x})$ that fulfills the conditions in (i), i.e., which is a symmetry with respect to a line $\mathcal{L}$ with the direction of V_1, the eigenspace of A associated with $\lambda = 1$. Since $\bar{u}_1 \in V_1$, $\mathcal{L}$ is parallel to $\bar{u}_1$. Additionally, observe that the vector $(b'_1, 0)$ is written in the basis $B = \{\bar{u}_1, \bar{u}_{-1}\}$, so this vector is parallel to $\bar{u}_1$. Thus, f is the composition of a reflection on a line $\mathcal{L}$ and a translation by a vector parallel to $\mathcal{L}$. This transformation is called a *glide*, or *sliding, reflection* (see Fig. 1.9).

We summarize the classification of planar Euclidean symmetries in Table 1.3.

1.4.2 Classification of space Euclidean symmetries

We move now to Euclidean symmetries $f(\boldsymbol{x}) = A\boldsymbol{x} + \boldsymbol{b}$ in $\mathbb{A}_3$. In order to carry out the classification, we start from the results in Section 1.2.2, where space vector isometries are classified from the eigenvalues of A. Recall that if $\lambda = 1$ is not an eigenvalue of A, then we have just one fixed point. Otherwise, we may or we may not have fixed points, and this depends on the ranks of $I - A$ and $I - A|\boldsymbol{b}$.

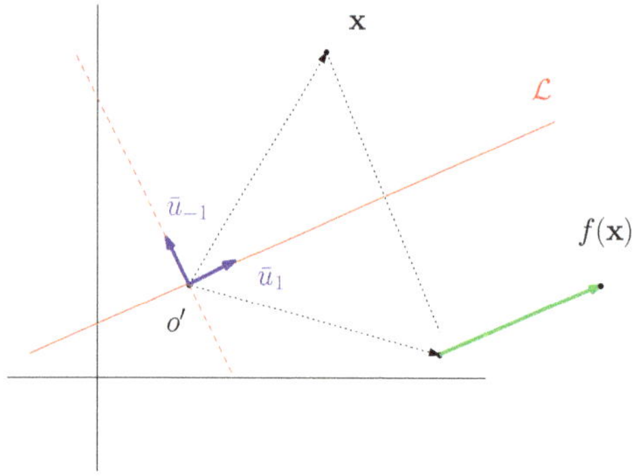

FIGURE 1.9 Sliding reflection

TABLE 1.3 Classification of planar Euclidean symmetries

$\lvert\mathbf{A}\rvert$	rank($\mathbf{I}-\mathbf{A}$)	rank($\mathbf{I}-\mathbf{A}\lvert\boldsymbol{b}$)	**Euclidean Symmetry**
1	0	0	Trivial symmetry (identity)
1	0	1	Translation by a vector $\boldsymbol{b}$
−1	1	1	Reflection on a line $\mathcal{L}$
−1	1	2	Glide reflection
1	2	2	Rotation about a point $\mathbf{c}$

$\lambda_1 = 1$ (simple), λ_2, λ_3 complex conjugate

In this case A represents a rotation of angle $\theta \neq \pi$ about the vector line defined by V_1, the eigenspace associated with $\lambda = 1$. We distinguish two cases:

(i) $\text{rank}(I - A) = \text{rank}(I - A|\boldsymbol{b})$: Since $\lambda = 1$ is simple, $\text{rank}(I - A) = 2$ and therefore the solution set of Eq. (1.15) is a line, i.e., we have a line $\mathcal{L}$ of fixed points which by Lemma 15 is parallel to V_1. By moving the origin to any point on $\mathcal{L}$, we notice that here we have a *rotation about a line* $\mathcal{L}$. The line $\mathcal{L}$ is called the *axis of rotation*.

(ii) $\text{rank}(I - A) \neq \text{rank}(I - A|\boldsymbol{b})$. There are no fixed points. Since from (i) we have $\text{rank}(I - A) = 2$, here $\text{rank}(I - A) = 3$. By decomposing the transformation as we did in the case of planar glide reflections, we observe that the transformation is the composition of a rotation of angle $\theta \neq \pi$ about a line

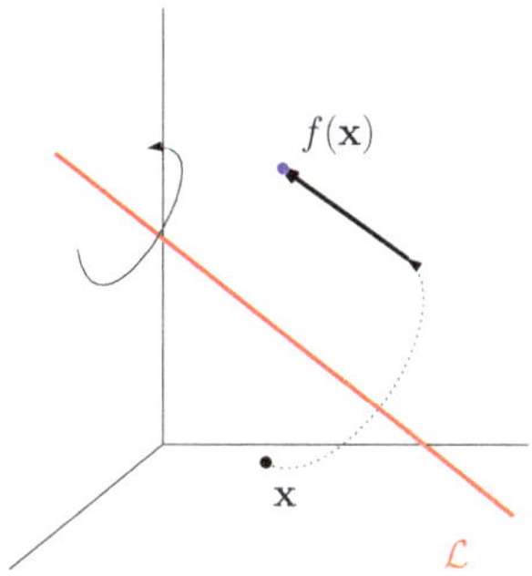

FIGURE 1.10 Screw motion

$\mathcal{L}$, which by Lemma 15 is parallel to V_1, and a translation by a vector parallel to $\mathcal{L}$. This transformation is called a *screw motion*, also called *helical motion* (see Fig. 1.10).

$\lambda_1 = 1$ (simple), $\lambda_2 = -1$ (double)

In this case A represents a reflection on the vector line defined by V_1, the eigenspace associated with $\lambda = 1$, i.e., an axial symmetry about V_1, which can also be described as a rotation of angle π about V_1, or a reflection on V_1. We distinguish two cases:

(i) $\text{rank}(I - A) = \text{rank}(I - A|\boldsymbol{b})$: Since $\lambda = 1$ is simple, $\text{rank}(I - A) = 2$ and therefore the solution set of Eq. (1.15) is a line, i.e., we have a line $\mathcal{L}$ of fixed points which by Lemma 15 is parallel to V_1. By moving the origin to any point on $\mathcal{L}$, we notice that here we have an *axial symmetry about a line* $\mathcal{L}$, i.e., a *reflection on a line* $\mathcal{L}$, also called *half-turn* or *mirror symmetry* with respect to $\mathcal{L}$. The line $\mathcal{L}$ is called the *symmetry axis*.

(ii) $\text{rank}(I - A) \neq \text{rank}(I - A|\boldsymbol{b})$: there are no fixed points. Since from (i) we have $\text{rank}(I - A) = 2$, here we get $\text{rank}(I - A) = 3$. This is just a special case of *screw motion*, where the rotation angle $\theta = \pi$. It can also be described as the composition of an axial symmetry about a line $\mathcal{L}$, parallel to V_1, and a translation by a vector parallel to $\mathcal{L}$.

$\lambda_1 = 1$ (double), $\lambda_2 = -1$ (simple)

The matrix A represents a reflection on the vector plane defined by V_1.

(i) $\text{rank}(I - A) = \text{rank}(I - A|\boldsymbol{b})$: Since $\lambda = 1$ is double and A is diagonalizable (see Section 1.2.2), $\text{rank}(I - A) = 1$ and therefore the solution set of Eq. (1.15) is a plane, i.e., we have a plane Π of fixed points which by Lemma 15 is parallel to V_1. By moving the origin to any point on Π, we notice that here we have a *reflection on a plane* Π, or *mirror symmetry* with respect to Π. The plane Π is called the *symmetry plane.*

(ii) $\text{rank}(I - A) \neq \text{rank}(I - A|\boldsymbol{b})$: there are no fixed points. Since from (i) we have $\text{rank}(I - A) = 1$, in this case $\text{rank}(I - A) = 2$. By appropriately decomposing the transformation, we identify that this transformation is the composition of a reflection on a plane Π, parallel to V_1, and a translation by a vector parallel to Π. This transformation is called a (space) *glide*, or *sliding*, *reflection.*

$\lambda_1 = 1$ (triple)

Since from Section 1.2.2 we know that A is diagonalizable, $A = I$.

(i) $\text{rank}(I - A) = \text{rank}(I - A|\boldsymbol{b})$: since $A = I$, $\text{rank}(I - A) = 0$, and therefore $\boldsymbol{b} = \boldsymbol{0}$. Thus, $f(\boldsymbol{x}) = \boldsymbol{x}$, i.e., the transformation is the *trivial symmetry.*

(ii) $\text{rank}(I - A) \neq \text{rank}(I - A|\boldsymbol{b})$: in this case $\boldsymbol{b} \neq \boldsymbol{0}$, and $f(\boldsymbol{x}) = \boldsymbol{x} + \boldsymbol{b}$, so we have a *translation* by a vector.

$\lambda_1 = -1$ (simple), λ_2, λ_3 complex conjugate

Since $\lambda = 1$ is not an eigenvalue of A, there is a unique fixed point $\mathbf{c}$. By moving the origin to $\mathbf{c}$ and from the results in Section

1.2.2, we identify that we have a *rotoreflection*, i.e., the composition of a rotation about a line $\mathcal{L}$, which is the line through $\mathbf{c}$ with the direction of the eigenspace V_{-1} associated with $\lambda = -1$, and a reflection on the plane through $\mathbf{c}$ parallel to $V_{-1}^{\perp}$.

$\boxed{\lambda_1 = -1 \text{ (triple)}}$

Since $\lambda = 1$ is not an eigenvalue of A, there is a unique fixed point $\mathbf{c}$. By moving the origin to $\mathbf{c}$ and from the results in Section 1.2.2, we identify that we have a *central symmetry* with respect to the point $\mathbf{c}$. The point $\mathbf{c}$ is called the *center of symmetry.*

We summarize the classification of space Euclidean symmetries in Table 1.4; here we write the multiplicity of the eigenvalues within the bracket, i.e., $\lambda_1 = 1$ (2) means that $\lambda = 1$ has algebraic multiplicity 2.

Furthermore, before finishing this chapter let us introduce the following definition.

Definition 16. *Let $f : \mathbb{A}_n \to \mathbb{A}_n$ be a Euclidean symmetry. We say that f is an* involution, *if $f \circ f = \mathrm{id}_{\mathbb{A}_n}$.*

Therefore, in $\mathbb{A}_2$ reflections on a line and central symmetries, which are rotations about a point of angle π, are involutions. In $\mathbb{A}_3$, reflections on a plane, axial symmetries and central symmetries are also involutions.

1.4.3 Final remarks and notation

For future reference, let us include the following two results on the composition of different Euclidean symmetries. We refer to [46, §7.3], [71, §1], [57, §19] for proofs. The first result concerns planar symmetries, and the second result, space symmetries.

Lemma 17. *(1) The composition of two planar reflections with parallel axes is a translation by a vector parallel to the axes.*

(2) The composition of two planar reflections with intersecting axes is a rotation about the intersection point, where the rotation angle is equal to twice the angle between the axes.

TABLE 1.4 Classification of space Euclidean symmetries

$\lvert\mathbf{A}\rvert$	**Eigenvalues**	rank$(\mathbf{I}-\mathbf{A})$	rank$(\mathbf{I}-\mathbf{A}\mid\boldsymbol{b})$	**Euclidean Symmetry**
1	$\lambda_1 = 1\ (1),\ \lambda_2, \lambda_3 \in \mathbb{C}$	2	2	Rotation about a line $\mathcal{L}$
1	$\lambda_1 = 1\ (1),\ \lambda_2, \lambda_3 \in \mathbb{C}$	2	3	Screw motion
1	$\lambda_1 = 1\ (1),\ \lambda_2 = -1\ (2)$	2	2	Axial symmetry about a line $\mathcal{L}$
1	$\lambda_1 = 1\ (1),\ \lambda_2 = -1\ (2)$	2	3	Screw motion of angle $\theta = \pi$
−1	$\lambda_1 = 1\ (2),\ \lambda_2 = -1\ (1)$	2	2	Reflection on a plane Π
−1	$\lambda_1 = 1\ (2),\ \lambda_2 = -1\ (1)$	2	3	Glide reflection
1	$\lambda = 1\ (3)$	0	0	Trivial symmetry (Identity)
1	$\lambda = 1\ (3)$	0	1	Translation by a vector $\boldsymbol{b}$
−1	$\lambda_1 = -1\ (1),\ \lambda_2, \lambda_3 \in \mathbb{C}$	3	3	Rotoreflection
−1	$\lambda_1 = -1\ (3)$	3	3	Central symmetry about a point $\mathbf{c}$

Lemma 18. *(1) The composition of two central symmetries of centers* $\mathbf{c}_1 \neq \mathbf{c}_2$ *is a translation by a vector parallel to the vector connecting* $\mathbf{c}_1, \mathbf{c}_2$.

(2) The composition of two rotations with parallel axes $\mathcal{L}_1, \mathcal{L}_2$ *is a rotation with axis parallel to* $\mathcal{L}_1, \mathcal{L}_2$.

(3) The composition of two rotations $\mathcal{R}_{\mathcal{L}_1,\alpha}$, $\mathcal{R}_{\mathcal{L}_2,\beta}$, *with concurrent axes* $\mathcal{L}_1 \neq \mathcal{L}_2$ *such that the angle between* $\mathcal{L}_1, \mathcal{L}_2$ *is* Φ, *is another rotation* $\mathcal{R}_{\mathcal{L}_3,\gamma}$, *where* $\mathcal{L}_3$ *is concurrent with* $\mathcal{L}_1, \mathcal{L}_2$ *and perpendicular to* $\mathcal{L}_1, \mathcal{L}_2$, *and*

$$\cos\left(\frac{\gamma}{2}\right) = \cos\left(\frac{\alpha}{2}\right) \cdot \cos\left(\frac{\beta}{2}\right) - \sin\left(\frac{\alpha}{2}\right) \cdot \sin\left(\frac{\beta}{2}\right) \cdot \cos(\Phi). \tag{1.16}$$

(4) The composition of two rotations with skew axes is a screw motion.

(5) The composition of two reflections with respect to two planes $\Pi_1 \neq \Pi_2$ *is a translation when* Π_1, Π_2 *are parallel, and a rotation about the line* $\Pi_1 \cap \Pi_2$, *of angle equal to double the angle formed by* Π_1, Π_2, *when* Π_1, Π_2 *intersect.*

To conclude, in this chapter we have completely classified planar and space Euclidean symmetries, i.e., the affine isometries of the plane $\mathbb{A}_2$ and the space $\mathbb{A}_3$, which are summarized in Tables 1.3 and 1.4. From now on we will mostly consider different types of subsets of $\mathbb{A}_2$ and $\mathbb{A}_3$, and we will focus on studying, and providing constructive methods to find, the symmetries of those subsets, i.e., the transformations studied in this section that leave those subsets invariant.

The notation $\mathbb{A}_n, \mathbb{V}_n$ for the affine Euclidean n-space and the underlying vector space was useful in this chapter to distinguish between vector isometries and affine isometries, that we called Euclidean symmetries. From now on, unless we believe that it can cause any misunderstanding, we will just use the notation $\mathbb{A}_n = \mathbb{R}^n$, and $\mathbb{R}^n$ will also denote, when necessary, $\mathbb{V}_n$. The elements of $\mathbb{R}^n$ will be denoted by $\boldsymbol{x} = (x_1, \ldots, x_n)$; when $n = 2$ we will write $\boldsymbol{x} = (x, y)$, and when $n = 3$ we will write $\boldsymbol{x} = (x, y, z)$. Similarly for the affine, Euclidean, complex n-space, $\mathbb{C}^n$.

Although we will be mostly working in an affine setting, occasionally we will also work in a *projective* setting. In this sense, recall that the *projective n-space*, $\mathbb{P}^n(\mathbb{R})$, is the set of $(n+1)$-tuples

$$[x_1 : x_2 : \cdots : x_n : x_{n+1}]$$

where $x_i \in \mathbb{R}$ for $i = 1, \ldots, n+1$, not all the x_i are zero, and

$$[x_1 : x_2 : \cdots : x_n : x_{n+1}] = [y_1 : y_2 : \cdots : y_n : y_{n+1}]$$
$$\Leftrightarrow x_i = \lambda y_i,\ \lambda \neq 0,\ i = 1, \ldots, n+1.$$

The projective points where $x_{n+1} = 0$ are called *points at infinity*, and can be identified with directions in n-space. The projective points where $x_{n+1} \neq 0$ can be identified with the affine points

$$\left(\frac{x_1}{x_{n+1}}, \frac{x_2}{x_{n+1}}, \ldots, \frac{x_n}{x_{n+1}}\right).$$

Conversely, given an affine point $\boldsymbol{x} = (x_1, x_2, \ldots, x_n) \in \mathbb{R}^n$ we can embed it into $\mathbb{P}^n(\mathbb{R})$ by identifying it with the projective point $[x_1 : x_2 : \cdots : x_n : 1]$. Thus, we can see $\mathbb{P}^n(\mathbb{R})$ as the union of the real affine n-space, plus the *hyperplane at infinity*, defined by $x_{n+1} = 0$; for $n = 2$ we speak about the *line at infinity*, and for $n = 3$ we speak about the *plane at infinity*. The complex projective n-space $\mathbb{P}^n(\mathbb{C})$ is defined analogously.

Furthermore, a *projective transformation* in $\mathbb{P}^n(\mathbb{R})$ is a mapping $\widehat{f} : \mathbb{P}^n(\mathbb{R}) \to \mathbb{P}^n(\mathbb{R})$ where $\widehat{f}(\mathbf{x}) = M\mathbf{x}$, with M an $(n+1) \times (n+1)$ matrix, and $\mathbf{x} \in \mathbb{P}^n(\mathbb{R})$. Since points in $\mathbb{P}^n(\mathbb{R})$ are defined by tuples up to multiplication by nonzero constants, the same happens with projective transformations: M and μM, with $\mu \neq 0$, define the same $\widehat{f}$.

As explained at the beginning of the chapter, the group of interest in this book is the group of Euclidean symmetries. There are other groups that one can explore. For instance:

- *Similarities*, namely the transformations $f(x) = \lambda A\boldsymbol{x} + \boldsymbol{b}$ with $\lambda \neq 0$ and A orthogonal. These transformations preserve angles, and preserve the topology, but they do not preserve distances. Any similarity is the composition of a Euclidean symmetry plus a homothety, i.e., a scaling.

- *Affinities*, i.e., the transformations $f(x) = A\boldsymbol{x} + \boldsymbol{b}$ with A non-singular but non-necessarily orthogonal. Here distances and angles are not preserved, but the global topology is preserved.

- *Projectivities*, i.e., non-singular projective transformations $\widehat{f}(\mathbf{x}) = M\mathbf{x}$ with M non-singular, also called *projective isomorphisms.* Here the global topology may not be preserved, since some affine points can be mapped to infinity, and conversely.

- *Möbius transformations*, which are the mappings

$$f(\boldsymbol{x}) = \boldsymbol{b} + \alpha \frac{A(\boldsymbol{x} - \boldsymbol{a})}{\|\boldsymbol{x} - \boldsymbol{a}\|^{\epsilon}}, \tag{1.17}$$

where $\boldsymbol{a}, \boldsymbol{b} \in \mathbb{R}^n$, $\alpha \in \mathbb{R}$, A is an orthogonal matrix, and $\epsilon = 0$ or $\epsilon = 2$ (although $\epsilon = 0$ yields a similarity). This group includes similarities as a subgroup and also *inversions.* Again, here the topology might not be preserved.

Fig. 1.11 illustrates the effect of these transformations on an algebraic curve, the *deltoid*, shown in the upper row, at the left: also in the upper row we see the image of the deltoid under a similarity (middle) and an affinity (right). In the lower row, we have the image of a deltoid under a projective transformation, at the left (here the topology is lost, since an affine point of the deltoid is mapped to a point at infinity), and under a Möbius transformation of the plane, at the right. If two objects are related by a similarity, affinity, projectivity or Möbius transformation, we say that they are *similar*, *affinely equivalent*, *projectively equivalent* or *Möbius equivalent*, respectively. We will occasionally provide some comments or references regarding these other groups, but our primary interest will be what we called Euclidean symmetries.

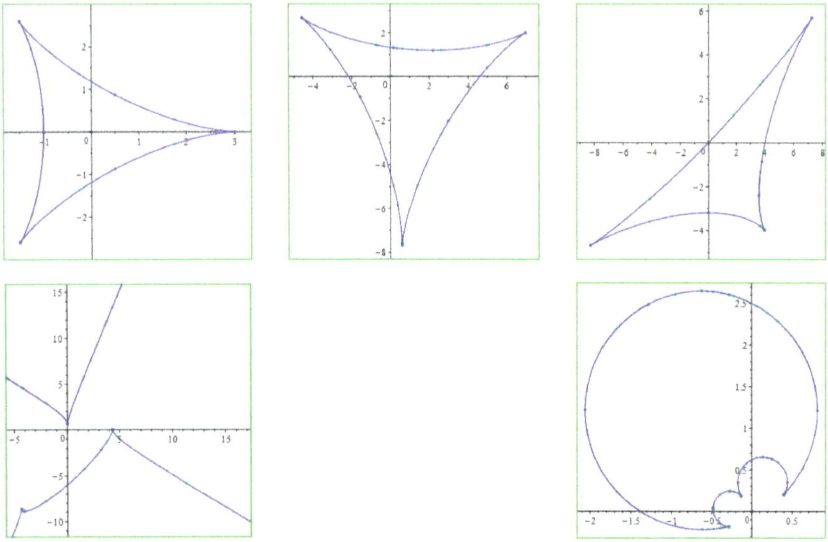

FIGURE 1.11 Image of a deltoid curve under different transformations

For other topics regarding Euclidean symmetries, e.g. finite subgroups of the group of Euclidean (planar and space) symmetries, generators of these groups and many other geometric questions, we recommend the interested reader the lovely classic and very readable [46]. Another good general reference is [36].

CHAPTER 2

Planar algebraic curves (I): implicit case

In this chapter we begin the analysis of the (Euclidean) symmetries of algebraic planar and space objects by considering algebraic planar curves, which are the zerosets of non-constant bivariate polynomials. For readers less familiar with planar algebraic curves we refer to [58, 117, 132] for a thorough introduction to the topic. In this chapter we explore the general properties of the symmetries of these objects, and present algorithms to compute these symmetries assuming that their implicit equation, i.e., the bivariate polynomial defining the curve, is given.

2.1 GENERAL PROPERTIES

(Affine) algebraic varieties are subsets of $\mathbb{K}^n$, with $\mathbb{K}$ a field, defined as the common zeroes of finitely many polynomials $F_i \in \mathbb{K}[x_1, \ldots, x_n]$, where $i = 1, \ldots, m$, i.e.,

$$\mathcal{V} = V(F_1(x_1, \ldots, x_n), \ldots, F_m(x_1, \ldots, x_n)).$$

We say that $\mathcal{U}$ is a *component* of $\mathcal{V}$ if $\mathcal{U}$ is also an algebraic variety, $\mathcal{U} \subset \mathcal{V}$, and there exists another algebraic variety $\mathcal{W} \subset \mathcal{V}$ such that $\mathcal{V} = \mathcal{U} \cup \mathcal{W}$. If $\mathcal{V}$ does not admit such a decomposition, we say that $\mathcal{V}$ is *irreducible*. Every algebraic variety can be written as

$$\mathcal{V} = \mathcal{V}_1 \cup \mathcal{V}_2 \cup \cdots \cup \mathcal{V}_p,$$

DOI: 10.1201/9781003648710-2

where each $\mathcal{V}_i$ is irreducible. We say that the $\mathcal{V}_i$ are the *irreducible components* of $\mathcal{V}$.

Furthermore, *projective algebraic varieties* are subsets of $\mathbb{P}^n(\mathbb{K})$ defined as the common (projective) zeroes of finitely many *homogenous* polynomials $\widehat{F}_i(x_1, \ldots, x_n, x_{n+1})$. We can embed any variety $\mathcal{V}$ into $\mathbb{P}^n(\mathbb{K})$ by considering its *projective closure* $\widehat{\mathcal{V}}$, which is the smallest projective algebraic variety containing $\mathcal{V}$. One might (wrongly) think that in order to get $\widehat{\mathcal{V}}$ from $\mathcal{V}$ one just needs to homogenize the polynomials F_i defining $\mathcal{V}$, using a homogenization variable x_{n+1}, which amounts to performing the substitution

$$x_i \to \frac{x_i}{x_{n+1}}$$

in $F_1, \ldots, F_n$, and clear denominators. This will be the case when $\mathcal{V}$ is defined by just one polynomial, which happens for planar algebraic curves, which we treat in this chapter, and for algebraic surfaces, treated in Chapter 6. But in general the situation is not so simple, as we will show in Chapter 5. Anyway, for now the affine setting will suffice. The notions of reducibility and irreducibility extend naturally to $\mathbb{P}^n(\mathbb{K})$.

For us, $\mathbb{K} = \mathbb{R}$ or $\mathbb{K} = \mathbb{C}$. Most of the time we will stay in the case $\mathbb{K} = \mathbb{R}$, so from now on in this chapter we will assume this; in Subsection 2.3, where a certain complex representation will be used, this will change. Now in the plane $\mathbb{R}^2$, a *planar algebraic curve* $\mathcal{C}$ is the zero set of a non-constant polynomial $F(x, y)$, so $\mathcal{C} = V(F(x, y))$. We say that $F(x, y) = 0$ is the *implicit equation* of $\mathcal{C}$. The following are all of them examples of implicit algebraic curves (see Fig. 2.1):

(i) A *lemniscate*: $(x^2 + y^2)^2 - (x^2 - y^2) = 0$.

(ii) A *three-leaved rose*: $x^4 + 2x^2y^2 + y^4 + 3x^2y - y^3 = 0$

(iii) *Descartes' Folium*: $8x^3 + 27y^3 - 36xy = 0$.

(iv) *Deltoid*: $(x^2 + y^2)^2 + 18(x^2 + y^2) - 27 - 8(x^3 - 3xy^2) = 0$

Furthermore, the projective closure $\widehat{\mathcal{C}}$ of $\mathcal{C}$ is defined by the polynomial $\widehat{F}(x, y, \omega)$, which results from homogenizing $F(x, y)$ with respect to a homogenization variable ω.

FIGURE 2.1 From left to right: a lemniscate, a three-leaved rose, Descartes' Folium, deltoid curve

The *degree* of $\mathcal{C}$ is the maximum power appearing in the polynomial $F(x, y)$ defining the curve. We can always write $F(x, y)$ as

$$F(x, y) = F_1(x, y)^{\ell_1} \cdots F_r(x, y)^{\ell_r}, \tag{2.1}$$

where the F_i cannot be written themselves as products of polynomials of smaller degree, in which case we say that they are *irreducible* . Thus, the curves $\mathcal{C}_i$ defined by $F_i(x, y) = 0$ are the irreducible components of $\mathcal{C}$. We say that $F(x, y)$ is *square-free* if in Eq. (2.1) we have $\ell_i = 1$ for $i = 1, \ldots, r$; thus, $\mathcal{C}$ is irreducible if the polynomial $F(x, y)$ is square-free and $r = 1$. Note that the implicit equation of $\mathcal{C}$ is unique except for multiplication by constants.

Any component of $\mathcal{C}$ is defined by a factor of $F(x, y)$. Related to this, and in order to use it in the future, we recall here *Bezout's Theorem*, well-known in Algebraic Geometry (see for instance [132]); what we present is an affine version of the theorem, enough for our purposes.

Theorem 19. *Let $\mathcal{C}, \mathcal{D} \subset \mathbb{R}^2$ be two planar algebraic curves of degrees m, n. If $\mathcal{C}, \mathcal{D}$ do not have any common component, then $\mathcal{C}, \mathcal{D}$ intersect in at most mn points.*

Remark 3. *In fact, when formulated in a projective setup, Bezout's Theorem ensures that two projective planar algebraic curves of degrees m, n either intersect in* exactly *mn projective points, counted with multiplicity, or have a common component.*

Remark 4. *If $\mathcal{C}, \mathcal{D}$ are defined by polynomials $F(x, y), G(x, y)$, of degrees m, n, Bezout's Theorem is saying that whenever $\mathcal{C}, \mathcal{D}$ have no common component, the polynomial system*

$$F(x, y) = 0, \ G(x, y) = 0$$

has at most mn solutions. A consequence is that if $\mathcal{C}, \mathcal{D}$ have more than mn points in common, in particular if they share infinitely many points, then $\mathcal{C}, \mathcal{D}$ must have a common component, implying that the polynomials $F(x, y), G(x, y)$ have a nontrivial common factor; if additionally, say, $F(x, y)$, is irreducible, then $F(x, y)$ must be a factor of $G(x, y)$.

Coming back to Definition 1 in Chapter 1, we will say that an algebraic curve $\mathcal{C}$ is *symmetric* if there is some non-trivial Euclidean planar symmetry, i.e., some transformation of the list in Table 1.3, leaving $\mathcal{C}$ invariant. For instance, all the curves in Fig. 2.1 show *reflectional symmetry*, since they all have axes of symmetry, i.e., are invariant under certain non-trivial reflections. Also, the 3-leaved rose and the deltoid show *rotational symmetry*, since they are invariant under non-trivial rotations, for instance rotations of angle $\theta = 2\pi/3$ around a certain point. Additionally, the lemniscate exhibits central symmetry as well.

We denote the set of symmetries of an algebraic curve $\mathcal{C}$ by $\mathrm{Sym}(\mathcal{C})$, which has the structure of a group and is in fact a subgroup of the group $\mathbb{E}(2)$ of planar Euclidean symmetries. In order to discover the properties of $\mathrm{Sym}(\mathcal{C})$, we will assume that $\mathcal{C}$ is defined by a square-free polynomial. The reason for this assumption is the following result.

Lemma 20. *Let $f(\boldsymbol{x}) = A\boldsymbol{x} + \boldsymbol{b}$ be a symmetry of $\mathcal{C} \subset \mathbb{R}^2$, defined by a square-free polynomial $F(x, y)$. Then there exists $\lambda \in \mathbb{R}$ such that $F \circ f = \lambda F$.*

Proof. Since f leaves $\mathcal{C}$ invariant, $f(\mathcal{C}) = \mathcal{C}$. Thus, $F \circ f$ also defines $\mathcal{C}$. Since $F, F \circ f$ are both square-free and both define $\mathcal{C}$, the result follows. □

The condition in Lemma 20 of $F(x, y)$ being square-free is necessary: for instance, $F(x, y) = x^2y^3$ defines a curve symmetric with respect to the line $y = x$, i.e., $f(x, y) = (y, x)$ is a symmetry of the curve. However, $(F \circ f)(x, y) = y^2x^3 \neq \lambda F$. Observe that Lemma 20 does not require irreducibility, though.

Lemma 20 is useful because it translates a geometric property, i.e., being symmetric, into an algebraic condition. To make good use of Lemma 20 we need to find out more about the constant $\lambda \in \mathbb{R}$ in Lemma 20. In order to do that, we first need some properties of $\mathrm{Sym}(\mathcal{C})$.

Lemma 21. *Let $\mathcal{C}$ not contain any line, and let $f \in \mathrm{Sym}(\mathcal{C})$, $f(\boldsymbol{x}) = A\boldsymbol{x} + \boldsymbol{b}$. Then f is neither a translation nor a glide reflection.*

Proof. Let $\mathbf{p} \in \mathcal{C}$. Assume first that f is a translation by the vector $\boldsymbol{b}$, i.e., $A = I$. Since $f(\mathcal{C}) = \mathcal{C}$, $\mathbf{p}_n = f^{(n)}(\mathbf{p})$, $f^{(n)} = f \circ \overset{n}{\cdots} \circ f$, is a sequence of different points of $\mathcal{C}$. Let L be the line through $\mathbf{p}$ parallel to $\boldsymbol{b}$. Then $\mathbf{p}_n \in L$ for all $n \in \mathbb{N}$; thus $\mathcal{C} \cap L$ consists of infinitely many points, and therefore, by Bezout's Theorem, $\mathcal{C}$ and L have a common component. Since L is a line, and therefore irreducible, $L \subset \mathcal{C}$. But this cannot happen, because by hypothesis $\mathcal{C}$ does not contain any line. Now assume that f is a glide reflection, where $\mathbf{v}$ is the vector defining the translation part. Then the sequence $f^{(2n)}(\mathbf{p})$ is also a sequence of different points of $\mathcal{C}$ belonging both to $\mathcal{C}$ and the line through $\mathbf{p}$ parallel to $\mathbf{v}$, and we can argue as before. □

Remark 5. *In fact, one can argue as in Lemma 21 to prove that the only algebraic curves invariant under some translation or some glide reflection are unions of parallel lines.*

Corollary 22. *Assume that $\mathcal{C}$ is not a union of parallel lines. Then all the axes of symmetry of $\mathcal{C}$ are concurrent.*

Proof. Suppose that $\mathcal{C}$ is invariant under two different reflections f_1, f_2 with parallel axes ℓ_1, ℓ_2. Since Sym($\mathcal{C}$) is a group, $\mathcal{C}$ is invariant under the composition $f_1 \circ f_2$. However, the composition of two reflections with parallel axes is a translation (see Lemma 17), and therefore the result follows from Lemma 21 and Remark 5. □

Next we address rotations. Properties of rotational symmetries of planar algebraic curves are addressed in [83, 84]. The following result is proven in [84], see Theorem 5.3 there.

Lemma 23. *Assume that $\mathcal{C}$ is not a union of lines. Then all rotations leaving $\mathcal{C}$ invariant have the same center.*

Furthermore, the rotation angle, except for circles, has some restrictions as well (see Section 5.1.1 of [84]).

Lemma 24. *Assume that $\mathcal{C}$, defined by $F(x, y) = 0$, is not a union of concentric circles, and let $\mathcal{R}_{\mathbf{c},\theta}$ be a rotation of center $\mathbf{c}$ and angle $\theta \in (0, 2\pi)$ leaving $\mathcal{C}$ invariant.*

(i) θ is a rational multiple of 2π, i.e., $\theta = \dfrac{2m\pi}{n}$, with $\gcd(m, n) = 1$.

(ii) There exists a rotation $\mathcal{R}_{\mathbf{c},\psi}$ leaving $\mathcal{C}$ invariant such that $\psi = 2\pi/N$, with $N \in \mathbb{N}$. Furthermore, any other θ is a multiple of ψ.

(iii) $N \leq 2\deg(F)$.

Proof. Let $\mathbf{p} \in \mathcal{C}$, not belonging to any circle contained in $\mathcal{C}$, and let C be the circle through $\mathbf{p}$ centered at $\mathbf{c}$. Since $\mathcal{R}_{\mathbf{c},\theta}$ leaves $\mathcal{C}$ invariant, for all $k \in \mathbb{N}$ the points in the sequence $\mathbf{p}_k = \mathcal{R}^{(k)}_{\mathbf{c},\theta}(\mathbf{p})$ are points of both $\mathcal{C}$ and C. Since $\mathcal{C}$ does not contain C, by Bezout's Theorem the number of intersection points of $\mathcal{C}$ and C is finite. Thus, there exists a minimum $n \in \mathbb{N}$, $n > 1$, such that $\mathcal{R}^{(n)}_{\mathbf{c},\theta}(\mathbf{p}) = \mathbf{p}$, so $n\theta = 2\pi m$ for some $m \in \mathbb{N}$; since we can always assume $\gcd(m, n) = 1$, (i) follows. In order to see (ii), for a given $\theta = 2m\pi/n$ and since $\gcd(m, n) = 1$, by Bezout's Identity [135, §3.1] there exist $a, b \in \mathbb{Z}$ such that $am + bn = 1$. Now if θ is a rotation angle, $a\theta + 2\pi\ell$ is also a rotation angle for every $\ell \in \mathbb{Z}$; taking $\ell = b$, we get

$$a\theta + 2\pi b = a\frac{2m\pi}{n} + 2\pi\frac{n}{n}b = \frac{am+bn}{n}2\pi = \frac{1}{n}2\pi = \frac{2\pi}{n},$$

so $\psi = 2\pi/n$ is also a rotation angle. Next by (i) there are finitely many values for θ, that we can write as $\theta_1, \ldots, \theta_\ell$. Let $\psi_1, \ldots, \psi_\ell$ be the corresponding values for ψ, and let $\psi_i = 2\pi/N_i$. Taking N as the least common multiple of the N_i, (ii) follows. Finally, let us see (iii). Since for $\psi = 2\pi/N$ we get $\mathcal{R}^{(N)}_{\mathbf{c},\psi}(\mathbf{p}) = \mathbf{p}$, and since all the iterations $\mathcal{R}^{(k)}_{\mathbf{c},\psi}(\mathbf{p})$ belong to the same circle C, the intersection of C and the curve $\mathcal{C}$ contains at least N points. However, by Bezout's Theorem the number of intersection points of C and $\mathcal{C}$, counting multiplicities, is at most $2\deg(F)$. Thus, $N \leq 2\deg(F)$. □

The following corollary follows from Lemma 23 and Lemma 24.

Corollary 25. *Assume that $\mathcal{C}$ is not a union of concentric circles. Then $\mathcal{C}$ has finitely many rotational symmetries, all of them sharing the same center. Furthermore, the subgroup of rotations leaving $\mathcal{C}$ invariant is cyclic and corresponds either to the trivial group or to the group of rotational symmetries of a k-gon (i.e., a regular polygon with k edges).*

Finally, we have the following theorem, which characterizes the planar algebraic curves with finitely many symmetries.

Theorem 26. *The group* $\text{Sym}(\mathcal{C})$ *is finite if and only if $\mathcal{C}$ is not a union of parallel lines or a union of concentric circles.*

Proof. ($\Rightarrow$) Unions of parallel lines are invariant under all translations by vectors parallel to the lines, and unions of circles are invariant under any rotation centered at their common center. Thus, the curves which are unions of parallel lines or unions of concentric lines have infinitely many symmetries. ($\Leftarrow$) By Lemma 21, the only possible symmetries of $\mathcal{C}$ are reflections and rotations. By Corollary 25, the number of rotations is finite. So let us see that the number of reflections is also finite. By Corollary 22 any two reflections of $\mathcal{C}$ have concurrent symmetry axes. Additionally, the composition of two reflections with concurrent axes is a rotation $\mathcal{R}_{\mathbf{c},\theta}$ about the intersection point of both axes (see Lemma 17), where the angle θ is twice the angle between the axes. Since by Corollary 25 the angle θ has just finitely many possible values, the number of axes of symmetry must also be finite. □

Theorem 26 has the following consequence on the value of the constant $\lambda \in \mathbb{R}$ appearing in Lemma 20.

Corollary 27. *Let $f(\boldsymbol{x}) = A\boldsymbol{x} + \boldsymbol{b}$ be a symmetry of $\mathcal{C} \subset \mathbb{R}^2$, not a union of parallel lines or a union of concentric circles, defined by $F(x, y)$ where F is square-free. Then*

$$F \circ f = \pm F. \tag{2.2}$$

Proof. From Lemma 20 we have $F \circ f = \lambda F$ for some $\lambda \in \mathbb{R}$. Since $\mathrm{Sym}(\mathcal{C})$ is a group, $f^{(k)} \in \mathrm{Sym}(\mathcal{C})$ for all $k \in \mathbb{N}$. Furthermore, since by Theorem 26 the group $\mathrm{Sym}(\mathcal{C})$ is finite, there must be a smallest k_0 such that $f^{(k_0)} = \mathrm{id}_{\mathbb{R}^2}$. Iterating $F \circ f = \lambda F$ we get $F \circ f^k = \lambda^k F$. Taking $k = k_0$ we have that $\lambda^{k_0} = 1$. Since $\lambda \in \mathbb{R}$, we deduce that $\lambda = \pm 1$. □

As we will see in the following section, quite often we will find the symmetries of $\mathcal{C}$ from a finite set containing them. In this sense, Eq. (2.2) will be useful in order to quickly test whether or not a given mapping f is a symmetry of $\mathcal{C}$.

Now we are ready to address how to effectively compute $\mathrm{Sym}(\mathcal{C})$, assuming that the curve is given by means of its implicit equation. Since we want to work with algebraic curves $\mathcal{C}$ such that $\mathrm{Sym}(\mathcal{C})$ is finite, we will be excluding from our study the curves in Corollary 27, i.e., unions of parallel lines and unions of concentric circles.

2.2 COMPUTATION OF THE SYMMETRIES

The fact that $\mathrm{Sym}(\mathcal{C})$ is finite and the relationship in Eq. (2.2) provides a brute-force method to find $\mathrm{Sym}(\mathcal{C})$: indeed, Eq. (2.2) yields a polynomial system in the degree of the curve for the entries of $A, \boldsymbol{b}$ in $f(\boldsymbol{x}) = A\boldsymbol{x} + \boldsymbol{b}$, whose (finitely many) solutions correspond to the symmetries of $\mathcal{C}$. However, here we will see a more efficient method to find $\mathrm{Sym}(\mathcal{C})$, developed in [13], that takes advantage of certain properties of *harmonic polynomials.*

Let us recall first the *Laplacian* operator. The Laplacian of a function $u(x_1, \ldots, x_n)$ admitting first and second derivatives is

$$\Delta u = \frac{\partial^2 u}{\partial x_1^2} + \cdots + \frac{\partial^2 u}{\partial x_n^2}.$$

A function $u(x_1, \ldots, x_n)$ is *harmonic* if Δu is identically zero. In our case, taking the implicit equation $F(x, y)$ of $\mathcal{C}$,

$$\Delta F = \frac{\partial^2 F}{\partial x^2} + \frac{\partial^2 F}{\partial y^2}.$$

If $F(x, y)$ is not harmonic, $\Delta F(x, y)$ is another polynomial whose degree is $\deg(F) - 2$. The reason why the Laplacian is useful in our context is the following property (see Chapter 1 of [24]), that ensures that the Laplacian commutes with orthogonal transformations.

Theorem 28. *Let $u(x_1, \ldots, x_n) \in C^2(\Omega)$, $\Omega \subset \mathbb{R}^n$, and let T be an orthogonal transformation on Ω. Then*

$$\Delta(u \circ T) = (\Delta u) \circ T. \tag{2.3}$$

Since every $f \in \mathrm{Sym}(\mathcal{C})$ is, in particular, an orthogonal transformation, Eq. (2.3) holds for the symmetries of $\mathcal{C}$. Now every symmetry f satisfies Eq. (2.2). Taking Laplacians in Eq. (2.2) and using Theorem 28, we have

$$F \circ f = \pm F \Rightarrow \Delta(F \circ f) = \pm \Delta F \Rightarrow (\Delta F) \circ f = \pm \Delta F.$$

Thus, every symmetry of $\mathcal{C}$, i.e., of the curve defined by $F(x, y)$, is also a symmetry of the curve defined by $\Delta F(x, y)$. Writing $\mathrm{Sym}(F)$ for the set of symmetries of $\mathcal{C}$, something that because of Eq. (2.2) makes perfect sense, we have $\mathrm{Sym}(F) \subset \mathrm{Sym}(\Delta F)$, where the gain is that the degree of ΔF is equal to the degree of F minus two. There is no reason why we cannot continue this process. If we do, we get a *Laplacian chain*

$$F \to \Delta F \to \Delta^2 F \to \cdots \Delta^k F \to \cdots \tag{2.4}$$

and in turn a chain of symmetry groups

$$\mathrm{Sym}(F) \subset \mathrm{Sym}(\Delta F) \subset \mathrm{Sym}(\Delta^2 F) \subset \cdots \mathrm{Sym}(\Delta^k F) \subset \cdots \tag{2.5}$$

Since every time we are dropping the degree of the polynomial by two, unless the chain in Eq. (2.4) hits a (possibly zero) constant, the chain ends up with a line, which requires $F(x,y)$ to have odd degree, or a conic, which requires $F(x,y)$ to have even degree. In this last case, whenever the conic is not either a circle or a product of parallel lines, we can compute the symmetry group of the conic, which is finite, by Linear Algebra methods and extract the symmetries of $\mathcal{C}$ from there by using Eq. (2.2). So let us analyze the remaining cases. We will see that these cases can be reduced to be first one, and in fact to the case of the Laplacian chain abruptly stopping at zero, meaning that in the previous step we have run into a harmonic polynomial.

2.2.1 Harmonic polynomials

Let $H(x,y)$ denote a harmonic polynomial of degree $\deg(H) > 1$, and let $\mathcal{H}$ represent the algebraic curve defined by $H(x,y)$. Let us see first that $\mathrm{Sym}(\mathcal{H})$ is finite. In order to do this, it is useful to recall the formula for the Laplacian in polar coordinates: if $F = F(r,\theta)$, then

$$\Delta F = \frac{1}{r}\frac{\partial}{\partial r}\left(r\frac{\partial f}{\partial r}\right) + \frac{1}{r^2}\frac{\partial^2 F}{\partial \theta^2}. \tag{2.6}$$

Lemma 29. *If* $H(x,y)$ *is harmonic and* $\deg(H) > 1$*, then* $\mathrm{Sym}(\mathcal{H})$ *is finite.*

Proof. From Theorem 26, we need to see that $\mathcal{H}$ is not either a union of parallel lines or a union of concentric circles. Assume that $\mathcal{H}$ is a union of parallel lines. Since the Laplacian operator commutes with orthogonal transformations, we can rotate the lines so that the lines are parallel to the y-axis, i.e., we can assume that $H(x,y) = H(x) = \sum_{i=0}^{n} a_i x^i$. However, in this case $\Delta H = 0$ implies that $H(x,y) = a_0 + a_1 x$, i.e., $\deg(H) = 1$, which cannot happen by hypothesis. Now assume that $\mathcal{H}$ is a union of concentric circles. Using polar coordinates, then H transforms into $\sum_{i=}^{n} b_{2i} r^{2i}$, and from Eq. (2.6) we see that the Laplacian vanishes if and only if H is either linear in r or a constant, which cannot happen because $\deg(H) > 1$. □

In order to compute $\mathrm{Sym}(\mathcal{H})$, we will make use of a certain planar *vector field*,

$$\boldsymbol{V}(x,y) = (H_x(x,y), -H_y(x,y)), \tag{2.7}$$

where H_x, H_y are the partial derivatives of $\boldsymbol{V}(x,y)$ with respect to x, y. A planar vector field can simply thought of as a mapping between planar points and planar vectors, so that at each point of the plane we attach a planar vector (if you like, with its origin linked to the point, as forces acting on precise points of a planar structure). Here we will consider *algebraic* planar vector fields, meaning that the components of the mapping are bivariate polynomials.

Definition 30. *We say that* $\mathbf{p}$ *is a* singular point *of a vector field* $\boldsymbol{V}(x,y)$ *iff* $\boldsymbol{V}(\mathbf{p}) = \mathbf{0}$. *The set of singular points of* $\boldsymbol{V}(x,y)$ *is denoted by* $\mathcal{S}_{\boldsymbol{V}}$.

An example of a vector field like Eq. (2.7) together with its singular points is provided in Example 1. Next by Eq. (2.2) we know that $H \circ f = \pm H$ for all $f \in \mathrm{Sym}(\mathcal{H})$. Using the Chain Rule, we deduce that

$$(\nabla H) \circ f = \pm J_f^{-1} \cdot \nabla H,$$

where ∇H is the gradient of H, and J_f is the Jacobian of f. Thus $(\nabla H) \circ f$ vanishes if and only if ∇H vanishes too. However, the points where ∇H vanishes are exactly the points of $\mathcal{S}_{\boldsymbol{V}}$. Thus, any symmetry f is also a bijection of the set of singular points, and therefore a symmetry of $\mathcal{S}_{\boldsymbol{V}}$, i.e., $\mathrm{Sym}(\mathcal{H}) \subset \mathrm{Sym}(\mathcal{S}_{\boldsymbol{V}})$. If

$\mathrm{Sym}(\mathcal{S}_{\mathbf{V}})$ is finite, then we can compute $\mathrm{Sym}(\mathcal{H})$ by first computing $\mathrm{Sym}(\mathcal{S}_{\mathbf{V}})$, and then testing whether each element of $\mathrm{Sym}(\mathcal{S}_{\mathbf{V}})$ belongs to $\mathrm{Sym}(\mathcal{H})$ with the help of Eq. (2.2).

In order to do this, we will make use of some tools from Complex Analysis and *holomorphic* functions; see for instance [35] for further reference on this topic. Thus, we first identify $\mathbb{R}^2$ with $\mathbb{C}$ via $(x, y) \leftrightarrow x + \mathbf{i}y$, where $\mathbf{i}^2 = -1$. Then, the vector field in Eq. (2.7) can be replaced by a complex function

$$g(x, y) = H_x(x, y) - \mathbf{i}H_y(x, y).$$

Since H is harmonic, $H_{xx} + H_{yy} = 0$; thus, denoting the partial derivatives with respect to x, y as ∂_x, ∂_y, we get that $\partial_x(H_x) = \partial_y(-H_y)$. Additionally, by Schwartz's rule, $\partial_y(H_x) = -\partial_x(-H_y)$. This means that the components of $g(x, y)$ satisfy *Cauchy-Riemann's conditions*, so g is *holomorphic*. Thus, using the change of variables

$$x = \frac{z + \overline{z}}{2},\ y = \frac{z - \overline{z}}{2\mathbf{i}}, \tag{2.8}$$

we can write $g(x, y)$ as a complex function $g(z, \overline{z})$, which in fact, because $g(x, y)$ is holomorphic, must be only a function of z. Thus, we can write

$$g(z) = \sum_{j=0}^{m} \alpha_j z^j = \alpha_m \Pi_{j=1}^{m}(z - \xi_j). \tag{2.9}$$

We say that $g(z)$ is *associated* with $H(x, y)$. Furthermore, $\mathcal{S}_{\mathbf{V}}$ coincides, via Eq. (2.8), with the set of roots of $g(z)$.

Proposition 31. *If $H(x, y)$ is harmonic and* $\deg(H) > 1$, *then: (i) $\mathcal{S}_{\mathbf{V}}$ is nonempty, and entirely consisting of finitely many real points; (ii) $H(x, y)$ is square-free.*

Proof. Since $\deg(H) > 1$ we deduce that $g(z)$ is not a constant and therefore it has finitely many roots. Additionally, the points of $\mathcal{S}_{\mathbf{V}}$ can be identified with the roots of $g(z)$; but every root of $g(z)$ corresponds to a real point by Eq. (2.8), so (i) holds. For (ii), notice that $H(x, y)$ is non-square-free if and only if $\gcd(H, H_x, H_y)$ is non-constant (see Theorem 4.4.2 in [135, §4.4]), in which case H_x, H_y share a non-trivial factor. However, in that case $\mathcal{S}_{\mathbf{V}}$ contains infinitely many points, namely all the points of the algebraic curve defined by the common factor of H_x, H_y. But this contradicts (i), so $H(x, y)$ must be square-free. □

Since $g(z)$ cannot be a constant by Proposition (31), unless $g(z) = \alpha_m (z - z_0)^m$ the set $\mathcal{S}_{\mathbf{V}}$ consists of at least two points, and therefore $\mathrm{Sym}(\mathcal{S}_{\mathbf{V}})$ is finite. Furthermore, if we can find $\mathrm{Sym}(\mathcal{S}_{\mathbf{V}})$, by statement (ii) of Proposition (31) we can check whether or not an element of $\mathrm{Sym}(\mathcal{S}_{\mathbf{V}})$ belongs to $\mathrm{Sym}(\mathcal{H})$ using, once again, Eq. (2.2). The case $g(z) = \alpha_m (z - z_0)^m$ can happen, though. The next lemma shows that this occurs for certain special forms of H, that can be guessed.

Lemma 32. *Let $g(z)$ be a nonzero polynomial, and let $H(x, y)$ be a harmonic polynomial such that $g(z)$ is associated with $H(x, y)$. Then there exists a primitive $G(z)$ of $g(z)$ such that $H(x, y) =$* $\mathrm{Re}(G(z))$.

Proof. Let $\tilde{G}(z)$ be a primitive of $g(z)$, and let $\tilde{G}(x + \mathbf{i}y) = U(x, y) + \mathbf{i}V(x, y)$. Since $\tilde{G}(z)$ is holomorphic, we have $\tilde{G}'(z) = U_x + \mathbf{i}V_x = U_x - \mathbf{i}U_y$. On the other hand, by definition $g(z) = H_x - \mathbf{i}H_y$. Therefore we have $U_x = H_x$, $U_y = H_y$ and thus $H = U + c$, where c is a real constant. Hence $H(x, y) = \mathrm{Re}(\tilde{G}(z) + c)$, where $G(z) = \tilde{G}(z) + c$ is also a primitive of $g(z)$. □

Thus, using Lemma 32 with $g(z) = \alpha_m (z - z_0)^m$ provides the form of the corresponding $H(x, y)$ where this special case occurs.

We focus now on the symmetries of $\mathcal{S}_{\mathbf{V}}$; the following lemma, in fact a particular case of Theorem 179 in Appendix A, is useful.

Lemma 33. *Let $\{\mathbf{p}_1, \ldots, \mathbf{p}_N\} \subset \mathbb{R}^2$ be a finite set of points. Then the center of any rotational (all the axes of reflectional) symmetry is (pass through) the* barycenter

$$\frac{1}{N} \sum_{i=1}^{N} \mathbf{p}_i. \tag{2.10}$$

Taking into account that the elements of $\mathcal{S}_V$ are the (complex) roots of the polynomial $g(z)$ in Eq. (2.9), the barycenter can be efficiently found using Cardano-Vieta formulae (see also Appendix A).

Theorem 34. *Let $H(x, y)$ be a harmonic polynomial defining a curve $\mathcal{H}$, and let $g(z)$ be the expression in Eq. (2.9). If $\mathcal{R}_{\mathbf{c},\theta}$ is a rotation leaving $\mathcal{H}$ invariant, then $\mathbf{c}$ corresponds to the complex number*

$$\mathbf{c} = -\frac{\alpha_{m-1}}{m\alpha_m}. \tag{2.11}$$

If $\mathcal{S}_V$ consists of at least two points, i.e., if $g(z) \neq \alpha_m(z - z_0)^m$, the rotation angle θ must be $\theta = \frac{2\pi}{k}$, with k a divisor of the number of elements in $\mathcal{S}_V$. Another criterion for θ will be given in Chapter 4, see Proposition 79 there, where vector fields will be considered again.

Furthermore, Cardano-Vieta formulae can also be used to find a closed expression for the reflectional axes, which is given in the following theorem (see Theorem 19 of [13] for a proof of this theorem). Here we assume that we have performed a translation so that the point $\mathbf{c}$ in Eq. (2.11) is moved to the origin; in that case, the polynomial defining the curve $\mathcal{H}$ remains harmonic, and the second highest coefficient of the new associated polynomial $g(z)$ vanishes.

Theorem 35. *Let $H(x, y)$ be a harmonic polynomial such that the associated polynomial $g(z)$ in Eq. (2.9) satisfies that $\alpha_{m-1} = 0$. Then any possible axis of symmetry of $H(x, y) = 0$ passes through the origin. Let ℓ be the index such that $\alpha_\ell \neq 0$ but $\alpha_i = 0$ for each $i < \ell$. If $\ell < m$, then the angles θ between the tentative axes of symmetry of $H(x, y) = 0$ and the x-axis satisfy the relationship*

$$(m - \ell)\theta \equiv \arg\left((-1)^{m-\ell}\frac{\alpha_\ell}{\alpha_m}\right) \mod 2\pi. \tag{2.12}$$

Otherwise, $g(z) = \alpha_m z^m$ and the tentative axes of symmetry of $H(x, y) = 0$ possess the directions

$$\theta = \frac{1}{2\pi}(\kappa\pi - 2\arg(\alpha_m)) \tag{2.13}$$

for $\kappa = 1, \ldots, 2m + 2$.

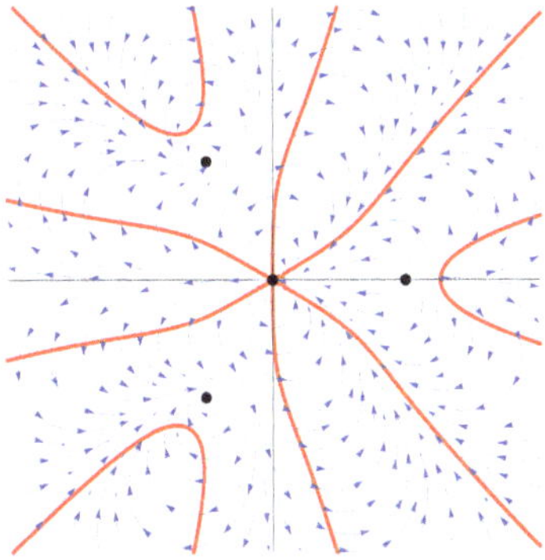

FIGURE 2.2 Curve $H(x, y) = 0$ in Example 1 (red), Vector field $(H_x, -H_y)$ (blue) and its singularities (thick black dots).

Example 1. *Consider the harmonic sextic*

$$H(x, y) = x^6 - 15x^4y^2 - 2x^3 + 15x^2y^4 + 6xy^2 - y^6. \tag{2.14}$$

The complex function associated with the vector field then has the form $g(x, y) = g_1(x, y) - \mathbf{i}g_2(x, y)$, *with*

$$\begin{array}{rcl} g_1(x, y) & = & 6x^5 - 60x^3y^2 - 6x^2 + 30xy^4 + 6y^2 \\ g_2(x, y) & = & -30x^4y + 60x^2y^3 + 12xy - 6y^5 \end{array},$$

which after the substitution in Eq. (2.8) gives rise to

$$g(z) = 6z^5 - 6z^2. \tag{2.15}$$

Therefore, $g(z) = z^2(6z^3-6)$ *and* $\mathcal{S}_{\mathbf{V}}$ *consists of four points (one of them, the origin, with multiplicity two), shown in Fig. 2.2. Formula (2.11) reveals that the only possible center of rotational symmetry of the curve* $H(x, y) = 0$ *is the point* $\mathbf{c} = 0 + 0\mathbf{i} \simeq (0, 0)$. *The only non-trivial rotation angle for the curve (see also Fig. 2.2) is* $\theta = 2\pi/3$; *therefore the curve* $H(x, y) = 0$ *is invariant under the rotations around the origin, by angles* $2k\pi/3$ *with* $k \in \mathbb{Z}$. *As for reflections, since the coefficient of* z^4 *in* $g(z)$ *is already zero, we can directly apply Eq. (2.13), which yields*

$$3\theta \equiv \arg(1) \bmod 2\pi,$$

so the tentative reflection axes are the lines through the origin forming angles of $2\pi/3$, 0 and $-2\pi/3$ with the x-axes. All these lines are, in fact, reflection axes of the curve $H(x, y) = 0$.

2.2.2 Reduction to the harmonic case

Let us go back to Eqs. (2.4) and (2.5), and let $\ell \in \mathbb{N}$ such that $\Delta^\ell F$ is not constant, but $\Delta^{\ell+1} F$ is a constant c. If $\deg(\Delta^\ell F) > 2$ and $c = 0$, then $\Delta^\ell F$ is harmonic. Thus we can compute $\mathrm{Sym}(\Delta^\ell F)$ with the results in the previous subsection, and use later that $\mathrm{Sym}(F) \subset \mathrm{Sym}(\Delta^\ell F)$ to find $\mathrm{Sym}(F)$. Let us analyze now the remaining cases, which can be essentially reduced to the case of harmonic polynomials.

$\boxed{\deg(\Delta^\ell F) = 1}$

In this case $\Delta^\ell F(x, y) = c_0 + c_1 x + c_2 y$, and therefore $\mathrm{Sym}(\Delta^\ell F)$ has infinitely many symmetries. So we need some strategy to reduce the number of tentative symmetries to finitely many. In order to present this strategy, for simplicity we will assume that $\Delta^\ell F(x, y) = x$; we will provide the algorithm for the general case later. Next let $k \leq \ell$ be the smallest natural number such that $\Delta^k F(x, y) = p(x)$ only depends on x, so that $\widehat{F} = \Delta^{k-1} F$ does depend on both x, y. Also, let $q \geq 0$ such that $k + q = \ell$ and let $\tilde{p}(x) \in \mathbb{R}[x]$ satisfy that $\Delta\tilde{p} = \tilde{p}''(x) = p$, $\tilde{p}'(0) = \tilde{p}(0) = 0$. Furthermore, observe that $p(x)$ is a polynomial in x of odd degree $2q + 1$, and therefore $\tilde{p}(x)$ is a polynomial of degree $2q + 3$. Now calling $\xi(x, y) = x$, the polynomials $\widehat{F}, p, \xi$, on one hand, and the polynomials $\tilde{p}, p, \xi$, on the other hand, belong to two Laplacian chains:

$$\widehat{F} \to p \to \overbrace{\cdots}^{q} \to \xi \tag{2.16}$$

$$\tilde{p} \to p \to \overbrace{\cdots}^{q} \to \xi \tag{2.17}$$

Recall also that $\widehat{F}, p, \xi$ belong to the Laplacian chain starting with F. Thus, by Theorem 28, if f is a symmetry of F, then

$\widehat{F} \circ f = \alpha \widehat{F}$, $p \circ f = \alpha p$, and $\xi \circ f = \alpha \xi$, where $\alpha = \pm 1$ has the same sign for $\widehat{F}, p, \xi$. Let us consider then the polynomial $\tilde{F} = \widehat{F} - \tilde{p}$, which is harmonic: indeed,

$$\Delta \tilde{F} = \Delta \widehat{F} - \Delta \tilde{p} = p - p = 0.$$

Then we have the following result.

Proposition 36. $\mathrm{Sym}(F) \subset \mathrm{Sym}(\tilde{F})$.

Proof. If suffices to see that $\mathrm{Sym}(\widehat{F}) \subset \mathrm{Sym}(\tilde{F})$. Thus, let $f \in \mathrm{Sym}(\widehat{F})$. Since $\widehat{F}, p$ belong to the same Laplacian chain, $f \in \mathrm{Sym}(p)$ too; furthermore, if $\widehat{F} \circ f = \alpha \widehat{F}$, with $\alpha = \pm 1$, then $p \circ f = \alpha p$ as well. Let us write $f = f(x, y) = (f_1(x, y), f_2(x, y))$. Taking into account the general structure of planar symmetries (see Eq. (1.10) in Section 1.2.1), because $p \circ f = \alpha p$ and $p = p(x)$ we have $f_1(x, y) = \kappa x$, where $\kappa = \pm 1$; and since p has odd degree and $p \circ f = \alpha p$, $\kappa = \alpha$. Since $\tilde{p}$ is univariate and has also odd degree, $p \circ f = \alpha p$ if and only if $\tilde{p} \circ f = \alpha \tilde{p}$. Thus,

$$\tilde{F} \circ f = \widehat{F} \circ f - \tilde{p} \circ f = \alpha \widehat{F} - \alpha \tilde{p} = \alpha(\widehat{F} - \tilde{p}) = \alpha \tilde{F},$$

and we conclude that $f \in \mathrm{Sym}(\tilde{F})$. □

Thus, we have reduced the problem to computing $\mathrm{Sym}(\tilde{F})$, where $\tilde{F}$ is harmonic. Notice that $\tilde{F}$ is not constant because $\widehat{F}$ depends on y, and p does not. It may, however, have degree one. In that case, it suffices to compute the symmetries of the product of $\xi(x, y) = \Delta^{\ell} F$ and $\tilde{F}$, which is a conic, not a circle and not a product of parallel lines. The whole algorithm (for a linear bivariate polynomial $\xi(x, y)$, not necessarily $\xi(x, y) := x$) is provided in Algorithm 1; notice that Step 4 of Algorithm 1 amounts to solving a linear system of equations in the coefficients of $\tilde{p}(x, y)$.

Example 2. *Let $F(x, y) = \frac{1}{3}x^3 - y + 1$. Here $\triangle F = x$. Since $\tilde{p}(x) = \frac{1}{3}x^3$, we get $\tilde{F} = \widehat{F} - \tilde{p} = F - \tilde{p} = -y + 1$. Therefore, the possible symmetries of F are the simultaneous symmetries of $x = 0$ and $y = 1$. However, one can check that none of these are, in fact, symmetries of $F(x, y) = 0$, so we certify that $F(x, y) = 0$ has no Euclidean symmetries.*

Algorithm 1 Case $\deg(\Delta^\ell F) = 1$

Require: A square-free polynomial $F(x, y)$ defining a real algebraic curve $\mathcal{C}$, not a union of parallel lines, such that the case $\deg(\Delta^\ell F) = 1$ occurs.

Ensure: A polynomial $\tilde{F}$ such that $\mathrm{Sym}(F) \subset \mathrm{Sym}(\tilde{F})$, which is either harmonic, or a conic, where $\mathrm{Sym}(\tilde{F})$ is finite.

1: let $\xi(x, y) = \Delta^\ell F$ [notice that $\xi(x, y)$ is linear in x, y.]
2: let k be the smallest number such that $\Delta^k F$ is the composition of a univariate function, and $\xi(x, y)$
3: let $\widehat{F} := \Delta^{k-1} F$
4: let $\tilde{p}(x, y)$ be the polynomial of degree $\deg(\Delta^k F)+2$ such that: (i) $\Delta\tilde{p} = \Delta^k F$; (ii) $\tilde{p}(0, 0) = 0$; (iii) $\tilde{p}_x(0, 0) = \tilde{p}_y(0, 0) = 0$
5: let $\tilde{F} := \widehat{F} - \tilde{p}$ [notice that $\tilde{F}$ is harmonic.]
6: **if** $\deg(\tilde{F}) = 1$ **then**
7: $\quad\tilde{F} := \tilde{F} \cdot \xi(x, y)$. [this polynomial is not necessarily harmonic, but defines a conic with finitely many symmetries, containing $\mathrm{Sym}(F)$].
8: **end if**

$\deg(\Delta^\ell F) = 2$

In this case $\Delta^\ell F(x, y) = 0$ is a conic section. If this conic is not a pair of parallel lines or a circle, it has finitely many symmetries which can be computed by elementary Linear Algebra methods. If the conic is a pair of parallel lines $(c_0+c_1x+c_2y)(d_0+c_1x+c_2y)$, we can reduce the problem to the previous subcase calling $\xi(x, y) = c_1x + c_2y$.

So we are left with the case when $\Delta^\ell F(x, y) = 0$ is a circle. Thus $\Delta^\ell F(x, y)$ can be written as a univariate function of $x^2 + y^2$, i.e., of r^2 where r, θ are the polar coordinates. Then we can proceed as in the subcase $\deg(\Delta^\ell F) = 1$: we look for $k \leq \ell$ the smallest natural number such that $\Delta^k F - p(r^2)$ only depends on r^2, we define $\widehat{F} = \Delta^{k-1} F$, and we find $\tilde{p} = \tilde{p}(r^2)$ satisfying that $\Delta\tilde{p} = p$, where $\Delta\tilde{p}$ is computed in in polar coordinates using Eq. (2.6). Finally, $\tilde{F} = \widehat{F} - \tilde{p}$ does the trick; one can check Section 3.2 of [13] for details.

$\boxed{\deg(\Delta^\ell F) > 2}$

The case $\Delta^{\ell+1}F = 0$ is already covered, so we just need to consider the case $\Delta^{\ell+1}F = c \neq 0$. Let us call $G(x,y) = \Delta^\ell F(x,y)$, and let us write

$$G(x,y) = G_m(x,y) + G_{m-1}(x,y) + \cdots + G_0(x,y), \qquad (2.18)$$

where $G_i(x,y)$ is a homogeneous polynomial of degree $i \in \{0,1,\ldots,m\}$, and $m = \deg(\Delta^\ell F)$. Since the Laplacian of a homogeneous polynomial of degree k is a homogeneous polynomial of degree $k-2$ (maybe identically zero), the fact that $\Delta^{\ell+1}F = c$ implies that the Laplacian of $G_m(x,y)$, the form of highest degree, must vanish. Thus, $G_m(x,y)$ is a harmonic polynomial. Furthermore, if $f(\boldsymbol{x}) = A\boldsymbol{x} + \boldsymbol{b}$ satisfies that $f \in \mathrm{Sym}(G)$, using that $G \circ f = \pm G$ and taking Eq. (2.18) into account, we get

$$G_m(A\boldsymbol{x}) = \pm G_m(\boldsymbol{x}),$$

so $\tilde{f}(\boldsymbol{x}) = A\boldsymbol{x}$ is a symmetry of $G_m(x,y) = 0$. Since G_m is harmonic, we can find the tentative matrices A by reducing the problem to the harmonic case, and then find $\boldsymbol{b}$, if any, for each matrix A by directly imposing $F \circ f = \pm f$, with $f(\boldsymbol{x}) = A\boldsymbol{x} + \boldsymbol{b}$. This amounts to solving a bivariate polynomial system in the coordinates of $\boldsymbol{b}$.

2.3 COMPLEX REPRESENTATION

In this section, we will consider another implicit representation for planar algebraic curves, which is an alternative to the implicit equation employed so far. Before introducing this new representation, let us observe that if $F(x,y)$ has degree d, in which case we have

$$F(x,y) = \sum_{i=0}^{d} \sum_{j=0}^{d-i} a_{i,j} x^i y^j,$$

then we can write $F(x, y)$ in matrix form as

$$F(x,y) = \begin{bmatrix} 1 & x & x^2 & \cdots & x^d \end{bmatrix} \mathbf{A}_F \begin{bmatrix} 1 \\ y \\ y^2 \\ \vdots \\ y^d \end{bmatrix} \tag{2.19}$$

with

$$\mathbf{A}_F = \begin{bmatrix} a_{0,0} & a_{0,1} & \cdots & a_{0,d-1} & a_{0,d} \\ a_{1,0} & a_{1,1} & \cdots & a_{1,d-1} & 0 \\ \vdots & \vdots & \ddots & \vdots & \vdots \\ a_{d-1,0} & a_{d-1,1} & \cdots & 0 & 0 \\ a_{d,0} & 0 & \cdots & 0 & 0 \end{bmatrix}. \tag{2.20}$$

Additionally, since the implicit equation is unique up to multiplication by constants, we can identify the curve $\mathcal{C}$ defined by $F(x, y) = 0$ with an element of the projective space $\mathbb{P}^{N-1}(\mathbb{R})$ whose coordinates are the $a_{i,j}$, where

$$N = (d+1) + d + (d-1) + \cdots + 1 = \frac{(d+2)(d+1)}{2} = \binom{d+2}{2}. \tag{2.21}$$

Now if we perform in $F(x, y)$ the substitution in Eq. (2.8) we get a polynomial which, abusing notation, we will write as $F(z, \overline{z})$, with

$$F(z, \overline{z}) = \sum_{i=0}^{d} \sum_{j=0}^{d-i} b_{i,j} z^i \overline{z}^j, \tag{2.22}$$

where $b_{i,j} = \overline{b}_{j,i} \in \mathbb{C}$ for all i, j, and therefore $b_{i,i} = \overline{b}_{i,i} \in \mathbb{R}$. We will call Eq. (2.22) the *complex representation* of the curve $\mathcal{C}$. This representation was introduced in [124, 125], and used also in [83, 84, 127], to solve problems related to curve recognition and symmetry computation. Eq. (2.22) also admits a matrix expression, namely

$$F(z,\overline{z}) = \begin{bmatrix}1 & z & z^2 & \cdots & z^d\end{bmatrix} \mathbf{B}_F \begin{bmatrix}1\\ \overline{z}\\ \overline{z}^2\\ \vdots\\ \overline{z}^d\end{bmatrix}, \tag{2.23}$$

with

$$\mathbf{B}_F = \begin{bmatrix} b_{0,0} & b_{0,1} & \cdots & b_{0,d-1} & b_{0,d}\\ b_{1,0} & b_{1,1} & \cdots & b_{1,d-1} & 0\\ \vdots & \vdots & \ddots & \vdots & \\ b_{d-1,0} & b_{d-1,1} & \cdots & 0 & 0\\ b_{d,0} & 0 & \cdots & 0 & 0 \end{bmatrix}. \tag{2.24}$$

Unlike the matrix $\mathbf{A}_F$ in Eq. (2.19), the matrix $\mathbf{B}_F$ in Eq. (2.23) is Hermitian. Furthermore, the relationship between the entries of $\mathbf{A}_F, \mathbf{B}_F$ is [83]

$$b_{k,\ell} = \frac{1}{2^{k+\ell}} \sum_{i=0}^{k+\ell} \sum_{j=0}^{i} \binom{k+\ell-i}{\ell-j} \binom{i}{j} \mathbf{i}^{2j-i} a_{k+\ell-i,i}.$$

In particular, we can also identify $\mathcal{C}$ with an element of the projective space $\mathbb{P}^{N-1}(\mathbb{C})$, with N as in Eq. (2.21).

The reason why this representation is useful here is the fact, observed in [83, 124, 125, 127] and thoroughly analyzed in [30], that the matrix $\mathbf{B}_F$ encodes all the necessary information about rotational symmetries of $\mathcal{C}$. This matrix can also be used to analyze reflectional symmetries, which is discussed to a certain extent in [84], but here we will focus on rotations following the ideas and notation of [30].

Let us consider first rotations around the origin. A rotation around the origin by an angle θ amounts to changing

$$z \to e^{\mathrm{i}\theta} z,\ \overline{z} \to e^{-\mathrm{i}\theta}\overline{z}.$$

Thus, the complex representation of any curve invariant under such a rotation with $\theta = 2\pi/m$ satisfies the condition, which follows from Eq. (2.2),

$$F(z, \overline{z}) = \pm F(e^{2i\pi/m} z, e^{-2i\pi/m} \overline{z}).$$

In turn, this leads to two families of linear conditions on the $b_{i,j}$,

$$b_{i,j} = \pm e^{2i\pi(i-j)/m} b_{i,j}, \ \overline{b}_{i,j} = \pm e^{-2i\pi(i-j)/m} \overline{b}_{i,j}. \tag{2.25}$$

The equations with + imply $b_{i,j} = 0$ for $(i - j) \neq km$, $k \in \mathbb{Z}$, for an arbitrary m, while the equations with $-$ only have a solution for even m, and imply $b_{i,j} \neq (2k + 1)m/2$. Hence, $\mathbf{B}_F$ is a sparse matrix. Following [30], we will refer to the curves where + (resp. $-$) holds as s_+ (resp. s_-) curves. Let us illustrate this with an example.

Example 3. *Consider the algebraic curves of degree $d = 6$ which are rotationally symmetric about the origin, by an angle $\theta = \pi/2$, i.e., $m = 4$. From the previous discussion, for s_+ curves the only terms in the complex representation of $F(z, \overline{z})$ are the terms $z^i \overline{z}^j$ where $0 \leq i + j \leq 6$ and $i - j$ is a multiple of 4, i.e., where the pair (i, j) is*

$$(0, 0), \ (0, 4), \ (1, 1), \ (1, 5), \ (2, 2), \ (3, 3), \ (4, 0), \ (5, 1).$$

In turn, the corresponding terms are

$$1, \ \overline{z}^4, \ z\overline{z}, \ z\overline{z}^5, \ z^2\overline{z}^2, \ z^3\overline{z}^3, \ z^4, \ z^5\overline{z}.$$

For s_- curves, the nonzero terms of $F(z, \overline{z})$ correspond to pairs (i, j) where $i - j$ is an odd multiple of 2,

$$(0, 2), \ (0, 6), \ (1, 3), \ (2, 0), \ (2, 4), \ (3, 1), \ (4, 2), \ (6, 0),$$

correspondinq to

$$\overline{z}^2, \ \overline{z}^6, \ z\overline{z}^3, \ z^2, \ z^3\overline{z}, \ z^4\overline{z}^2, \ z^6.$$

We visualize the structure of the matrices $\mathbf{B}_F$ in each of these cases in Fig. 2.3; the nonzero elements are shown with a red bullet.

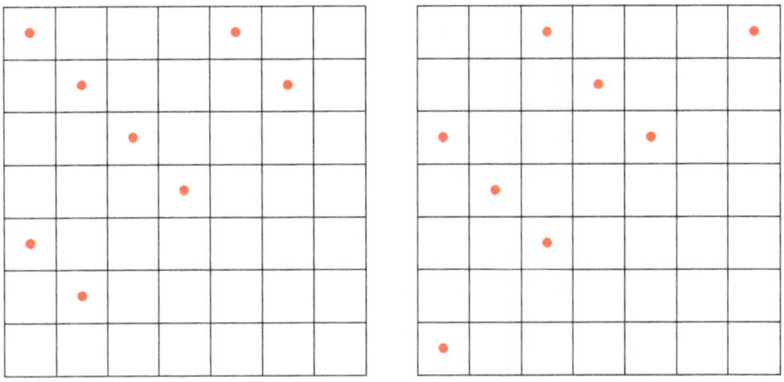

FIGURE 2.3 Matrices $\mathbf{B}_F$ in Example 3: s_+ curves (left); s_- curves (right)

The structure of the matrices is symmetric and sparse, as expected, and we observe that the conditions for the possibly nonzero $b_{i,j}$ induce a band-like structure for $\mathbf{B}_F$, where the bands are parallel to the main diagonal.

In more detail, for s_+ curves the only terms which may be nonzero correspond to

$$\left\{(z\overline{z})^i(z^{km}+\overline{z}^{km})\right\}_{i=0,k-0}^{\lfloor\frac{d}{2}\rfloor,\lfloor\frac{d-2}{m}\rfloor} \tag{2.26}$$

and for s_- curves, the only possibly nonzero terms correspond to

$$\left\{(z\overline{z})^i(z^{\frac{(2k+1)}{2}m}+\overline{z}^{\frac{(2k+1)}{2}m})\right\}_{i=0,k=0}^{\lfloor\frac{d}{2}\rfloor,\lfloor\frac{2d-4i-m}{2m}\rfloor}. \tag{2.27}$$

From here, it follows that the matrices $\mathbf{B}_F$ have a band-like structure, as shown in Example 3; for instance, for s_+ curves, the bands start at the entries corresponding to the terms $z^i\overline{z}^j$ where (i,j) is $(0,0)$, $(0,m)$, $(m,0)$, $(0,2m)$, $(2m,0)$, etc. with $km \leq d$, and stop when the secondary diagonal of $\mathbf{B}_F$ is reached.

Next plugging Eq. (2.26) into Eq. (2.22) yields

$$F(z,\overline{z}) = \sum_{i=0}^{\lfloor\frac{d}{2}\rfloor}\sum_{k=0}^{\lfloor\frac{d-2i}{2m}\rfloor}\left((\alpha_{ik}+\mathbf{i}\beta_{ik})(z\overline{z})^i z^{km}+(\alpha_{ik}-\mathbf{i}\beta_{ik})(z\overline{z})^i\overline{z}^{km}\right)$$

for s_+ curves, which can be written as

$$\begin{aligned}F(z,\overline{z}) = \textstyle\sum_{i=0}^{\lfloor\frac{d}{2}\rfloor}\sum_{k=0}^{\lfloor\frac{d-2i}{2m}\rfloor}\Big[\alpha_{ik}\left((z\overline{z})^i\left(z^{km}+\overline{z}^{km}\right)\right)+\\ +\mathbf{i}\beta_{ik}\left((z\overline{z})^i\left(z^{km}-\overline{z}^{km}\right)\right)\Big],\end{aligned}$$

and analogously for s_- curves. Thus, for s_+ curves with the rotational symmetries of an m-gon about the origin one can use the basis

$$\mathcal{B}^+_{d,m} = \left\{ (z\overline{z})^i \left(z^{km} + \overline{z}^{km} \right), \mathbf{i}(z\overline{z})^i \left(z^{km} - \overline{z}^{km} \right) \right\}_{i=0,k=0}^{\lfloor \frac{d}{2} \rfloor, \lfloor \frac{d-2i}{m} \rfloor}, \quad (2.28)$$

and for s_- curves with the rotational symmetries of an m-gon about the origin one can use the basis

$$\begin{aligned} \mathcal{B}^-_{d,m} &= \left\{ (z\overline{z})^i \left(z^{\frac{(2k+1)}{2}m} + \overline{z}^{\frac{(2k+1)}{2}m} \right), \right. \\ & \left. \mathbf{i}(z\overline{z})^i \left(z^{\frac{(2k+1)}{2}m} - \overline{z}^{\frac{(2k+1)}{2}m} \right) \right\}_{i=0,k=0}^{\lfloor \frac{d}{2} \rfloor, \lfloor \frac{2d-4i-m}{2m} \rfloor}. \end{aligned} \quad (2.29)$$

By using these bases, one just needs real coefficients α_{ik}, β_{ik}. This means that even though we are using complex representations, we can identify the curves with the rotational symmetry of an m-gon about the origin with elements of $\mathbb{P}^{N-1}(\mathbb{R})$ with N as in Eq. (2.21). Moreover, we observe that the s_+ (resp. s_-) curves correspond to a projective subspace of $\mathbb{P}^{N-1}(\mathbb{R})$.

Next let us move to rotations about an arbitrary center $\mathbf{c}$. In order to do this, observe that any curve $\mathcal{C}$ with the rotational symmetry of an m-gon about an arbitrary point $\mathbf{c}$ can be considered as the result of translating a curve $\mathcal{C}_0$, with the same symmetry about the origin, by the vector $-\mathbf{c}$. So let us explore the effect of a translation on such a curve $\mathcal{C}_0$, whose structure we have just described. We will denote by $\mathbf{B}_F$ and $\mathbf{B}^0_F$ the matrices associated with the complex representations of $\mathcal{C}$ and $\mathcal{C}_0$. When working with complex representations, a translation amounts to changing

$$z \to z + p, \ \overline{z} \to \overline{z} + \overline{p},$$

where here p represents a complex number that we identify with the translation vector. Now the coefficients $c_{k,\ell}$ of the complex representation of the translated curve, say $\mathcal{C}$, have the following relationship with the coefficients $b_{i,j}$ of the original curve, say $\mathcal{C}_0$ [30, 83]:

$$c_{k,\ell} = \sum_{i=0}^{d-k} \sum_{j=0}^{d-k-\ell-i} \binom{k+i}{i} \binom{\ell+j}{j} p^i \overline{p}^j b_{k+i,\ell+j}. \quad (2.30)$$

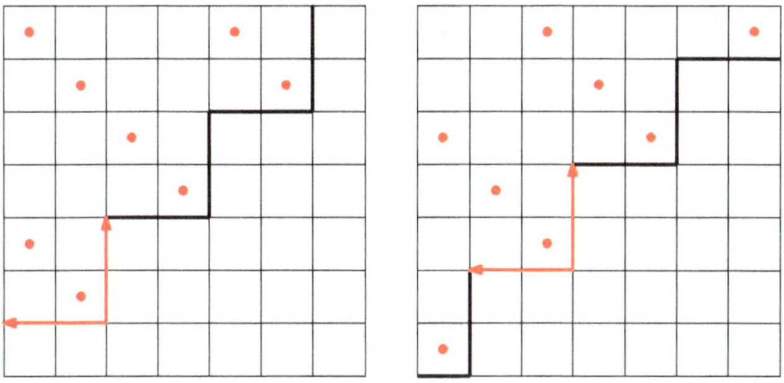

FIGURE 2.4 Effect of a translation in the matrix of a curve with the symmetry of an m-gon about the origin (I)

This means that if $c_{2,3}$ in $\mathbf{B}_F$, for instance, is nonzero, there must be nonzero elements in $\mathbf{B}_F^0$ of the type $b_{2+i,3+j}$, i.e., occupying positions $(2+i, 3+j)$ with $i \geq 0$, $j \geq 0$, generating $c_{2,3}$: in other words, there must be nonzero elements in the second row of $\mathbf{B}_F^0$, or higher, and/or in the third column of $\mathbf{B}_F^0$, or in a column situated more to the left. So when undergoing a translation, the elements in $\mathbf{B}_F^0$ "expand" up and to the left. Fig. 2.4 illustrates this, continuing Example 3: the red arrows show the directions by which the matrix $\mathbf{B}_F^0$ expands when a translation is applied. The expansion itself is shown in Fig. 2.5: while in $\mathbf{B}_F^0$ the only possibly nonzero elements are the grey ones, in $\mathbf{B}_F$ the possibly nonzero elements are the grey ones and the yellow ones. Notice that the number of "echelons" corresponding to the possibly nonzero elements in $\mathbf{B}_F^0$ and $\mathbf{B}_F$ is the same. The reason is that the positions corresponding to the "corners", i.e., the last elements in the echelons (assuming that echelons grow to the right), which are also the elements in the bands of $\mathbf{B}_F^0$ furthest from the row/column where they start, are not changed when moving from $\mathbf{B}_F^0$ to $\mathbf{B}_F$. This is, again, a consequence of how $\mathbf{B}_F^0$ expands.

Now let us see how we can find the center of rotation $\mathbf{c}$, when we are given $\mathbf{B}_F$. To do this, we consider the translation taking back $\mathcal{C}$ to $\mathcal{C}_0$, i.e., moving $\mathbf{B}_F$ to $\mathbf{B}_F^0$, which corresponds to $p = \mathbf{c}$. Since $\mathbf{B}_F^0$ has a very special structure, we are looking for a translation that creates in the resulting matrix a band-like structure. Since

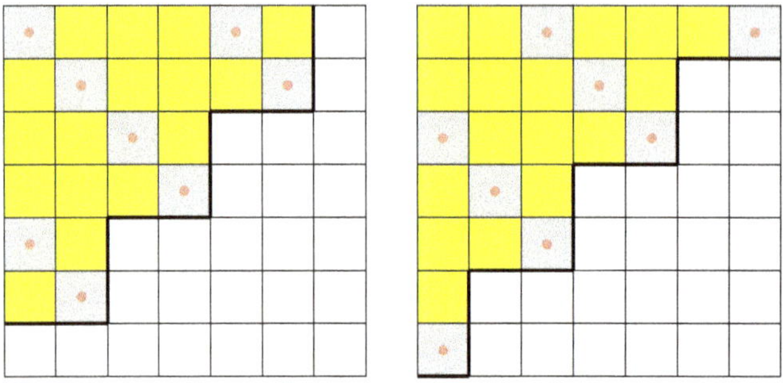

FIGURE 2.5 Effect of a translation in the matrix of a curve with the symmetry of an m-gon about the origin (II)

now we are moving from $\mathcal{C}$ to $\mathcal{C}_0$, in Eq. (2.30) the $c_{k,\ell}$ correspond to $\mathcal{C}_0$, and the $b_{i,j}$, to $\mathcal{C}$. The main result is the following.

Theorem 37. *Let $b_{k,\ell}$ be an element in the matrix $\mathbf{B}_F$, associated with an algebraic curve $\mathcal{C}$, such that for $i, j \in \{0, 1, \ldots\}$ all $b_{k+i,\ell+j}$ except at least one of $b_{k+1,\ell}$, $b_{k,\ell+1}$, are zero. If the curve $\mathcal{C}$ is invariant under a rotation about a point $\mathbf{c}$, then*

$$\mathbf{c} = \frac{(\ell+1)b_{k,\ell+1}\overline{b}_{k,\ell} - (k+1)b_{k,\ell}\overline{b}_{k+1,\ell}}{(k+1)^2|b_{k+1,\ell}|^2 - (\ell+1)^2|b_{k,\ell+1}|^2}. \tag{2.31}$$

In particular, if $b_{k,\ell+1} = 0$, then

$$\mathbf{c} = -\frac{b_{k,\ell}}{(k+1)b_{k+1,\ell}} \tag{2.32}$$

and if $b_{k+1,\ell} = 0$, then

$$\mathbf{c} = -\frac{\overline{b}_{k,\ell}}{(\ell+1)\overline{b}_{k,\ell+1}} \tag{2.33}$$

Before proving Theorem 37, let us clarify the hypotheses. The $b_{k,\ell}$ mentioned in the statement of the theorem correspond, in Example 3, to the crossed out elements of $\mathbf{B}_F$ in Fig. 2.6. These elements have something special: because of the way $\mathbf{B}_F^0$ expands when moving from $\mathcal{C}_0$ to $\mathcal{C}$, only one element of $\mathbf{B}_F^0$ generates them. Now when moving back from $\mathcal{C}$ to $\mathcal{C}_0$, i.e., when passing from $\mathbf{B}_F$ to

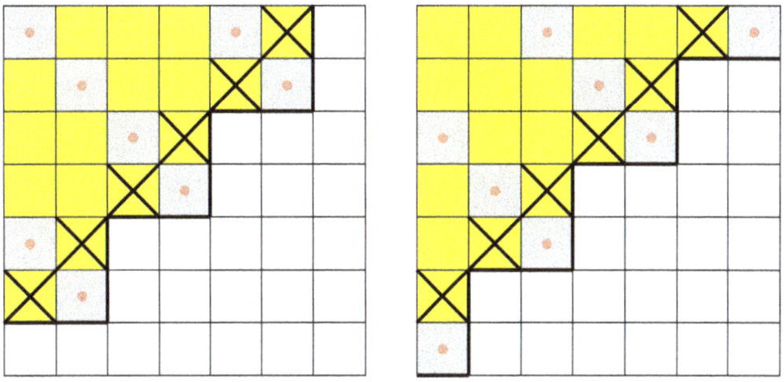

FIGURE 2.6 Illustrating Theorem 37

$\mathbf{B}_F^0$, the matrix $\mathbf{B}_F$ expands also up and to the left, so there is just one element (a corner) of $\mathbf{B}_F$ that generates such an element of $\mathbf{B}_F^0$, too. Furthermore, this element in $\mathbf{B}_F^0$ must be a zero because of the band-like structure of $\mathbf{B}_F^0$. Since the translation taking $\mathcal{C}$ to $\mathcal{C}_0$ satisfies that $p = \mathbf{c}$, we need to use Eq. (2.30), with $c_{k,\ell} = 0$, $p = \mathbf{c}$, plus the fact that all the $b_{k+i,\ell+j}$ for $i, j \geq 1$, $i + j \geq 2$, are zero.

Proof. (of Theorem 37) Plugging $c_{k,\ell} = 0$, $p = \mathbf{c}$, $b_{k+i,\ell+j} = 0$ for $i, j \geq 1$, $i + j \geq 2$ in Eq. (2.30) yields

$$0 = b_{k,\ell} + (k+1)b_{k+1,\ell}\mathbf{c} + (\ell+1)b_{k,\ell+1}\overline{\mathbf{c}}.$$

Conjugating this equation we get a linear system in $\mathbf{c}, \overline{\mathbf{c}}$, whose solution yields the desired result. Eqs. (2.32) and (2.33) are just particular instances of the general formula. □

Once that $\mathbf{c}$ is found, it remains to compute the value of m for the rotation angle $\theta = 2\pi/m$; however, the combinatorial information to find m is contained in $\mathbf{B}_F^0$ and somehow preserved in $\mathbf{B}_F$ via the fact that the echelons do not change in $\mathbf{B}_F^0$ and $\mathbf{B}_F$. This can be exploited to find m; see [30] for details. Observe, additionally, that by plugging $z \to z - \mathbf{c}$, $\overline{z} \to \overline{z} - \overline{\mathbf{c}}$ in Eqs. (2.28) and (2.29), we get bases $\mathcal{B}_{d,m}^{\pm}(\mathbf{c})$ for the projective subspaces of curves with the symmetry of an m-gon about the point $\mathbf{c}$.

But [30] does something more, which deserves commenting. The last part of [30] is devoted to *approximate* symmetries, meaning the following: assume that we are given a curve $\mathcal{C}$ defined by a polynomial $F(x, y) = 0$ whose coefficients are known only up to a certain precision, e.g. with floating point coefficients, so that $\mathcal{C}$ is not rotationally symmetric, but is "very close" to another, unknown, curve $\tilde{\mathcal{C}}$ (probably not unique) which *is* rotationally symmetric. The problem is to detect whether we are in this situation, and find the approximate center of rotation $\tilde{\mathbf{c}}$ and rotation angle $\tilde{\theta} = 2\pi/m$ of $\mathcal{C}$ as the *exact* center of rotation and rotation angle of $\tilde{\mathcal{C}}$.

The underlying idea is that $F(x, y) = \tilde{F}(x, y) + \epsilon h(x, y)$, with $\|h\|_\infty \leq 1$ and $0 < \epsilon \ll 1$, where $\tilde{F}(x, y)$ defines $\tilde{\mathcal{C}}$ and $h(x, y)$ defines the perturbation undergone by $\tilde{F}$. Thus, $\mathbf{B}_F = \mathbf{B}_{\tilde{F}} + \mathbf{B}_\epsilon$, where $\mathbf{B}_\epsilon$ corresponds to the perturbation. First, a criterion, based on how perturbations are transmitted to the associated matrix of the complex representation, is given to recognize some "fake" nonzero entries of $\mathbf{B}_F$ to be actual zeros, and correct $\mathbf{B}_F$ accordingly. Then, using Eq. (2.31), the approximate center of rotation $\tilde{\mathbf{c}}$ and m are computed from the (corrected) matrix $\mathbf{B}_F$.

Finally, after finding $\tilde{\mathbf{c}}, m$, one computes a suitable $\tilde{\mathcal{C}}$ with the symmetry of an m-gon about $\tilde{\mathbf{c}}$ by projecting, as it is usual in Linear Algebra, the curve $\mathcal{C}$, identified with a point of $\mathbb{P}^{N-1}(\mathbb{R})$, onto the subspaces corresponding to the curves with the symmetry of an m-gon about the point $\mathbf{c}$, whose bases are $\mathcal{B}^{\pm}_{d,m}(\mathbf{c})$.

2.4 REALIZATION OF SYMMETRIES

One can wonder whether, given a subgroup $G \subset \mathbb{E}(2)$, where $\mathbb{E}(2)$ denotes the group of Euclidean planar symmetries, one can construct algebraic curves $\mathcal{C}$ such that $\forall f \in G$, f is a symmetry of $\mathcal{C}$. An example of this, and an affirmative answer in a particular case, was already given in the previous section: Eqs. (2.28) and (2.29) provide the bases $\mathcal{B}^{\pm}_{m,d}$, for the complex representation of all curves of degree d with the rotational symmetry of the m-gon about the origin: since the rotational symmetries of the m-gon form a subgroup G_m of $\mathbb{E}(2)$, certainly we can *realize* G_m using $\mathcal{B}^{\pm}_{m,d}$, for appropriate degrees.

Here we will present an approach to the problem which takes us back to *invariant theory*, an active field of research in Mathematics in the last part of the XIXth and the first part of the XXth centuries, with contributions by mathematicians like Emmy Noether and David Hilbert, among many others. We will follow the ideas and notation in Chapter 7 of [45]; in turn, the authors of [45] acknowledge [121] as their main reference for this topic. The approach in [45] works for polynomials in any number of variables; here we will restrict ourselves to just two variables, in accordance with the rest of the chapter, but the results are completely generalizable to an arbitrary number of variables.

In Chapter 7 of [45], there are two requirements for G: the first one is that G is a *matrix group*, i.e., a subgroup of the *linear group* GL, in our case of GL(2), which is the set of invertible 2×2 matrices; in particular, the elements of G fix the origin. The second one is that G is finite; we will represent the cardinal of G, i.e., the number of elements of G, by $|G|$. An example of such G is, again, the group G_m of the rotational symmetries of the m-gon: indeed, G_m is generated by the matrix

$$A = \begin{bmatrix} \cos(2\pi/m) & -\sin(2\pi/m) \\ \sin(2\pi/m) & \cos(2\pi/m) \end{bmatrix}$$

and its elements are

$$G = \{A, A^2, \ldots, A^{m-1}, A^m = I\},$$

so $|G_m| = m$. Next the following definition is needed.

Definition 38. *Let* $\boldsymbol{x} = (x, y)$*, let* $F(x, y) \in \mathbb{R}[x, y]$*, and let* $G \subset$ GL(2) *be a finite group. We say that* $F(x, y)$ *is* invariant *under* G *if*

$$F(A\boldsymbol{x}) = F(\boldsymbol{x}) \tag{2.34}$$

for all $A \in G$*.*

Although Definition (38) does not require the matrices $A \in G$ to be orthogonal, certainly Eq. (2.34) reminds us of Eq. (2.2). Thus, if G is a subgroup of $\mathbb{E}(2)$ and $F(x, y)$ is invariant under G, then all the transformations in G define symmetries of the algebraic curve $F(x, y) = 0$. Notice that all these symmetries have the form $f(\boldsymbol{x}) = A\boldsymbol{x}$, i.e., the origin is a fixed point. A first observation is that if F_1, F_2 are invariant under G, so are $F_1 + F_2$ and $F_1 \cdot F_2$, which means that the set of polynomials invariant under G is a ring, that we will denote by $\mathbb{R}[x, y]^G$ and will refer to as the *ring of invariants of* G.

So we want to describe $\mathbb{R}[x, y]^G$. To do this, given $F_1, \ldots, F_\ell \in \mathbb{R}[x, y]^G$ we denote by $\mathbb{R}[F_1, \ldots, F_\ell]$ the set of all polynomial expressions in $F_1, \ldots, F_\ell$ with coefficients in $\mathbb{R}$. Thus, an element of $\mathbb{R}[F_1, \ldots, F_\ell]$ looks like

$$M(F_1, \ldots, F_\ell),$$

with M a multivariate polynomial. Since $F_i = F_i(x, y)$ for $i = 1, \ldots, \ell$, the elements of $\mathbb{R}[F_1, \ldots, F_\ell]$ are bivariate polynomials. And since the F_i are invariant by G, so is any other element of $\mathbb{R}[F_1, \ldots, F_\ell]$. Therefore, if we have some invariants for G, we can have many more. This brings up the question of how to generate all of them, which requires a new ingredient, called the *Reynolds operator.*

Definition 39. *Let $G \subset \mathrm{GL}(2)$ be a finite group. The* Reynolds operator *of G is the map $R_G : \mathbb{R}[x, y] \to \mathbb{R}[x, y]$ defined by*

$$R_G(F(\boldsymbol{x})) = \frac{1}{|G|} \sum_{A \in G} F(A\boldsymbol{x}). \tag{2.35}$$

Thus, the Reynolds operator "averages", as suggested in [45], the effect of G on a polynomial. One can easily prove that if $F \in \mathbb{R}[x, y]^G$, then $R_G(F) \in \mathbb{R}[x, y]^G$, and in fact that $R_G(F) = F$. It turns out that the Reynolds operator is the key to generate $\mathbb{R}[x, y]^G$. The reason is the following theorem, due to Emmy Noether; see Chapter 7 of [45] for a proof. Here we represent the monomials of total degree β by $\mathbf{x}^\beta$; thus, in $\mathbb{R}^2$ we have $\mathbf{x}^\beta = x^i y^j$ with $i + j = \beta$.

Theorem 40. *Let $G \subset \mathrm{GL}(2)$ be a finite group, and let $\mathbf{x}^{\beta_1}, \ldots, \mathbf{x}^{\beta_\ell}$ be all monomials of total degree $\beta \leq |G|$. Then*

$$\mathbb{R}[x,y]^G = \mathbb{R}[R(\mathbf{x}^{\beta_1}), \ldots, R(\mathbf{x}^{\beta_\ell})].$$

In particular, $\mathbb{R}[x,y]^G$ is finitely generated in the sense that it consists of the polynomials that can be written as polynomial expressions of the (finitely many) $R(\mathbf{x}^\beta)$ with $\beta \leq |G|$.

Let us illustrate all these ideas with an example.

Example 4. *Let G_4 be the group of rotational symmetries of the 4-gon, i.e., the square. This is a cyclic group generated by*

$$A = \begin{bmatrix} 0 & -1 \\ 1 & 0 \end{bmatrix}$$

with $|G_4| = 4$. The elements of the group are

$$\begin{bmatrix} 0 & -1 \\ 1 & 0 \end{bmatrix}, \begin{bmatrix} -1 & 0 \\ 0 & -1 \end{bmatrix}, \begin{bmatrix} 0 & 1 \\ -1 & 0 \end{bmatrix}, \begin{bmatrix} 1 & 0 \\ 0 & 1 \end{bmatrix}. \tag{2.36}$$

From Definition 39, given $F(x,y)$ we have

$$R_{G_4}(F(x,y)) = \frac{1}{4}[F(-y,x) + F(-x,-y) + F(y,-x) + F(x,y)].$$

Moreover, according to Theorem 40, all the algebraic curves admitting the symmetries of G_4 are defined by polynomial expressions of the $R_{G_4}(\mathbf{x}^\beta)$s, where the $\mathbf{x}^\beta$ are the monomials in x, y of degree $\leq |G_4| = 4$,

$$x,\ y,\ x^2,\ xy,\ y^2,\ x^3,\ x^2y,\ xy^2,\ y^3,\ x^4,\ x^3y,\ x^2y^2,\ xy^3, y^4. \tag{2.37}$$

For instance,

$$R_{G_4}(x^3y) = \frac{1}{4}[(-y)^3x + (-x)^3(-y) + y^3(-x) + x^3y] = \frac{1}{2}(x^3y - xy^3).$$

If we proceed this way for all the monomials in Eq. (2.37) (see

Chapter 7 of [45]), leaving multiplying constants aside, we will see only four polynomials arising in the list of $R_{G_4}(\mathbf{x}^\beta)$*s, namely*

$$x^2+y^2,\ x^2y^2,\ x^3y-xy^3,\ x^4+y^4.$$

In fact, we can remove x^4+y^4 *since it can be generated from* x^2+y^2 *and* x^2y^2*,*

$$x^4+y^4=(x^2+y^2)^2-2x^2y^2.$$

This means that the elements of $\mathbb{R}[x,y]^{G_4}$ *are the polynomials* $F(x,y)$ *corresponding to polynomial expressions of* $B_1(x,y)=x^2+y^2$*,* $B_2(x,y)=x^2y^2$ *and* $B_3(x,y)=x^3y-xy^3$*. The curve* $F(x,y)=0$ *exhibits the symmetries of* G_4*. For instance, an option to generate a polynomial of that kind of degree 12 could be*

$$F(x,y)=B_3^3(x,y)+B_2(x,y)^2-B_1(x,y).$$

Notice that Theorem 40 allows us to generate as many $\mathcal{C}$s as we want such that $G\subset \mathrm{Sym}(\mathcal{C})$; however, the inclusion may be strict. For instance, in the above example it is clear that the curve $x^2+y^2=0$ has more symmetries than just those of G_4; in fact, $x^2+y^2=0$ is invariant under any rotation about the origin, by any angle θ, and is also invariant under the reflections on both coordinate axes.

To end this short section, let us observe that if $|G|$ is big, the process can be cumbersome, and the decision on which invariants are algebraically independent, not trivial. However, there are methods to optimize the process; we refer to [45] and [121] for details.

2.5 FINAL REMARKS AND OPEN PROBLEMS

Let us finish this chapter with some remarks on the generalization of the problems treated here to other groups of interest, and an enumeration of open problems. The method in Section 2.2 to compute the symmetries of an implicit algebraic curve can be extended to find the similarities between two different algebraic curves, if any; this is also done in [13]. The problem of checking whether or not two given implicit algebraic curves are affinely equivalent is considered in [29], where the problem is reduced to solving a polynomial system derived from the forms of high degree of the curves, and later in [6], taking advantage of the properties of the *Hessian* operator and using polynomial factoring, instead of polynomial system solving. Projective equivalences between two implicit planar algebraic curves are harder, though the case of elliptic and hyperelliptic curves is considered in [8], taking advantage of the rich theory behind these curves.

Additionally, we can mention other related questions related to this chapter that up to our knowledge have not been solved so far, and that we pose here as pending problems.

Open Problem 1. *Extend the ideas in [30] to the case of reflectional symmetries.* Although some aspects of this question are addressed in [84], in light of [30] it seems that more can be said, and done, about this.

Open Problem 2. *Find an efficient, exact method to check whether or not two given, implicit algebraic curves are projectively equivalent.* So far, this problem has only been considered for elliptic and hyperelliptic curves.

Open Problem 3. *Similarity detection for curves with parameters.* A reason why similarity detection is useful is to be able to identify, or "label", a given algebraic curve as, say, a lemniscate, a cardioid, a deltoid, an n-leaved rose, etc. This is relevant in Pattern Recognition, or in Dynamic Geometry, where the user might be interested in identifying a curve obtained as a geometric locus, or by whatever process. However, quite often notable curves correspond to *families* depending on parameters. For instance, Cassini's ovals are the algebraic curves defined by

$$(x^2 + y^2)^2 - 2a^2(x^2 - y^2) + a^4 - b^4 = 0,$$

for $a, b \in \mathbb{R}$. Thus, to identify an algebraic curve as a Cassini's oval implies to find a, b such that the given curve is, up to a similarity (because the curve to be identified might be in a different position/size), like the curve above. This requires to adapt and implement an algorithm to detect similarity to the case when one of the curves depends on parameters.

Open Problem 4. *Approximate symmetries/similarities/affine equivalences/projective equivalences.* First, let us mention that the notion of "approximate", in the context of Computational Algebraic Geometry, is used in a special way that we should clarify. In the wider context of Applied Mathematics, and in particular in applied fields like Pattern Recognition or Computer Vision, "approximate" can be identified with "reasonable". In these fields, in many cases algebraic objects are just an approximation of the object that one is analyzing, and therefore the notion of symmetry, affine or projective equivalence is, in a way, "fuzzy". Good examples of this are papers like [124, 126, 134], but there are many more (see also the Bibliography of [2]). There is certainly nothing wrong about this, which in applications can lead to good results.

However, in Computational Algebraic Geometry the notion of "approximate" often means the following [114]: let us say that we are given an algebraic object $\mathcal{E}$ (say, an algebraic curve), and that we want to investigate a certain property $\mathcal{P}$ (say, having rotational symmetry). Especially if $\mathcal{E}$ is known only up to a certain precision ϵ, for instance because it is represented by mathematical expressions with floating point coefficients, it is possible that $\mathcal{E}$, due to the inaccuracy of its representation, does not have the property $\mathcal{P}$. However, there might exist another object $\tilde{\mathcal{E}}$, "close" to $\mathcal{E}$, which can be considered as an ϵ-deformation of $\mathcal{E}$, and that *has* the property $\mathcal{P}$. One might then consider the problem of finding such an $\tilde{\mathcal{E}}$ (probably not unique, if it exists), and computing from $\tilde{\mathcal{E}}$ whatever information could be derived regarding $\mathcal{P}$ (e.g. center of rotation, rotation angle). If such $\tilde{\mathcal{E}}$ exists, then we say that $\mathcal{E}$ has, "approximately", property $\mathcal{P}$, and for practical purposes regarding $\mathcal{P}$, we can consider $\tilde{\mathcal{E}}$ as a more reliable version of $\mathcal{E}$.

A good example of this approach, as already observed in Section 2.3, is [30]; other good examples, although not related to the computation of symmetries or equivalences but to other problems on planar algebraic curves, are [99, 101]. For many algebraic or geometric problems regarding algebraic varieties, one can consider its approximate, in this sense, version. The computation, in this sense, of approximate symmetries, affine equivalences and projective equivalences of implicit algebraic curves is, with the exception of [30], completely open.

Open Problem 5. *Möbius equivalences.* The equivalence of planar algebraic curves under transformations belonging to the Möbius group are completely open. A first, unsolved, problem here is to recognize whether a given curve is the *inverse* of another curve, i.e., the image of the latter curve under an inversion (to be found); a particular case is the recognition of curves which are their own inverses, called *anallagmatic* curves. For instance, Cassini's ovals belong to this class.

CHAPTER 3

Planar algebraic curves (II): rational curves

Some planar algebraic curves, certainly not all of them, admit a *rational parametrization*, i.e., a parametrization whose components are quotients of polynomial functions in one variable. This is an advantage that makes these curves especially suitable in fields like Computer Aided Geometric Design. In this chapter we continue, and finish, our study of planar algebraic curves examining the computation of the symmetries of this type of curves, assuming that they are defined by means of a rational parametrization.

3.1 GENERALITIES

We begin with the following definition. Here $\mathbb{K}$ is a field; for us, either $\mathbb{K} = \mathbb{R}$ or $\mathbb{K} = \mathbb{C}$.

Definition 41. *Let $\mathcal{C} \subset \mathbb{K}^2$ be a planar algebraic curve. We say that $\mathcal{C}$ is* rational *if $\mathcal{C}$ admits a* parametrization $\boldsymbol{p}(t) = (p_1(t), p_2(t))$ *where for $i = 1, 2$, $p_i(t)$ is a rational function, i.e., a quotient of polynomial functions. The functions $p_1(t), p_2(t)$ are the* components *of $\boldsymbol{p}(t)$.*

If $\mathcal{C}$ is a rational, planar algebraic curve with implicit equation $F(x, y)$, then $F(\boldsymbol{p}(t)) = F(p_1(t), p_2(t))$ is identically zero. We will mostly work with $\mathcal{C} \subset \mathbb{R}^2$ and real parametrizations $\boldsymbol{p}(t)$, so that the coefficients in the numerators and denominators of the $p_i(t)$

DOI: 10.1201/9781003648710-3

are real numbers; however, complex (space) rational curves will come up in Chapter 9. Regardless of whether or not $\boldsymbol{p}(t)$ is real, the *parameter* t lives, in general, in $\mathbb{C}$. If $\boldsymbol{p}(t)$ is real, though, the *real part* of the curve $\mathcal{C}$ defined by $\boldsymbol{p}(t)$ will be obtained for $t \in \mathbb{R}$, with possibly the exception of finitely many points. Moreover, we say that $\boldsymbol{p}(t)$ is *regular* at $t = t_0$ if $\boldsymbol{p}'(t_0)$ exists and is nonzero; if $\boldsymbol{p}'(t_0) = 0$ we say that $\boldsymbol{p}(t)$ is *singular* at $t = t_0$, or that $P = \boldsymbol{p}(t_0)$ is a singular point of $\boldsymbol{p}(t)$. For real parametrizations $\boldsymbol{p}(t)$, *isolated real singularities*, i.e., real singular points $P \in \mathcal{C}$ not containing any other real point of $\mathcal{C}$ other than P in a small neighborhood of P, provide an example of real points of $\mathcal{C}$ generated by complex, non-real, values of the parameter.

Certainly not every planar algebraic curve is rational. A curve is rational if and only if its *genus* is zero, a condition that can be algorithmically checked; for the notion of genus and many other questions regarding planar rational curves one can check the monograph [117]. If the genus of $\mathcal{C}$ is zero and its implicit equation $F(x, y)$ is known, then there are efficient algorithms to compute a rational parametrization $\boldsymbol{p}(t)$ of $\mathcal{C}$; see again [117]. Furthermore, if $\mathcal{C}$ has a rational parametrization $\boldsymbol{p}(t)$, then it has infinitely many, since composing $\boldsymbol{p}$ with any rational function $\xi : \mathbb{C} \to \mathbb{C}$ provides another parametrization $(\boldsymbol{p} \circ \xi)(t)$ of $\mathcal{C}$, which is said to be a *reparametrization* of $\boldsymbol{p}(t)$. Let us also mention that every rational curve (in whatever dimension) is irreducible.

Parametrizations allow us to visualize $\mathcal{C}$ easily, and are very convenient, for instance, for intersecting the curve with other implicit curves. But some parametrizations are better than others. For instance, $\boldsymbol{p}(t) = (t, t^2)$ and $\boldsymbol{q}(t) = (t^2, t^4)$ both parametrize the parabola $y = x^2$. However, the second parametrization is not injective because $\boldsymbol{q}(t_0) = \boldsymbol{q}(-t_0)$ for all $t_0 \in \mathbb{C}$, which means that we are covering the curve twice. The algebraic consequences are that $\boldsymbol{q}^{-1}(x, y)$ does not exist, and that the degrees that we observe in the components of $\boldsymbol{q}(t)$ are unnecessarily big. Because of this, we introduce the following definition.

Definition 42. *Let $\mathcal{C} \subset \mathbb{K}^2$ be a rational, planar algebraic curve, and let $\mathbf{p}(t)$ be a parametrization of $\mathcal{C}$. We say that $\mathbf{p}(t)$ is* proper

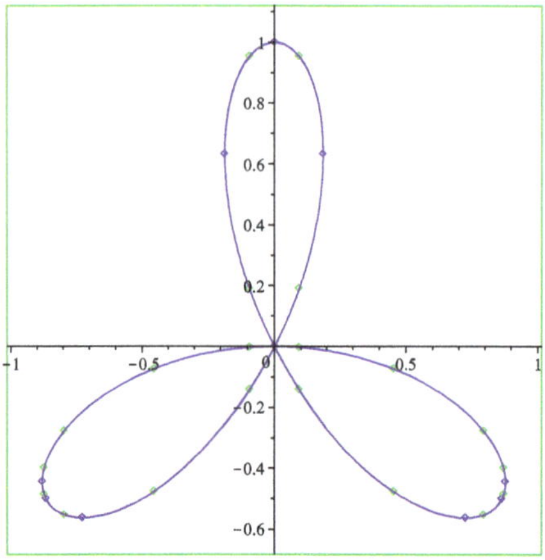

FIGURE 3.1 Three-leaved rose

if $\boldsymbol{p}(t)$ *is generically injective, i.e., if for almost all points* $P \in \mathcal{C}$ *(in other words, except perhaps for finitely many points) there is just one* $t_0 \in \mathbb{K}$ *satisfying that* $P = \boldsymbol{p}(t_0)$*. If* $\boldsymbol{p}(t)$ *is not proper, then we say that* $\boldsymbol{p}(t)$ *is* improper.

Thus, $\boldsymbol{p}(t) = (t, t^2)$ is a proper parametrization of the parabola $y = x^2$, while $\boldsymbol{q}(t) = (t^2, t^4)$ is improper. Observe that the fact that the parameter t lives in $\mathbb{C}$ matters, e.g., $\boldsymbol{p}(t) = (t, t^3)$ is proper but $\boldsymbol{q}(t) = (t^3, t^6)$ is not: the reason is that in general there are three complex values of t, although just one of them is real, generating each real point of the curve parametrized by $\boldsymbol{q}(t)$. Also, notice that even if $\boldsymbol{p}(t)$ is proper, there might be some points of the curve generated by more than one value of t. For instance, one can check that

$$\boldsymbol{p}(t) = \left(\frac{t(t^2-3)}{t^4+2t^2+1}, \frac{t^2(t^2-3)}{t^4+2t^2+1}\right) \tag{3.1}$$

is a proper parametrization of the *three-leaved rose* (see Fig. 3.1). However, this curve has a self-intersection at $(0,0)$, so $\boldsymbol{p}(t)$ crosses the origin three times, when $t = 0$, $t = \sqrt{3}$, and $t = -\sqrt{3}$, i.e., $\boldsymbol{p}(0) = \boldsymbol{p}(\sqrt{3}) = \boldsymbol{p}(-\sqrt{3})$.

Proper parametrizations satisfy that the equation $\boldsymbol{p}(t) = \boldsymbol{p}(s)$, for a generic value of t, only has one solution, namely $t = s$. This is the idea behind the following lemma; see [117] for a proof. For $i = 1, 2$ we denote by $G_i(t, s)$ the numerator of $p_i(t) - p_i(s)$.

Lemma 43. *Let $\boldsymbol{p}(t)$ be a rational parametrization of a planar algebraic curve $\mathcal{C} \subset \mathbb{K}^2$, and let $\mathcal{G}(t, s) = \gcd(G_1(t, s), G_2(t, s))$. Then $\boldsymbol{p}(t)$ is proper if and only if $\mathcal{G}(t, s) = t - s$.*

When $\mathcal{G}(t, s)$ is not equal to $t - s$, $\boldsymbol{p}(t)$ has undergone an unnecessary reparametrization of bigger degree, as it happened with the parabola and (t^2, t^4). In that case, $\mathcal{G}(t, s)$ somehow shows the reparametrization that we need to remove to make the parametrization proper: for instance, if $\boldsymbol{p}(t) = (t^2, t^4)$, then $\mathcal{G}(t, s) = t^2 - s^2$, which relates to the fact that $\xi(t) = t^2$ is the reparametrization undergone by the parabola. This idea can be made into an algorithm [117], so we can always check whether or not a parametrization is proper, and reparametrize it properly in case it is not. Furthermore, if $\boldsymbol{p}$ is proper, then $\boldsymbol{p}^{-1}$ exists, and the following result holds.

Lemma 44. *Let $\boldsymbol{p}(t)$ be a rational parametrization of a planar algebraic curve $\mathcal{C} \subset \mathbb{K}^2$. Then $\boldsymbol{p}^{-1}$ exists and is a rational function.*

Lemma 44 essentially follows by applying the Euclidean Algorithm to find the gcd of $H_1(x, t) = \text{num}(x - p_1(t))$, $H_2(y, t) = \text{num}(y - p_2(t))$, where $\text{num}(\bullet)$ denotes the numerator of the rational function between brackets, seen as polynomials in t with coefficients in the field of rational functions of the curve defined by $\boldsymbol{p}(t)$; see [117] for a detailed proof, which is constructive. Thus, when $\boldsymbol{p}$ is proper, $\boldsymbol{p}^{-1}$ can be algorithmically computed.

The notion of properness makes sense not only for planar parametrizations, but for real or complex parametrizations $\boldsymbol{p}(t) = (p_1(t), \ldots, p_n(t))$ in any dimension. Lemma 43 can be generalized for any dimension and the same holds for proper reparametrization algorithms, so we can always get proper parametrizations of curves, regardless of the dimension. Lemma 44 can also be generalized to arbitrary dimension.

3.2 COMPUTATION OF SYMMETRIES

Given a planar rational algebraic curve $\mathcal{C} \subset \mathbb{R}^2$, defined by a rational, real proper parametrization $\boldsymbol{p}(t)$, we aim to find its symmetries. The method that we present here is the result of adapting the ideas of [11] to the case of planar curves. Furthermore, we will exclude lines and circles (which are rational curves); thus, by Theorem 26, we can be sure that $\text{Sym}(\mathcal{C})$ is finite.

We begin with the following result, where a more general problem is considered. Before providing this result, let us recall *Möbius transformations*. A Möbius transformation $\varphi(t)$ is a mapping

$$\varphi : \mathbb{C} \to \mathbb{C}, \ \varphi(t) = \frac{at+b}{ct+d}, \ ad - bc \neq 0. \tag{3.2}$$

For us, the most important property of Möbius transformations is that they are the only birational mappings of the real and complex line, see [128]. One can go even further to prove that they are the only biholomorphic transformations of the complex line (see Remark 2 in [19]).

The following result is valid for *birational* mappings $f : \mathbb{C}^n \to \mathbb{C}^n$, i.e., for mappings in $\mathbb{C}^n$ of the type

$$f(x_1, \ldots, x_n) = (f_1(x_1, \ldots, x_n), \ldots, f_n(x_1, \ldots, x_n))$$

where the f_i are rational functions of $x_1, \ldots, x_n$, that have an also rational inverse f^{-1}. Moreover, the result is valid for whatever n, not only $n = 2$. We present it this way for future reference. Notice that planar or space symmetries are, in particular, birational mappings. Furthermore, the theorem is valid for mappings between two different curves $\mathcal{C}_1, \mathcal{C}_2$, real or complex, which certainly includes the case when $\mathcal{C}_1 = \mathcal{C}_2 = \mathcal{C}$. We state the result over $\mathbb{C}$ because this will be needed in Chapter 9; for the case treated in this chapter and in most chapters, it would suffice to state it over $\mathbb{R}$, which certainly holds too.

Theorem 45. *Let $\mathcal{C}_i \subset \mathbb{C}^n$, with $i = 1, 2$, be two rational curves defined by rational, proper parametrizations $\mathbf{p}_i(t)$. If the curves $\mathcal{C}_i$*

are related by a birational mapping f so that $f(\mathcal{C}_1) = \mathcal{C}_2$, then there exists a Möbius transformation φ such that

$$f \circ \boldsymbol{p}_1 = \boldsymbol{p}_2 \circ \varphi, \tag{3.3}$$

i.e., making commutative the following diagram:

$$\begin{array}{ccc} \mathcal{C}_1 & \xrightarrow{\ f\ } & \mathcal{C}_2 \\ \uparrow \scriptstyle{\boldsymbol{p}_1} & & \uparrow \scriptstyle{\boldsymbol{p}_2} \\ \mathbb{C} & \dashrightarrow_{\varphi} & \mathbb{C} \end{array} \tag{3.4}$$

Furthermore, if f_1, f_2, $f_1 \neq f_2$, satisfy Eq. (3.3) for Möbius transformations φ_1, φ_2, then $\varphi_1 \neq \varphi_2$.

Proof. Assume that there exists $f : \mathbb{C}^n \to \mathbb{C}^n$, birational, such that $f(\mathcal{C}_1) = \mathcal{C}_2$. Since for $i = 1, 2$ the parametrization $\boldsymbol{p}_i(t)$ is proper, $\boldsymbol{p}_1^{-1}, \boldsymbol{p}_2^{-1}$ exist and are rational. Thus, $\varphi : \mathbb{C} \to \mathbb{C}$, $\varphi = \boldsymbol{p}_2^{-1} \circ f \circ \boldsymbol{p}_1$ is well-defined and is rational, and $\varphi^{-1} = \boldsymbol{p}_1^{-1} \circ f^{-1} \circ \boldsymbol{p}_2$ is also well-defined and is also rational. Thus, φ is a birational mapping of the complex line, and therefore it must be a Möbius transformation. For the last part, since $\varphi_i = \boldsymbol{p}_2^{-1} \circ f_i \circ \boldsymbol{p}_1$, $\varphi_i = \varphi_j$ if and only if $f_i = f_j$. □

Notice that if the parametrizations $\boldsymbol{p}_1, \boldsymbol{p}_2$ and the mapping f in Theorem 45 are real, then φ must also be real by Eq. (3.3). We have the following corollary for the case of symmetries.

Corollary 46. *Let $\mathcal{C} \subset \mathbb{C}^n$ be a rational curve defined by a rational, proper parametrization $\mathbf{p}(t)$. If there exists a birational mapping $f : \mathbb{R}^n \to \mathbb{R}^n$, $f(\boldsymbol{x}) = A\boldsymbol{x} + \boldsymbol{b}$, $A^T A = I$, $\boldsymbol{b} \in \mathbb{R}^n$, such that $f(\mathcal{C}) = \mathcal{C}$, then there exists also a Möbius transformation φ such that*

$$A\boldsymbol{p}(t) + \boldsymbol{b} = (\boldsymbol{p} \circ \varphi)(t) \tag{3.5}$$

Furthermore, $f^k = \mathrm{id}_{\mathbb{R}^n}$, $k \in \mathbb{N}$, if and only if $\varphi^k = \mathrm{id}_{\mathbb{C}}$. In particular, f is an involution if and only if φ is an involution of $\mathbb{C}$.

Proof. We just need to prove the last part: since $\varphi = \boldsymbol{p}^{-1} \circ f \circ \boldsymbol{p}$, $\varphi^k = \boldsymbol{p}^{-1} \circ f^k \circ \boldsymbol{p}$, and the result follows. □

Therefore, from Theorem 45 (resp. Corollary 46), in order to find the birational mappings f between two curves $\mathcal{C}_i$, properly parametrized (resp. the symmetries f of a curve $\mathcal{C}$, properly parametrized), the idea is to find, first, the Möbius transformation φ at the bottom of the diagram Eq. (3.4), and derive f from φ. We will refer to φ as the Möbius transformation *associated* with f. Notice that the hypothesis on the properness of the parametrization(s) is crucial, since otherwise we cannot guarantee the birationality of φ.

Theorem 45 provides a brute-force approach to solve the problem of computing the symmetries of $\mathcal{C}$: we might derive a polynomial system in the parameters of φ and the entries of $A, \boldsymbol{b}$ from Eq. (3.5), and solve it. However, we will proceed in a different way, computing φ directly.

Let us go back to planar, real curves. From the previous paragraph, our goal is to find the Möbius transformations φ "behind" the symmetries of the curve. To do this, we invoke the (signed) *curvature* (see [53, 65] for references on Differential Geometry); for convenience, here we will write $\boldsymbol{p}(t) = (x(t), y(t))$, where $x(t), y(t)$ are real rational functions, and we will also write $\det(\bullet)$ for the determinant of the matrix whose column vectors appear in $\bullet$. Then, the curvature of $\boldsymbol{p}(t)$ is

$$\kappa_{\boldsymbol{p}} = \frac{\det(\boldsymbol{p}', \boldsymbol{p}'')}{\|\boldsymbol{p}'\|^3} = \frac{x'y'' - y'x''}{(x'^2 + y'^2)^{3/2}}. \tag{3.6}$$

The curvature captures the local information of the curve in the vicinity of a point. Now if the curve is symmetric, at symmetric points the local behavior is essentially the same. Thus, intuitively the curvature κ must be preserved under a symmetry, except perhaps for the sign, which is related to the orientation of the basis $\{\mathbf{t}, \mathbf{n}\}$, with $\mathbf{t}, \mathbf{n}$ the unitary tangent and normal vectors. For instance, the curve parametrized by $\boldsymbol{p}(t) = (t^2, t^3)$ (see Fig. 3.2, left) is symmetric with respect to the x-axis; the points $\boldsymbol{p}(1) = (1, 1)$ and $\boldsymbol{p}(-1) = (1, -1)$ are symmetric, and $\kappa(1) = 6/\sqrt{13} = \kappa(-1)$. However, the curve parametrized by $\boldsymbol{p}(t) = (t, t^3)$ (Fig. 3.2, right) is symmetric with respect to the origin; the points $\boldsymbol{p}(1) = (1, 1)$

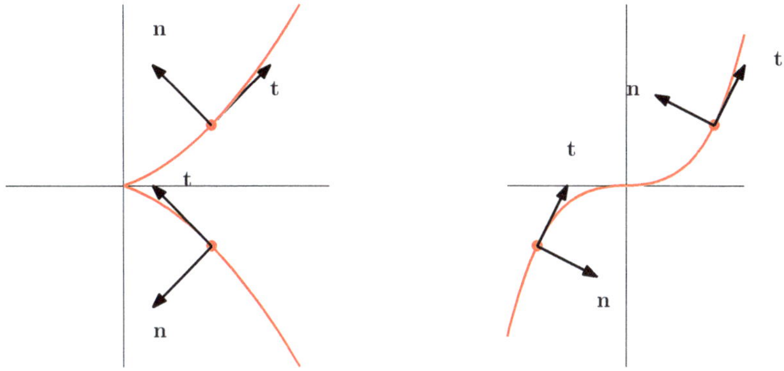

FIGURE 3.2 $\boldsymbol{p}(t) = (t^2, t^3)$ (left), $\boldsymbol{p}(t) = (t, t^3)$ (right)

and $\boldsymbol{p}(-1) = (-1, -1)$ are symmetric, and $\kappa(1) = 6/\sqrt{10}$, while $\kappa(-1) = -6/\sqrt{10}$. The reason is that the orientation of $\{\mathbf{t}, \mathbf{n}\}$ is the same at both points $\boldsymbol{p}(1), \boldsymbol{p}(-1)$ for the first curve, but not for the second curve.

However, the potential change of sign in κ will not be a problem for us, because we will be taking the square of κ. The reason is that, because of the denominator in Eq. (3.6), in general Eq. (3.6) does not provide a rational function, but a function involving a square-root. Thus, instead we will use

$$\mathrm{k}_{\boldsymbol{p}} = \kappa_{\boldsymbol{p}}^2 = \frac{(x'y'' - y'x'')^2}{(x'^2 + y'^2)^3} = \frac{\|\boldsymbol{p}' \times \boldsymbol{p}''\|^2}{\|\boldsymbol{p}'\|^6}, \tag{3.7}$$

which is a rational function.

Algebra tells us that $\mathrm{k}_{\boldsymbol{p}}$ is preserved by Euclidean symmetries: in order to see this, let us recall the following formula for the cross product,

$$M\boldsymbol{a} \times M\boldsymbol{b} = \det(M) M^{-T} (\boldsymbol{a} \times \boldsymbol{b}), \tag{3.8}$$

where $\boldsymbol{a}, \boldsymbol{b} \in \mathbb{R}^n$ and M is a nonsingular $n \times n$ matrix. If M is orthogonal, in which case $M^{-T} = M$ and $\det(M) = \pm 1$, we obtain

$$M\boldsymbol{a} \times M\boldsymbol{b} = \pm M(\boldsymbol{a} \times \boldsymbol{b}). \tag{3.9}$$

Thus, we get the following result.

Lemma 47. *Let $\mathcal{C} \subset \mathbb{R}^2$ be a planar curve properly parametrized by $\mathbf{p}(t)$, and let $f(\boldsymbol{x}) = A\boldsymbol{x} + \boldsymbol{b}$ be a symmetry of $\mathcal{C}$. Then*

$$\mathrm{k}_{\boldsymbol{p}} = \mathrm{k}_{f \circ \boldsymbol{p}}. \tag{3.10}$$

Proof. Let $\tilde{\mathbf{p}}(t) = (f \circ \mathbf{p})(t)$. Since $\tilde{\mathbf{p}}(t) = A\boldsymbol{p}(t) + \boldsymbol{b}$ we get that $\tilde{\mathbf{p}}'(t) = A\boldsymbol{p}'(t)$ and $\tilde{\mathbf{p}}''(t) = A\mathbf{p}''(t)$. Taking Eq. (3.9) into account with $M = A$, $\boldsymbol{a} = \mathbf{p}'(t)$, $\boldsymbol{b} = \mathbf{p}''(t)$ and the fact that orthogonal mappings preserve the norm of a vector, we deduce that $\mathrm{k}_{\tilde{\boldsymbol{p}}} = \mathrm{k}_{\boldsymbol{p}}$, i.e., $\mathrm{k}_{f \circ \boldsymbol{p}} = \mathrm{k}_{\boldsymbol{p}}$. □

Still, we need a bit more. Our goal is to obtain φ all at once, and to do that the computational tool that we will be using is *bivariate factoring*. But for this to work we need the following result; we will express this result by saying that $\mathrm{k}_{\boldsymbol{p}}$ *commutes* with Möbius transformations.

Lemma 48. *Let $\mathbf{p}(t)$ be a planar parametrization, and let φ be a Möbius transformation. Then*

$$\mathrm{k}_{\boldsymbol{p} \circ \varphi} = \mathrm{k}_{\boldsymbol{p}} \circ \varphi. \tag{3.11}$$

Proof. Let $\widehat{\mathbf{p}} = \boldsymbol{p} \circ \varphi$. Using the Chain Rule,

$$\widehat{\boldsymbol{p}}' = (\mathbf{p}' \circ \varphi)\varphi', \ \widehat{\boldsymbol{p}}'' = (\mathbf{p}'' \circ \varphi)(\varphi')^2 + (\mathbf{p}' \circ \varphi)\varphi''.$$

Thus,

$$\widehat{\boldsymbol{p}}' \times \widehat{\boldsymbol{p}}'' = [(\mathbf{p}' \circ \varphi) \times (\mathbf{p}'' \circ \varphi)](\varphi')^3,$$

so

$$\|\widehat{\boldsymbol{p}}' \times \widehat{\boldsymbol{p}}''\|^2 = \|(\mathbf{p}' \circ \varphi) \times (\mathbf{p}'' \circ \varphi)\|^2(\varphi')^6.$$

Furthermore,

$$\|\widehat{\boldsymbol{p}}'\|^6 = \|\boldsymbol{p}' \circ \varphi\|^6(\varphi')^6.$$

Therefore,

$$\mathrm{k}_{\boldsymbol{p} \circ \varphi} = \mathrm{k}_{\widehat{\boldsymbol{p}}} = \frac{\|\widehat{\boldsymbol{p}}' \times \widehat{\boldsymbol{p}}''\|^2}{\|\widehat{\boldsymbol{p}}'\|^6} = \frac{\|(\mathbf{p}' \circ \varphi) \times (\mathbf{p}'' \circ \varphi)\|^2 \cancel{(\varphi')^6}}{\|\boldsymbol{p}' \circ \varphi\|^6 \cancel{(\varphi')^6}} = \mathrm{k}_{\boldsymbol{p}} \circ \varphi$$

because the term $(\varphi')^6$ cancels in the numerator and the denominator. □

The following corollary follows from Lemma 48 and Lemma 47.

Corollary 49. *Let* $\mathcal{C} \subset \mathbb{R}^2$ *be a planar curve properly parametrized by* $\boldsymbol{p}(t)$*, and let* φ *be a Möbius transformation associated with a Euclidean symmetry of* $\mathcal{C}$*. Then*

$$\mathrm{k}_{\boldsymbol{p}} = \mathrm{k}_{\boldsymbol{p}} \circ \varphi. \tag{3.12}$$

Proof. Because of Eq. (3.3), $f \circ \boldsymbol{p} = \boldsymbol{p} \circ \varphi$. By Lemma 47, $k_{\boldsymbol{p}} = k_{f \circ \boldsymbol{p}}$, and therefore $k_{\boldsymbol{p}} = k_{\boldsymbol{p} \circ \varphi}$. By Lemma 48, $k_{\boldsymbol{p} \circ \varphi} = k_{\boldsymbol{p}} \circ \varphi$, so the result follows. □

The result in Corollary 49 leads to an algorithm to compute the symmetries using bivariate polynomial factoring, as promised. In order to do this, let

$$\mathrm{k}_{\boldsymbol{p}} = \mathrm{k}_{\boldsymbol{p}}(t) = \frac{P(t)}{Q(t)},$$

and let us consider the polynomial

$$K(t, \omega) = P(t)Q(\omega) - P(\omega)Q(t),$$

which is obtained by clearing denominators in $\mathrm{k}_{\boldsymbol{p}}(t) = \mathrm{k}_{\boldsymbol{p}}(\omega)$. Notice that since we are assuming that $\mathcal{C}$ is not either a circle or a line, $\mathrm{k}_{\boldsymbol{p}}(t)$ is not a constant, and therefore $K(t, \omega)$ is not the zero polynomial. Furthermore, we say that $F(t, \omega)$ is *Möbius-like* if

$$F(t, \omega) = (ct + d)\omega - (at + b), \tag{3.13}$$

i.e., it is the result of clearing denominators in $\omega = \varphi(t)$, where φ is a Möbius transformation as in Eq. (3.2); observe that $F(t, \omega)$ is an irreducible polynomial, because it is linear in ω, and $at + b, ct + d$ have a trivial common factor. Then we have the following theorem, which contains two ingredients: first, the polynomial $K(t, \omega)$ and its Möbius-like factors, and second, a condition on the *speed* of the curve, i.e., the norm of the first derivative of the parametrization. This second condition appears so that we can take advantage of the Fundamental Theorem of Planar Curves, which states that, when parametrized by *arc-length*, the non-signed curvature

$|\kappa|$ completely defines the curve up to rigid motions, i.e., Euclidean symmetries. Recall that if $\boldsymbol{p}(t)$ is a regular parametrization for $t \in I \subset \mathbb{R}$ and $t_0 \in I$, the arc-length parameter is

$$s = s(t) := \int_{t_0}^{t} \|\boldsymbol{p}'(t)\| dt,$$

which (locally) has a differentiable inverse $t = t(s)$.

Theorem 50. *Let $\mathcal{C} \subset \mathbb{R}^2$ be a planar rational curve, properly parametrized by $\mathbf{p}(t)$. The curve $\mathcal{C} \subset \mathbb{R}^2$ is invariant under a Euclidean symmetry $f(\boldsymbol{x}) = A\boldsymbol{x} + \boldsymbol{b}$ if and only if there exists a Möbius transformation $\varphi(t)$, as in Eq. (3.2), such that the polynomial $K(t,\omega)$ has a Möbius-like factor as in Eq. (3.13), and*

$$\|\boldsymbol{p}'\| = \|(\boldsymbol{p} \circ \varphi)'\|. \tag{3.14}$$

Furthermore, φ is associated with f as in Eq. (3.5).

Proof. ($\Rightarrow$) Let φ be the Möbius transformation associated with f according to Theorem 45. By Corollary 49, $\mathrm{k}_{\boldsymbol{p}}(t) = (\mathrm{k}_{\boldsymbol{p}} \circ \varphi)(t)$. Thus, seeing $K(t,\omega) = 0$ as the implicit equation of a planar algebraic curve in the (t,ω) plane, all the points (t,ω) satisfying that $\omega = \varphi(t)$, i.e., $F(t,\omega) = (ct+d)\omega - (at+b) = 0$, also satisfy $K(t,\omega) = 0$. But $F(t,\omega)$, which is Möbius-like, defines another planar algebraic curve that has infinitely many points in common with the curve defined by $K(t,\omega) = 0$ (because every point of the first curve is a point of the latter curve). Thus, by Bezout's Theorem (see Theorem 19 in Chapter 2, and see also Remark 4), $F(t,\omega)$ and $K(t,\omega)$ must share a component; since $F(t,\omega)$ is irreducible, $F(t,\omega)$ must be a factor of $K(t,\omega)$. As for the condition in Eq. (3.14), because of Eq. (3.5) we get that $A\boldsymbol{p}' = (\boldsymbol{p} \circ \varphi)'$. Since A is orthogonal, Eq. (3.14) follows.

($\Leftarrow$) Assume that $K(t,\omega)$ has a Möbius-like factor $F(t,\omega) = (ct+d)\omega - (at+b)$. By the definition of $K(t,\omega)$, $\mathrm{k}_{\boldsymbol{p}}(t) = (\mathrm{k}_{\boldsymbol{p}} \circ \varphi)(t)$ for $\varphi(t) = (at+b)/(ct+d)$, and therefore $\kappa_{\boldsymbol{p}} = \pm\kappa_{\boldsymbol{p}} \circ \varphi$, so $|\kappa_{\boldsymbol{p}}| = |\kappa_{\boldsymbol{p}} \circ \varphi|$. In turn, from Lemma 5.10 we get that $|\kappa_{\boldsymbol{p}}| = |\kappa_{\boldsymbol{p}\circ\varphi}|$. Because of the condition in Eq. (3.14), this equality still holds when reparametrizing (locally) by the arc-length. By the Fundamental

Theorem of Planar curves, $|\kappa_{\boldsymbol{p}}|$ and $|\kappa_{\boldsymbol{p}\circ\varphi}|$, after reparametrizing by the arc-length, define the same curve up to a Euclidean symmetry f, satisfying that $f\circ\boldsymbol{p}=\boldsymbol{p}\circ\varphi$. Since the reparametrization by the arc-length is, in general, local, the equality $f\circ\boldsymbol{p}=\boldsymbol{p}\circ\varphi$ holds for $t\in I$, with $I\subset\mathbb{R}$ a small interval where $\boldsymbol{p}(t)$ is regular. However, since the curves defined by $f\circ\boldsymbol{p}$ and $\boldsymbol{p}\circ\varphi$ are algebraic and irreducible, and have infinitely many points in common, namely the points in each curves generated for $t\in I$, the two curves must be equal, i.e., $f\circ\boldsymbol{p}=\boldsymbol{p}\circ\varphi$ holds for all t. Since $\boldsymbol{p}$ and $\boldsymbol{p}\circ\varphi$ both parametrize $\mathcal{C}$, we deduce that f is a Euclidean symmetry of $\mathcal{C}$. □

Therefore, in practice we compute φ from the Möbius-like factors of $K(t,\omega)$; this requires factoring $K(t,\omega)$, in general, over the reals, since φ needs not have rational coefficients. This can be efficiently done by several computer algebra systems (e.g. the command `AFactor` in Maple). After checking Eq. (3.14), we can compute f from Eq. (3.5); we can do this by directly solving the linear system provided by Eq. (3.5) in the entries of $A,\boldsymbol{b}$, or we can compute A from its action on $\boldsymbol{p}'(0),\boldsymbol{p}''(0)$ after differentiating Eq. (3.5) and substituting $t=0$, and then compute $\boldsymbol{b}$ from Eq. (3.5). Thus, we have the following algorithm `Rational Planar Symmetries`, to compute the symmetries of a planar rational curve, properly parametrized.

Example 5. *Let us consider the three-leaved rose, whose parametrization appears in Eq.* (3.1)*. One can check that this parametrization is proper. After computing $K(t,\omega)$ and factoring $K(t,\omega)$ over the reals, we get six Möbius-like factors,*

$$
\begin{aligned}
F_1(t,\omega) &= \left(t-\frac{\sqrt{3}}{3}\right)\omega - t\frac{\sqrt{3}}{3}-1,\\
F_2(t,\omega) &= \left(t+\frac{\sqrt{3}}{3}\right)\omega + t\frac{\sqrt{3}}{3}-1,\\
F_3(t,\omega) &= t-\omega,\\
F_4(t,\omega) &= \left(t-\frac{\sqrt{3}}{3}\right)\omega + t\frac{\sqrt{3}}{3}-1,
\end{aligned}
$$

Algorithm 2 Rational Planar Symmetries

Require: A proper parametrization $\boldsymbol{p}(t)$ of a planar curve $\mathcal{C} \subset \mathbb{R}^2$, not a line or a circle.

Ensure: $\mathrm{Sym}(\mathcal{C})$.

compute $K(t, \omega)$
factor $K(t, \omega)$ over the reals
for each Möbius-like factor $F(t, \omega)$ of $K(t, \omega)$ **do**
    solve $F(t, \omega) = 0$ for ω to find $\varphi(t)$
    check whether or not $\|\boldsymbol{p}'\| = \|(\boldsymbol{p} \circ \varphi)'\|$.
    if the above condition holds **then**
        compute $A, \boldsymbol{b}$ from Eq. (3.5)
        return $f(\boldsymbol{x}) = A\boldsymbol{x} + \boldsymbol{b}$
    end if
end for

$$
\begin{aligned}
F_5(t, \omega) &= \left(t + \frac{\sqrt{3}}{3}\right)\omega - t\frac{\sqrt{3}}{3} - 1, \\
F_6(t, \omega) &= t + \omega.
\end{aligned}
$$

The Möbius function $\varphi(t)$ corresponds, in each case, to solving for ω in $F_i(t, \omega) = 0$. Thus, we have

$$
\begin{aligned}
\varphi_1(t) &= -\frac{t\sqrt{3} + 3}{-3t + \sqrt{3}}, \quad & \varphi_2(t) &= -\frac{t\sqrt{3} - 3}{3t + \sqrt{3}}, \quad & \varphi_3(t) &= t \\
\varphi_4(t) &= \frac{t\sqrt{3} + 3}{-3t + \sqrt{3}}, \quad & \varphi_5(t) &= \frac{t\sqrt{3} - 3}{3t + \sqrt{3}}, \quad & \varphi_6(t) &= -t
\end{aligned}
$$

In all the cases the condition in Eq. (3.14) is satisfied, so all the $\varphi_i(t)$ correspond to symmetries of the curve. For instance, using Eq. (3.5) we get that $\varphi_1(t)$ corresponds to

$$
f_1(\boldsymbol{x}) = \begin{bmatrix} 1/2 & -\sqrt{3}/2 \\ -\sqrt{3}/2 & -1/2 \end{bmatrix} \boldsymbol{x},
$$

which describes a reflection on the line $y = -\sqrt{3}/3x$. The remaining $\varphi_i(t)$ correspond to a reflection on the line $y = \sqrt{3}/3x$, the identity, rotations around the origin by angles $\theta = \pm 2\pi/3$, and a reflection on the y-axis.

3.3 POLYNOMIALLY PARAMETRIZED CURVES

In this section we consider the special case when $\boldsymbol{p}(t)$ is a polynomial, proper, real, non-linear parametrization, so $\boldsymbol{p}(t) = (x(t), y(t))$ with $x(t), y(t)$ polynomials. Here we will also exploit Theorem 45 and Corollary 46. However, the fact that $\boldsymbol{p}(t)$ is polynomial implies that, because of Eq. (3.5), the Möbius transformation $\varphi(t)$ behind each symmetry must be a linear function. This advantage allows us to find closed expressions for the notable elements of the symmetries, i.e., the center of rotation, the rotation angle, and the axes of symmetry, if any. In this section, we will be following the ideas in [2].

It is useful here to consider complex numbers. Thus, we will identify $\mathbb{R}^2$ with $\mathbb{C}$ via $(x, y) \to z = x + \mathbf{i}y$, with $\mathbf{i}^2 = -1$. This way $\boldsymbol{p}(t) = (x(t), y(t))$, where now $x(t), y(t)$ are polynomials, is identified with the complex, polynomial parametrization $z(t) = x(t) + \mathbf{i}y(t)$. Since we want to work over $\mathbb{C}$, we observe that symmetries of $\mathbb{R}^2$ correspond to complex mappings

$$f(z) = \boldsymbol{a}z + \boldsymbol{b}, \;\; f(z) = \boldsymbol{a}\overline{z} + \boldsymbol{b}, \tag{3.15}$$

where the left-hand mapping describes orientation-preserving symmetries, i.e., rotations in our case, and the right-hand mapping describes orientation-reversing symmetries, i.e., reflections. Additionally, since by assumption $\mathcal{C}$ is not a line, $f^n = \mathrm{id}_{\mathbb{R}^2}$ for some $n \in \mathbb{N}$, so $\boldsymbol{a}$ must be a root of unity. For rotations we have $\boldsymbol{a} = e^{\mathrm{i}\alpha}$, with $\alpha = 2\pi/m$, $m \in \mathbb{N}$, and for reflections, which satisfy $f \circ f = \mathrm{id}_{\mathbb{R}^2}$, we have $\boldsymbol{a} = \pm 1$, $\boldsymbol{a}\overline{\boldsymbol{b}} + \boldsymbol{b} = 0$.

Let us separately analyze rotations, and reflections. We start with rotations.

Proposition 51. *Let $\mathcal{C} \subset \mathbb{R}^2$ be a planar algebraic curve parametrized by a proper and polynomial parametrization $\boldsymbol{p}(t) = (x(t), y(t))$.*

(1) If $\mathcal{C}$ is invariant under a non-trivial rotation f, then f must be a central symmetry.

(2) *If $\mathcal{C}$ exhibits central symmetry, then the degrees of $x(t), y(t)$ are odd.*

Proof. Let us see (1). Since $\boldsymbol{p}(t)$ is polynomial, the curve $\mathcal{C}$ only has one branch at infinity, i.e., the projective closure of $\mathcal{C}$ has just one projective point at infinity. Since any rotation sends branches at infinity to branches at infinity, we can have at most one rotation leaving $\mathcal{C}$ invariant, which preserves the branch at infinity of $\boldsymbol{p}(t)$; thus, the rotation angle θ can be either 0, i.e., the identity, or π, i.e., a central symmetry. Now let us see (2). Assume, after performing if necessary a change of parameters, that the center of symmetry is the origin. Since $\boldsymbol{p}(t)$ is polynomial and symmetric with respect to the origin, $\lim_{t\to\infty} x(t)$ and $\lim_{t\to-\infty} x(t)$, $\lim_{t\to\infty} y(t)$ and $\lim_{t\to-\infty} y(t)$, must have different signs. Thus, the degrees of both $x(t), y(t)$ must be odd. □

Next if $f(z)$ represents a central symmetry with respect to a point identified with $z_0 \in \mathbb{C}$, then the equation of the central symmetry must be (see Fig. 3.3)

$$f(z) = z_0 + [-(z - z_0)] = -z + 2z_0. \tag{3.16}$$

Thus, we have the following result, which follows from Theorem 45, taking Eq. (3.3) into account and the fact that, since we are working with polynomial parametrizations, the Möbius function associated with any symmetry must be linear.

Theorem 52. *Let $\mathcal{C} \subset \mathbb{R}^2$ be an algebraic planar curve properly and polynomially parametrized by $\mathbf{p}(t) = (x(t), y(t))$, written as $z(t) = x(t) + \mathbf{i}y(t)$ in complex form. Then $\mathcal{C}$ is symmetric with respect to the point z_0 if and only if there exists $\varphi(t) = at + b$ such that*

$$-z + 2z_0 = z(\varphi(t)). \tag{3.17}$$

Let us see that Theorem 52 provides a closed expression for the center of symmetry of $\mathcal{C}$, if any. In order to do this, let us write

$$z(t) = c_n t^n + c_{n-1} t^{n-1} + \cdots + c_1 t + c_0, \tag{3.18}$$

where $c_i \in \mathbb{C}$ for $i = 0, 1, \ldots, n$, $c_n \neq 0$. Then Eq. (3.17) leads to the triangular system $\mathcal{S}$ in Eq. (3.21), with $n+1$ equations, in a, b, z_0; we denote the equations of $\mathcal{S}$ by $(0), (1), \ldots, (n)$.

Theorem 53. *Let $\mathcal{C} \subset \mathbb{R}^2$ be an algebraic planar curve properly and polynomially parametrized by $\mathbf{p}(t) = (x(t), y(t))$, written as $z(t) = x(t) + \mathbf{i}y(t)$ in complex form. Then $\mathcal{C}$ exhibits central symmetry if and only if*

$$a = -1, \ b = -\frac{2c_{n-1}}{nc_n}, \tag{3.19}$$

with $b \in \mathbb{R}$, and

$$z_0 = c_0 + \frac{1}{2}(c_1 b + c_2 b^2 + \cdots + c_n b^n) \tag{3.20}$$

with b as in Eq. (3.19), are the solutions of the system $\mathcal{S}$ in Eq. (3.21). Furthermore, in that case Eq. (3.20) corresponds to the center of symmetry of $\mathcal{C}$.

Proof. The implication ($\Leftarrow$) follows from Theorem 52, so let us see ($\Rightarrow$). From the last equation of $\mathcal{S}$, i.e., equation (n) of Eq. (3.21), we get that $a^n = -1$. Since n is odd by statement (2) of Proposition 51, $a = -1$. Plugging $a = -1$ into the equation $(n-1)$ of Eq. (3.21) we get that $b = -2c_{n-1}/nc_n$. Since the Möbius function $\varphi(t) = at + b$ must be real, $-2c_{n-1}/nc_n \in \mathbb{R}$. Thus, the only possible solution of the equations $(1), \ldots, (n)$ of Eq. (3.21) is Eq. (3.19). Finally, Eq. (3.20) follows by substituting Eq. (3.19) into the first equation of $\mathcal{S}$. □

$$
\begin{array}{rcl}
-c_0 + 2z_0 &=& c_0 + c_1 b + c_2 b^2 + \cdots + c_{n-1} b^{n-1} + c_n b^n \\
-c_1 &=& c_1 a + c_2 \cdot 2ab + \cdots + c_{n-1}(n-1)ab^{n-2} + c_n n a b^{n-1} \\
-c_2 &=& c_2 a^2 + \cdots + c_{n-1}\frac{(n-1)(n-2)}{2}a^2 b^{n-3} + c_n \frac{n(n-1)}{2} a^2 b^{n-2} \\
&\vdots& \\
-c_{n-1} &=& c_{n-1} a^{n-1} + c_n n a^{n-1} b \\
-c_n &=& c_n a^n
\end{array}
\tag{3.21}
$$

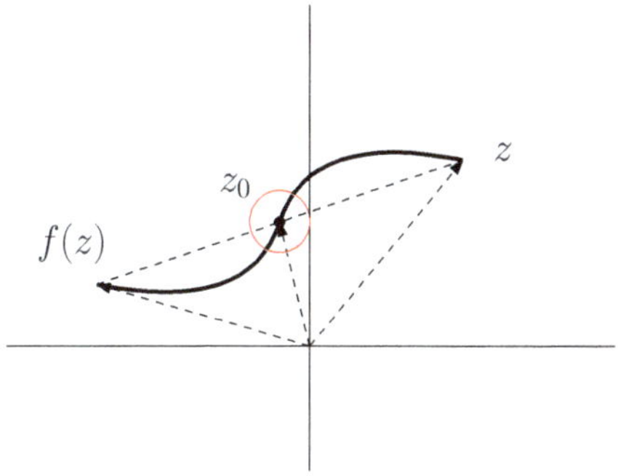

FIGURE 3.3 Complex equation of a central symmetry with respect to a point z_0

Now let us address reflections. A first observation is that we can have at most one axis of symmetry $\mathcal{L}$, which is also a consequence of the fact that $\mathcal{C}$ has just one point at infinity. In order to find $\mathcal{L}$, we observe that there is a rigid motion T, which is the composition of a translation and a rotation, that transforms $\mathcal{L}$ into the x-axis. In complex form, if $z_0 \in \mathcal{L}$ and α denotes the angle between $\mathcal{L}$ and the x-axis, the mapping

$$g(z) = (z - z_0)e^{\mathrm{i}\phi}, \ \phi = -\alpha,$$

transforms $\mathcal{L}$ into the x-axis (see Fig. 3.4). Then, the curve $T(\mathcal{C})$ is parametrized by $\omega(t) = (z(t) - z_0)e^{\mathrm{i}\phi}$. Since a symmetry with respect to the x-axis corresponds to the complex mapping $h(z) = \overline{z}$, we get the following result, which follows from Theorem 45.

Proposition 54. *Let $\mathcal{C} \subset \mathbb{R}^2$ be an algebraic planar curve properly and polynomially parametrized by $\mathbf{p}(t) = (x(t), y(t))$, written as $z(t) = x(t) + \mathbf{i}y(t)$ in complex form. Then $\mathcal{C}$ is symmetric with respect to a line $\mathcal{L}$ if and only if there exist $z_0 \in \mathbb{C}$, $\phi \in (-2\pi, 0]$, $\varphi(t) = at + b$, $a, b \in \mathbb{R}$ such that $\omega(t) = (z(t) - z_0)e^{\mathrm{i}\phi}$ satisfies that*

$$\overline{\omega(t)} = \omega(at + b), \tag{3.22}$$

where $z_0 \in \mathcal{L}$ and $\alpha = -\phi$ is the angle between the line $\mathcal{L}$ and the x-axis.

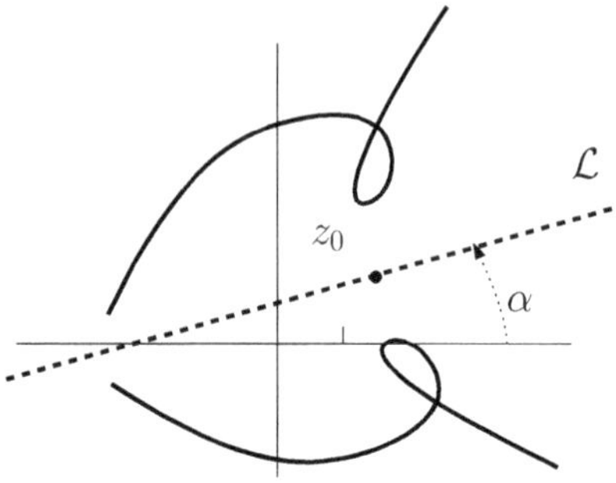

FIGURE 3.4 Reflections of polynomially parametrized curves

Proposition (54) leads to the following result, which follows from the fact that since $\omega(t)$ is symmetric with respect to the x-axis, the degree of the y-component of $\omega(t)$ must be odd; see [2] for details.

Lemma 55. *If Eq.* (3.22) *holds, then* $a = -1$.

Next we proceed as we did for central symmetries. First, Eq. (3.22) leads to a triangular system $\mathcal{W}$, shown in Eq. (3.24), resembling that in Eq. (3.21), with $n+1$ equations in a, b, z_0, ϕ; here we use again the notation in Eq. (3.18) for the complex parametrization $z(t)$, and we denote the equations of $\mathcal{W}$ by $(0), (1), \ldots, (n)$. The analysis of the system $\mathcal{W}$ is similar to that of Eq. (3.21) (see [2] for details); together with Lemma 55, the only possible solutions for a, b in $\mathcal{W}$ are

$$a = -1,\ b = -\frac{2\mathrm{Re}(c_n \overline{c}_{n-1})}{n|c_n|^2}, \tag{3.23}$$

where we observe that the expression for b is a real number. However, the existence of reflectional symmetry requires that the solution of $\mathcal{W}$ contains infinitely many values for z_0 defining a line, which is, precisely, the axis of symmetry $\mathcal{L}$. So we need that the first equation of $\mathcal{W}$, i.e., Eq. (0), that is the only equation of $\mathcal{W}$ containing z_0, defines a real line, which must be $\mathcal{L}$. In order to guarantee this (see again [2] for details), we define

$$Q(b) = c_1 b + c_b^2 + \cdots + c_n b^n,\ Q^\star(b) = Q(b)\frac{\overline{c}_n}{|\overline{c}_n|}\sqrt{(-1)^{n+1}}.$$

$$
\begin{array}{rcl}
(\overline{c}_0 - \overline{z}_0)e^{-2i\phi} & = & c_0 + c_1 b + c_2 b^2 + \cdots + c_{n-1} b^{n-1} + c_n b^n \\
\overline{c}_1 e^{-2i\phi} & = & c_1 a + c_2 \cdot 2ab + \cdots + c_{n-1}(n-1)ab^{n-2} + c_n nab^{n-1} \\
\overline{c}_2 e^{-2i\phi} & = & c_2 a^2 + \cdots + c_{n-1}\dfrac{(n-1)(n-2)}{2}a^2 b^{n-3} + c_n \dfrac{n(n-1)}{2}a^2 b^{n-2} \\
& \vdots & \\
\overline{c}_{n-1} e^{-2i\phi} & = & c_{n-1} a^{n-1} + c_n n a^{n-1} b \\
\overline{c}_n e^{-2i\phi} & = & c_n a^n
\end{array}
\tag{3.24}
$$

Then one can prove that Eq. (0) of $\mathcal{W}$ defines a real line if and only if $Q^\star(b) \in \mathbb{R}$ for the b in Eq. (3.23). Finally, one gets the following theorem.

Theorem 56. *Let $\mathcal{C} \subset \mathbb{R}^2$ be an algebraic planar curve properly and polynomially parametrized by $\mathbf{p}(t) = (x(t), y(t))$, written as $z(t) = x(t) + \mathbf{i}y(t)$ in complex form. Then $\mathcal{C}$ exhibits reflectional symmetry if and only if the following two conditions hold: (1) The a, b in Eq. (3.23) satisfy the equations (1), …, (n) of the system $\mathcal{W}$; (2) $Q^\star(b)$, with b as in Eq. (3.23), is a real number. Furthermore, in that case the reflection axis $\mathcal{L}$ of $\mathcal{C}$ is, in complex form,*

$$\overline{c}_n z - c_n(-1)^n \overline{z} + \overline{c}_0 c_n(-1)^n - \overline{c}_n c_0 - Q(b) = 0. \tag{3.25}$$

3.3.1 Bézier curves

Let us consider a special case within the already special case of polynomially parametrized curves, namely that of *Bézier curves* [56, 61], widely used in Computer Aided Geometric Design. A Bézier curve, to be precise, is a portion of a polynomially parametrized curve attached to a discrete object, called the *control polygon* of the curve. The control polygon $\mathcal{P}$ is defined from its vertices $\boldsymbol{b}_0, \boldsymbol{b}_1, \ldots, \boldsymbol{b}_N$, called the *control points*, so that the edges of the polygon are the segments connecting $\boldsymbol{b}_i, \boldsymbol{b}_{i+1}$ for $i = 0, 1, \ldots, N-1$; if $\boldsymbol{b}_0 = \boldsymbol{b}_N$ then $\mathcal{P}$, and also the corresponding Bézier curve, is closed. The Bézier curve itself is the set of points parametrized by

$$\boldsymbol{p}(t) = \sum_{i=0}^{N} \boldsymbol{b}_i B_{i,N}(t), \; t \in [0, 1], \tag{3.26}$$

where the $B_{i,N}(t)$ are the *Bernstein polynomials*,

$$B_{i,N} = \binom{N}{i} t^i (1-t)^{N-i}, \; i = 0, 1, \ldots, N.$$

Bernstein polynomials serve as a basis for the vector space of polynomials of any degree, alternative to the monomial basis $\{1, t, t^2, \ldots\}$, and with optimal numerical properties for evaluation or root computation, in particular with a better conditioning

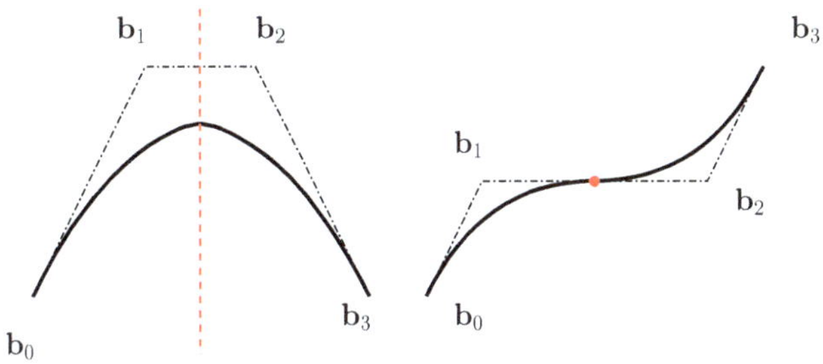

FIGURE 3.5 Bézier curves and their symmetries

compared to the monomial basis (see [56]). If the $\boldsymbol{b}_i$'s are points in $\mathbb{R}^2$ we have *planar* Bézier curves, but we can also consider Bézier curves in any dimension by taking $\boldsymbol{b}_i \in \mathbb{R}^n$.

Bézier curves interpolate the first and last control points $\boldsymbol{b}_0, \boldsymbol{b}_N$, and are tangent to the first and last edges of the control polygon (see Fig. 3.5, where the control polygon is plotted in dashed-dotted line); furthermore, the shape of the control polygon suggests the shape of the Bézier curve.

Symmetries of Bézier curves are studied in [112]. If the $\boldsymbol{p}(t)$ in Eq. (3.26), considered now as a polynomial parametrization of a rational curve, i.e., with t allowed to take values in $\mathbb{C}$, is proper, the Bézier curve defined by $\boldsymbol{p}(t)$, $t \in [0,1]$, exhibits reflectional/central symmetry if and only if the control polygon $\mathcal{P}$ exhibits reflectional/central symmetry. In other words, the axes of symmetry/center of symmetry of the Bézier curve, are the axes of symmetry/center of symmetry of $\mathcal{P}$. This follows from the fact that, whenever $\boldsymbol{p}(t)$ is proper, the control polygon is unique up to reversing the order of the control points; see [112] for more details. As a consequence, one can easily create Bézier curves with prescribed axes of symmetry, or with central symmetry: it suffices to choose control points exhibiting that kind of symmetry; examples of this are provided in Fig. 3.5, where the symmetry axis/center of symmetry are plotted in red.

Finally, the Bézier curve is part of a bigger curve, the curve defined by $\boldsymbol{p}(t)$ for $t \in \mathbb{C}$ (and not only over $[0,1]$, as it happens with the Bézier curve). One can wonder whether the symmetries of the Bézier curve correspond to symmetries of the bigger curve. The answer is affirmative, and the reason is, again, Bezout's Theorem, which ensures that two algebraic curves with infinitely many points in common must share a component. Indeed, let us take the algebraic curve C_1, defined by $\boldsymbol{p}(t)$ for $t \in \mathbb{C}$, and the algebraic curve C_2, defined by $(f \circ \boldsymbol{p})(t)$ for $t \in \mathbb{C}$, where f is a symmetry of the Bézier curve defined by $\boldsymbol{p}(t)$ for $t \in [0, 1]$. Since f is a symmetry of this Bézier curve, C_1, C_2 have infinitely many points in common, namely all the points of the Bézier curve. Thus, C_1, C_2 must share a component. However, C_1, C_2 are both rational and therefore irreducible, so $C_1 = C_2$ and f must be a global symmetry of C_1.

Let us observe that the same conclusion, i.e., that the symmetries of the curve are the symmetries of the control polygon, is reached for more elaborate curves also used in Computer Aided Geometric Design, namely *B-splines* and NURBS *curves* (see [12]).

3.4 TRIGONOMETRIC CURVES

In this section we treat another special type of rational curves. Here we will follow the ideas and notation of [31].

Definition 57. *Let $\mathcal{C} \subset \mathbb{R}^2$. We say that $\mathcal{C}$ is a* trigonometric curve *if $\mathcal{C}$ admits a parametrization $\boldsymbol{p}(t)$ of the type*

$$\begin{aligned} \boldsymbol{p}(t) &= \textstyle\sum_{k=0}^{N} [\mathbf{a}_k \cos(kt) + \mathbf{b}_k \sin(kt)] \\ &= \mathbf{a}_0 + \textstyle\sum_{k=1}^{N} [\mathbf{a}_k \cos(kt) + \mathbf{b}_k \sin(kt)] = \mathbf{a}_0 + \textstyle\sum_{k=1}^{N} \boldsymbol{p}_k(t), \end{aligned} \tag{3.27}$$

where $\mathbf{a}_k, \mathbf{b}_k \in \mathbb{R}^2$ and at least one of $\mathbf{a}_N, \mathbf{b}_N$ is nonzero.

Thus, trigonometric curves are parametric curves whose components are trigonometric polynomials of degree at most N, i.e., truncated Fourier series. Several algorithms for this type of curves are given in [73]. In particular, in [73] it is proven that any trigonometric curve admits either a polynomial parametrization, or a parametrization $\boldsymbol{p}(t)$ like Eq. (3.27) with $t \in [0, 2\pi)$ and with just finitely many points generated by more than one value of t; such a parametrization is called *simple* in [73], although we might as well call it *proper*, using the terminology introduced at the beginning of the chapter. We will assume here that $\mathcal{C}$ is not a circle, and that the parametrization in Eq. (3.27) is trigonometric and simple; in that case, we call N the *trigonometric degree* of $\boldsymbol{p}(t)$. Also, notice that each $\boldsymbol{p}_k(t)$, where $t \in [0, 2\pi)$, is a parametrization of a k-times traced ellipse centered at the origin. Fig. 3.6 shows several examples of trigonometric curves, some of them with different symmetries, others non-symmetric.

Trigonometric curves are rational: in order to see this, it suffices to consider the change $z = e^{\mathbf{i}t}$, where $\mathbf{i}^2 = -1$ and z belongs to the unit circle S^1. Since $e^{\mathbf{i}t} = \cos t + \mathbf{i} \sin t$, and taking into account that for $z \in S^1$ the conjugate $\overline{z}$ satisfies that $\overline{z} = \dfrac{1}{z}$, we deduce that

$$\begin{aligned} \cos t &= \frac{z^2+1}{2z}, \quad \sin t = \frac{z^2-1}{2\mathbf{i}z}, \\ \cos(kt) &= \frac{z^{2k}+1}{2z^k}, \quad \sin(kt) = \frac{z^{2k}-1}{2\mathbf{i}z^k}, \end{aligned} \tag{3.28}$$

where $k \in \mathbb{N}$. Substituting these relationships into Eq. (3.27) we get a rational parametrization

$$\tilde{\boldsymbol{p}}(z) = (\tilde{p}_1(z), \tilde{p}_2(z)), \tag{3.29}$$

where each component satisfies that

$$\tilde{p}_i(z) = \frac{P_i(z)}{z^{m_i}}, \; i = 1, 2, \tag{3.30}$$

with $P_i(z)$ a complex polynomial of degree $2m_i$, $N = \max\{m_1, m_2\}$, and $z \in S^1$. Thus, $\tilde{\boldsymbol{p}}(z)$ is a rational parametrization of $\mathcal{C}$ of degree $2N$, with complex coefficients.

The special form of Eq. (3.29) is exploited in [16] to solve the problem of computing the affine equivalences between two trigonometric curves in any dimension, where symmetry computation is a particular case. However, we will stick to the presentation in [31], where no conversion to rational form is required.

FIGURE 3.6 Some examples of trigonometric curves

For later use, we start introducing a special instance of trigonometric curves, named *higher cycloid curves* or *cycloids of Nth stage* [136]. While a cycloid is the curve traced by a point on a circle as it rolls along a straight line without slipping, higher cycloids are curves traced by a point that rolls without slipping on another circle that rolls without slipping on a further circle, etc. They correspond to Eq. (3.27) with the additional requirement that $\mathbf{a}_k, \mathbf{b}_k$ are orthogonal, so that denoting $\mathbf{a}_k = (a_{k1}, a_{k2})$, $\mathbf{b}_k = (b_{k1}, b_{k2})$,

$$(b_{k1}, b_{k2}) = \sigma_k(-a_{k2}, a_{k1}),\ \sigma_k \in \{-1, 1\}.$$

Thus, writing $\cos(\sigma_k kt) + \mathbf{i}\sin(\sigma_k kt) = e^{\sigma_k kt\mathbf{i}}$, $a_{k1} + a_{k2}\mathbf{i} = \lambda_k e^{\psi_k \mathbf{i}}$, the parametrization of a higher cycloid corresponds to

$$\boldsymbol{p}(t) = \mathbf{a}_0 + \sum_{k=1}^{N} \lambda_k e^{(\sigma_k kt + \psi_k)\mathbf{i}}. \tag{3.31}$$

One can check that $\boldsymbol{p}(t)$ has the form in Eq. (3.31) if and only if each $\boldsymbol{p}_k(t)$ is a circle. The blue curves in Fig. 3.6 are higher cycloids; the red curves are not.

We are now almost ready to begin our analysis. The following result, proven in [73], will be necessary.

Lemma 58. *Two simple trigonometric parametrizations of a same trigonometric curve are equal up to a linear change of parameters.*

Thus, if f is a Euclidean symmetry of a trigonometric curve $\mathcal{C}$ parametrized by $\boldsymbol{p}(t)$, where $\boldsymbol{p}(t)$ is simple, $(f \circ \mathbf{p})(t)$ is another simple parametrization of $\mathcal{C}$, and from Lemma 58 there exists a linear change of parameters $\varphi(t) = \alpha t + \beta$ such that

$$(f \circ \mathbf{p})(t) = \mathbf{p}(\alpha t + \beta). \tag{3.32}$$

Next let us consider rotations.

Theorem 59. *Let $\mathcal{C}$ be a trigonometric curve, not a circle, defined by a simple parametrization $\mathbf{p}(t)$ as in Eq. (3.27). If $\mathcal{C}$ has rotational symmetry, then:*

(1) *The rotation center is* $\mathbf{a}_0$, *and the rotation angle has the form* $\theta = 2\pi/m$, $m \in \mathbb{N}$.

(2) *If* $m > 2$, *the curve* $\mathcal{C}$ *is a higher cycloid.*

Proof. Assume, by applying if necessary a change of coordinates, that the rotation center $\mathbf{c}$ is the origin, and let $f(\boldsymbol{x}) = A\boldsymbol{x}$ define the rotation about the origin, of angle $\theta \in [0, 2\pi)$, such that $f(\mathcal{C}) = \mathcal{C}$. From Eq. (3.27) we get that $f(\boldsymbol{p}(t)) = \sum_{k=0}^{N} f(\boldsymbol{p}_k(t))$. Furthermore, since f is affine $f(\boldsymbol{p}_k(t)) = \tilde{\mathbf{a}}_k\cos(kt) + \tilde{\mathbf{b}}_k\sin(kt)$ for $k = 0, 1, \ldots, N$, and therefore $f(\boldsymbol{p}_k(t))$ must be a reparametrization of $\boldsymbol{p}_k(t)$ for each k. Thus, the curves defined by $\boldsymbol{p}_k(t)$, for $k = 0, 1, \ldots, N$, must be, each of them, invariant under f. For $k = 0$, we get that $A\mathbf{a}_0 = \mathbf{a}_0$, i.e., $\mathbf{a}_0$ is either the zero vector, or an eigenvector associated with the eigenvalue $\lambda = 1$. However, since A is a rotation matrix $\lambda = 1$ is not an eigenvalue of A. Thus, $\mathbf{a}_0$ must be the zero vector, i.e., $\mathbf{a}_0 = \mathbf{c}$, and the first part of statement (1) is proven. The second part of (1) follows from Lemma 24 because, by hypothesis, $\mathcal{C}$ is not a circle.

In order to see statement (2), notice that the curve $\mathcal{C}_k$, with $k \geq 1$, defined by $\boldsymbol{p}_k(t)$ must also be rotationally invariant by a rotation of angle $\theta = 2\pi/m$, where we are assuming that $m \neq 2$. However, $\boldsymbol{p}_k(t)$ parametrizes an ellipse: since the only ellipses with rotational symmetry by an angle $\theta = 2\pi/m$, $m \neq 2$, are circles, we deduce that $\boldsymbol{p}_k(t)$, for $k \geq 1$, must parametrize a circle, in which case $\mathcal{C}$ must be a higher cycloid. □

Theorem 59 implies that any trigonometric curve with the rotational symmetry of an m-gon must be a higher cycloid, but the converse is not true, as the right-most blue curve in the third row of Fig. 3.6 shows. Also, Theorem 59 does not include the case $m = 2$, i.e., central symmetry. In order to explore the question further, we can benefit from the rational form of these curves in Eq. (3.29). Recall that we achieve this form by applying the change of variables $z = e^{\mathbf{i}t}$. Thus, a linear reparametrization $t \to \alpha t + \beta$ corresponds to

$$e^{\mathbf{i}(\alpha t+\beta)} = e^{\mathbf{i}\beta}\left(e^{\mathbf{i}t}\right)^{\alpha},$$

so that the symmetry f behind is $f(z) = kz^{\alpha}$, in complex form, with $k = e^{\mathbf{i}\beta}$ a unitary complex number. Since the complex form of any Euclidean symmetry corresponds to one of the expressions in Eq. (3.15) (see Subsection 3.3), we deduce that $\alpha = \pm 1$, where $\alpha = 1$ corresponds to orientation-preserving symmetries, i.e., rotations, and $\alpha = -1$ corresponds to orientation-reversing symmetries, i.e., reflections; in this last case, since $z = e^{\mathbf{i}t} \in S^1$, $z^{-1} = \overline{z}$. Therefore, we deduce the following result. Here we work with rotations around the origin, since we can always reduce to that case by applying, if necessary, a change of coordinates making $\mathbf{a}_0 = 0$ (see also statement (1) of Theorem 59).

Lemma 60. *Let $\mathcal{C}$ be a trigonometric curve defined by a simple parametrization $\mathbf{p}(t)$, and let f be a rotation about the origin by an angle $2\pi/m$. Then f is a symmetry of $\mathcal{C}$ if and only if there exists $d \in \mathbb{Z}$ such that*

$$f(\mathbf{p}(t)) = \mathbf{p}\left(t + \frac{2\pi d}{m}\right). \tag{3.33}$$

Proof. ($\Rightarrow$) From Lemma 58 there exists a Möbius transformation $\varphi(t) = \alpha t + \beta$ satisfying Eq. (58). From the previous discussion we get that $\alpha = 1$, so $f(\mathbf{p}(t)) = \mathbf{p}(t + \beta)$. Since $f^m = \mathrm{id}$, we deduce that $\mathbf{p}(t) = \mathbf{p}(t + m\beta)$. Finally, since $\mathbf{p}(t)$ is 2π-periodic, there must exist $d \in \mathbb{Z}$ such that $m\beta = 2\pi d$.

($\Leftarrow$) The mapping $t \to t + \frac{2\pi d}{m}$ corresponds, via $z = e^{\mathbf{i}t}$, to $f(z) = e^{\mathbf{i}2\pi/m} z$, which is a rotation about the origin by an angle $2\pi/m$. Since Eq. (3.33) holds by hypothesis, $\mathbf{p}(t)$ and $(f \circ \mathbf{p})(t)$ parametrize the same curve, so $f(\mathcal{C}) = \mathcal{C}$ and thus f is a symmetry of $\mathcal{C}$. □

Now we can address central symmetry.

Corollary 61. *Let $\mathcal{C}$ be a trigonometric curve defined by a simple parametrization $\mathbf{p}(t)$. Then $\mathcal{C}$ is symmetric with respect to the origin if and only if $-\mathbf{p}(t) = \mathbf{p}(t + \pi)$; in particular, in this case $\mathbf{p}_k(t)$ must be identically zero for all even k.*

We can exploit Lemma 60 to characterize higher cycloids with the rotational symmetry of an m-gon, with $m > 2$, about the origin. To do that, let $\mathbf{p}(t)$ parametrize a higher cycloid with the symmetry of an m-gon about the origin. The statement of Theorem 59 applies to each $\boldsymbol{p}_k(t)$, so

$$\boldsymbol{p}_k(t) = \lambda_k e^{(\sigma_k kt+\psi_k)\mathbf{i}}.$$

By Lemma 60, $\boldsymbol{p}_k(t)$ has the symmetry of an m-gon about the origin if and only if

$$e^{\mathbf{i}\frac{2\pi}{m}} \cdot \lambda_k e^{(\sigma_k kt+\psi_k)\mathbf{i}} = \lambda_k e^{(\sigma_k k(t+\frac{2\pi d}{m})+\psi_k)\mathbf{i}}$$

for some $d \in \mathbb{Z}$. In turn, this yields

$$\frac{2\pi}{m}(1 - \sigma_k kd) \equiv 0 \bmod 2\pi.$$

This condition forces some $\mathbf{p}_k(t)$ to disappear from Eq. (3.27); in particular, we can just have terms corresponding to k where $\sigma_k kd - 1$ is divisible by m. So we get the following result, which helps to create trigonometric curves with rotational symmetry.

Theorem 62. *Let $\mathbf{p}(t)$ be a simple parametrization of a trigonometric curve $\mathcal{C}$. Then $\mathcal{C}$ has the rotational symmetry of an m-gon about the point $\mathbf{c}$ if and only if $\mathbf{a}_0 = \mathbf{c}$ and $\mathcal{C}$ is a higher cycloid where the only non-zero $\mathbf{p}_k(t)$ in Eq. (3.27) are those fulfilling $\sigma_k kd = jm + 1$ for some $j \in \mathbb{Z}$.*

Example 6. *Let us see that the curve parametrized by*

$$\mathbf{p}(t) = (\cos(t) + \cos(4t) + \cos(7t), \sin(t) + \sin(4t) + \sin(7t)),$$

i.e.,

$$\mathbf{p}(t) = \underbrace{(\cos(t), \sin(t))}_{\mathbf{p}_1(t)} + \underbrace{(\cos(4t) + \sin(4t))}_{\mathbf{p}_4(t)} + \underbrace{(\cos(7t) + \sin(7t))}_{\mathbf{p}_7(t)},$$

has the symmetry of a 3-gon. Indeed, from Theorem 62 if $m = 3$, choosing $d = 1$, then the non-zero $\mathbf{p}_k(t)$ correspond to $k = 3j + 1$, with $j \in \mathbb{Z}$. In our case, the non-zero $\mathbf{p}_k(t)$ are $1 = 3 \cdot 0 + 1$, $4 = 3 \cdot 1 + 1$, $7 = 3 \cdot 2 + 1$. The plotting of this curve appears in Fig. (3.6): it is the left curve in the third row; here we can check that certainly the curve has the symmetry of a 3-gon.

Further observations and examples of rotationally symmetric trigonometric curves can be found in [31]. But we will now move to reflections. An argument similar to the one used to prove statement (1) of Theorem 59 provides the following result.

Lemma 63. *Let $\mathcal{C}$ be a trigonometric curve, not a circle, defined by a simple parametrization $\mathbf{p}(t)$ as in Eq. (3.27). If $\mathcal{C}$ shows reflectional symmetry about a line $\mathcal{L}$, then $\mathbf{a}_0 \in \mathcal{L}$.*

As a consequence, we can always apply a translation so that $\mathbf{a}_0$ is moved to the origin, and consider only reflections on lines through the origin. Additionally, we have the following result.

Lemma 64. *Let $\mathcal{C}$ be a trigonometric curve, not a circle, defined by a simple parametrization $\mathbf{p}(t)$ as in Eq. (3.27). If $\mathcal{C}$ shows reflectional symmetry about a line $\mathcal{L}$ through the origin, then there exists t_0 such that $\mathbf{p}(t_0) \in \mathcal{L}$, i.e., $\mathcal{C}$ and $\mathcal{L}$ have at least one point $\mathbf{p}$ in common.*

Proof. Let f be a reflection on a line $\mathcal{L}$ through the origin such that $f(\mathcal{C}) = \mathcal{C}$. From Lemma 58 and the discussion after Theorem 59, we get that $f(\mathbf{p}(t)) - \mathbf{p}(-t+\beta)$. Making $t = \beta/2$, we deduce that $f(\mathbf{p}(\beta/2)) = \mathbf{p}(\beta/2)$, and therefore $\mathbf{p}(\beta/2)$ is a fixed point of $\mathcal{L}$; thus, $\mathbf{p}(\beta/2) \in \mathcal{L}$ and $\mathbf{p} = \mathbf{p}(\beta/2)$. □

Also using an analogous argument to that in the proof of Theorem 59, we deduce that if f is a reflection of $\mathcal{C}$, $\mathcal{L}$ must be a symmetry axis of every $\mathbf{p}_k(t)$, so the point $\mathbf{p}$ on Lemma 64 must be a vertex of each $\mathbf{p}_k(t)$ reached, in every $\mathbf{p}_k(t)$, for the same value of t, say t_0. Thus, we conclude that all the $\mathbf{p}_k(t)$, which in general are ellipses, must share the same axes and have a common vertex. So we get the following result.

Proposition 65. *Let $\mathcal{C}$ be a trigonometric curve, not a circle, defined by a simple parametrization $\mathbf{p}(t)$ as in Eq. (3.27). Assume that not all $\mathbf{p}_k(t)$ are circles.*

(1) The curve $\mathcal{C}$ shows reflectional symmetry about a line $\mathcal{L}$ through the origin if and only if $\mathcal{L}$ is a common axis of all the $\mathbf{p}_k(t)$, and there exists $\mathbf{p} \in \mathcal{L}$ and t_0 so that $\mathbf{p} = \mathbf{p}_k(t_0) \in \mathcal{L} \cap \mathcal{C}_k$ for all k, where $\mathcal{C}_k$ is the curve defined by $\mathbf{p}_k(t)$.

(2) *The curve $\mathcal{C}$ can have at most two axes of symmetry; furthermore, if $\mathcal{C}$ has two axes of symmetry, then the axes of symmetry are orthogonal.*

Proof. To see (1) we just need to prove ($\Leftarrow$), since ($\Rightarrow$) is a consequence of Lemma 64 and the discussion before the statement of this proposition. Now assume that $\mathcal{L}$ is a common axis of all the $\boldsymbol{p}_k(t)$, and let f denote the reflection on $\mathcal{L}$. Since $\mathcal{L}$ is a symmetry axis of $\boldsymbol{p}_k(t)$, from Lemma 58 we have $f(\boldsymbol{p}_k(t)) = \boldsymbol{p}_k(-t + \beta_k)$. Now making $t = t_0$, we deduce that $\beta_k = 2t_0$, so β_k is the same for all k. Since $\mathcal{L}$ passes through the origin $\mathbf{a}_0$, from Eq. (3.27) we get that $f(\boldsymbol{p}(t)) = \boldsymbol{p}(-t + 2t_0)$. Thus, f is a symmetry of $\mathcal{C}$. To see (2) we just need to observe that since not all the $\boldsymbol{p}_k(t)$ are circles, there exists k_0 such that $\boldsymbol{p}_{k_0}(t)$ has at most two orthogonal axes of symmetry. □

More details about how to find the axes of symmetry of the $\boldsymbol{p}_k(t)$, as well as an analysis of the case when all the $\boldsymbol{p}_k(t)$ are circles, can be found in [31]. The results on the symmetries of trigonometric curves are also used in [31] to address symmetries of discrete curves and point clouds transforming the problem via trigonometric interpolation.

3.5 FINAL REMARKS AND OPEN PROBLEMS

In [9, 18, 69, 29] we can find generalizations of the problem treated in this section. The paper [9] provides an algorithm to compute the similarities between two planar rational curves, using complex numbers. Similarities are also addressed in [18], although for a wider range of parametric curves, not just rational curves, using the notions of center of gravity and inertia tensor. In [69], the more general problem of computing the affine and projective equivalences between two given rational curves in arbitrary dimension is studied by exploiting the special structure of the polynomial system that follows from Eq. (3.3). In [29] the problem in [69] is also treated, but the approach is different: first, projective equivalences between finite sets of points in $\mathbb{P}^1(\mathbb{C})$ are studied using the notion of *cross-ratio*, which stays invariant under projective equivalences. From there the authors introduce a unifying method to address projective equivalences between rational curves in any dimension, ruled rational surfaces, and some similar problems on implicit plane algebraic curves and implicit algebraic surfaces. The special case of conic curves defined by a quadratic parametrization is addressed in [60], taking advantage, as we did here, of the fact that any proper reparametrization of a properly parametrized conic is a linear reparametrization. For conics in rational Bézier form, simple formulae, derived by using complex numbers, are found in [111]. Both in [60, 111] we can find simple algorithms to derive the main elements of a rational conic, from which the symmetries of the curve easily follow.

Let us also mention that equivalences (under different groups) have also been investigated in the context of *Classical Invariant Theory*, see for instance [92]. In this context, it is proven that the equivalence is characterized by the equality of the *signature curves* of the curves that we want to compare (see Theorem 8.53 in [92]), where the signature curve of $\mathcal{C}$ is defined as

$$\mathcal{S} = \{(\kappa(\mathbf{x}), \kappa_s(\mathbf{x}))|\mathbf{x} \in \mathcal{C}\},$$

with κ, κ_s being the curvature and the derivative of the curvature with respect to arc-length, respectively. This criterion has been used to solve problems in object recognition [37]. For rational curves, however, working with parametrizations by the arc-length is a problem, since in that case we cannot keep rationality (see Section 16.1 of [56]).

Additionally, we can pose several open problems.

Open Problem 6. *Compute affine and projective equivalences between B-splines/*NURBS *curves.* In Section 3.3.1, we observed that computing the symmetries of a Bézier curve, B-spline or NURBS curve is easy, since it suffices to compute the symmetries of a discrete object, the control polygon. Checking whether or not two Bézier curves are affinely equivalent also boils down to checking whether or not the control polygons are affinely equivalent. This is also true for projective equivalences and *rational* Bézier curves (see Chapter 13 of [55]). However, if we have two B-spline curves or two NURBS curves with different knot vectors or different weights, it is not so clear how to check affine/projective equivalence between the curves. Thus, the problem is to devise algorithms to find the affine and projective equivalences between these types of curves, exploiting the discrete structure behind (i.e., control polygons, knot vectors, weights).

Open Problem 7. *Similarity detection for rational curves depending on parameters.* This is just the rational version of Open Problem 3, in the previous chapter.

Open Problem 8. *Approximate symmetries/similarities/affine equivalences/projective equivalences of rational curves.* This is the rational version of Open Problem 4, in the previous chapter. Some contributions in this direction appear in [18, 16, 31], but this is still a very open question.

Open Problem 9. *Möbius equivalences of rational curves.* This is the rational version of Open Problem 5, in the previous chapter.

Open Problem 10. *Generalization to non-rational parametrizations.* Many classical curves, like the *cycloid*, the *tractix* or the planar spirals, are defined by parametrizations that are not rational. Thus, it is just natural to wonder how to extend the results and techniques of this chapter to the case of non-rational parametrizations. The problem is solved in [8] for planar elliptic and hyperelliptic curves, which admit parametrizations involving one square-root, taking advantage of the rich properties of these curves. Also, certain types of meromorphic parametrizations are addressed in [20], using the notion of *rational invariants*, that will be introduced in Section 5.6 of Chapter 5. However, the parametrizations in [20] need to satisfy strong hypotheses which imply the existence of a global, meromorphic, inverse: this guarantees the validity of Theorem 45 also in this setup, with a Möbius transformation at the bottom of the diagram in Eq. (3.4) (this is a consequence of the fact, already observed in Section 3.2, that the Möbius functions are the biholomorphic transformations of the complex line). Nevertheless, the conditions required in [20] are quite restrictive, so it would be good to find an alternative approach applicable to a wider class of curves. A crucial, difficult question here is to study the existence of a global inverse for non-rational parametrizations, and develop results and/or algorithms to check that condition.

CHAPTER 4

Planar algebraic vector fields

The last planar algebraic objects that we will consider are *vector fields*, which come up naturally, for instance, when analyzing systems of differential equations. This notion appeared briefly in Chapter 2, where the idea of *singular point* of a vector field was introduced and used to compute symmetries of implicit planar algebraic curves. However, the vector fields used in Chapter 2 were special, because they came from harmonic bivariate polynomials. Here we consider the question in full generality, following [14]. In particular, we will show that, in the generic case, the symmetries of algebraic vector fields are encoded in a univariate complex polynomial. There are also interesting special cases, which will lead us to study vector fields with infinitely many symmetries.

4.1 MOTIVATION AND GENERAL NOTIONS

To motivate the problem in this chapter, consider the following system of differential equations, which is a particular instance of a *Lotka-Volterra* model [39, 42], arising in Population Dynamics:

$$\begin{cases} x' &= x(1-x-\alpha y), \\ y' &= y(1-\beta x-y). \end{cases} \tag{4.1}$$

Here α, β are the interspecific interactions between the two species in competition whose sizes (over time) are x, y. This dynamical system can be written, in short, as

$$\boldsymbol{x}' = \boldsymbol{V}(\boldsymbol{x}),\ \boldsymbol{x} = (x,y),\ \boldsymbol{V}(\boldsymbol{x}) = (x(1-x-\alpha y), y(1-\beta x-y)),$$

DOI: 10.1201/9781003648710-4

where $\boldsymbol{V}(\boldsymbol{x})$ is a planar vector field. Now when $\alpha = \beta = 1$, the plot of $\boldsymbol{V}(\boldsymbol{x})$, i.e., the phase portrait of the system in Eq. (4.1) for $\alpha = \beta = 1$, is shown in Fig. 4.1. This plot shows symmetry with respect to the line $y = x$, which is easy to discover by hand since calling $\boldsymbol{V}(x, y) = (V_1(x, y), V_2(x, y))$, one observes that $\boldsymbol{V}(y, x) = (V_2(x, y), V_1(x, y))$. This symmetry implies that if we choose symmetric initial conditions with respect to the line $y = x$, the evolution of the system will be completely symmetric with respect to that line, as well.

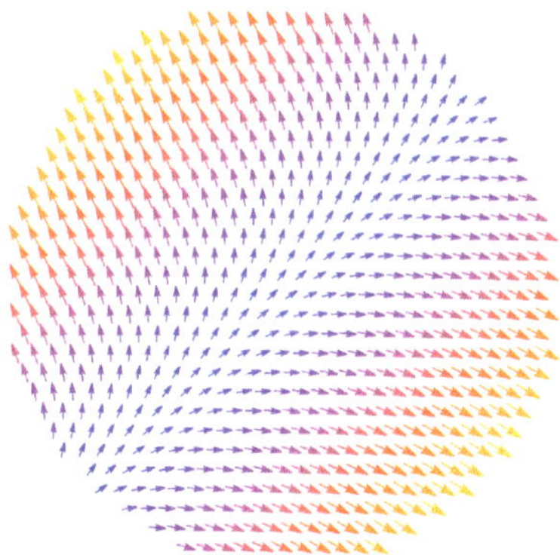

FIGURE 4.1 The vector field $\boldsymbol{V}(x, y) = (x(1-x-y), y(1-x-y))$, which appears as an instance of a Lotka-Volterra model.

In the previous example, the symmetry of the vector field was pretty clear both from Fig. 4.1, and from the expression of the vector field. However, in other cases the symmetries might be far from obvious, so we want to find algorithms to compute them. Of course, the first step is to make precise what a symmetry of a vector field is, which requires to take a closer look into the notion of vector field itself. The simplest way to understand a vector field in the affine n-space $\mathbb{A}_n$ (in this introductory section we recover the distinction between $\mathbb{A}_n$, for the set of points in the affine n-space, and $\mathbb{V}_n$, for the vectors in the underlying vector space) is to see it as a mapping $\boldsymbol{V} : \mathbb{A}_n \to \mathbb{V}_n$, i.e., as a mapping that attaches a vector of $\mathbb{V}_n$ to each point of $\mathbb{A}_n$. However, in order to give a formal definition of what a symmetry of a vector field

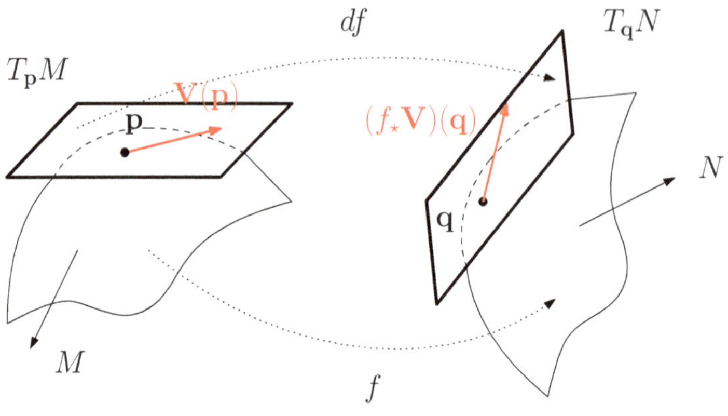

FIGURE 4.2 Pushforward

is, it is useful to consider a higher-level description. Indeed, vector fields are more generally introduced in the context of *differentiable manifolds* (see for instance [85]): given such a manifold M, that for simplicity we can roughly think of as a hypersurface of $\mathbb{A}_n$ (with good analytic properties, so that we can describe it by means of a suitable parametrization), a vector field is a mapping

$$\boldsymbol{V} : M \to \bigcup_{\mathbf{p}\in M} T_{\mathbf{p}}M,$$

where $T_{\mathbf{p}}M$ is the *tangent space* to M at $\mathbf{p} \in M$. If we have a differentiable mapping $f : M \to N$, where M, N are manifolds, and a vector field $\boldsymbol{V}$ defined in M, we can generate a vector field $f_\star\boldsymbol{V}$ in N via the *differential* df of f, so that

$$(f_\star\boldsymbol{V})(\mathbf{q}) = df_{f^{-1}(\mathbf{q})}(\boldsymbol{V}(f^{-1}(\mathbf{q})), \tag{4.2}$$

where $\mathbf{q} \in N$. The mapping $f_\star\boldsymbol{V}$ is called the *push-forward* of $\boldsymbol{V}$ by f (see p. 183 of [85]). Observe in particular that $df : T_{\mathbf{p}}M \to T_{f(\mathbf{p})}N$, so that $f^{-1}(\mathbf{q}) = \mathbf{p}$ implies that $\boldsymbol{V}(\mathbf{p})$ is mapped onto $(f_\star\boldsymbol{V})(\mathbf{q}) = df_{\mathbf{p}}(\boldsymbol{V}(\mathbf{p}))$ by df; see Fig. 4.2.

Now let us go back to the case of vector fields in $\mathbb{A}_n$. Here, $M = N = \mathbb{A}_n$, and for $\mathbf{p}, \mathbf{q} \in \mathbb{A}_n$, $T_{\mathbf{p}}M = T_{\mathbf{q}}N = \mathbb{V}_n$. Additionally $f(\boldsymbol{x}) = A\boldsymbol{x} + \boldsymbol{b}$, where $A \in \mathcal{M}_{n\times n}(\mathbb{R})$, A orthogonal, $\boldsymbol{b} \in \mathbb{R}^n$, in which case df is the linear mapping $df : \boldsymbol{v} \to A\boldsymbol{v}$. Furthermore, if $\mathbf{q} = A\mathbf{p} + \boldsymbol{b}$ and since f is a bijection, $f^{-1}(\mathbf{q}) = f^{-1}(A\mathbf{p} + \boldsymbol{b}) = \mathbf{p}$, so

$$(f_\star \boldsymbol{V})(A\mathbf{p} + \boldsymbol{b}) = df_{\mathbf{p}}(\boldsymbol{V}(\mathbf{p})) = A\boldsymbol{V}(\mathbf{p}). \tag{4.3}$$

Finally, we can reach a definition for symmetry in this context. The idea behind a symmetry is, somehow, that of invariance, so taking Eq. (4.2) into account, $\boldsymbol{V}$ being invariant under f is equivalent to $f_\star \boldsymbol{V} = \boldsymbol{V}$, in which case $(f_\star \boldsymbol{V})(A\mathbf{p} + \boldsymbol{b}) = \boldsymbol{V}(A\mathbf{p} + \boldsymbol{b})$ for all $\mathbf{p} \in \mathbb{A}_n$. Since $(f_\star \boldsymbol{V})(A\mathbf{p} + \boldsymbol{b}) = A\boldsymbol{V}(\mathbf{p})$ from Eq. (4.3), we get the following definition, where we also include the notion of an *antisymmetry*, which will be needed at the end of the chapter.

Definition 66. *Let $\boldsymbol{V} : \mathbb{A}_n \to \mathbb{V}_n$ be a vector field in $\mathbb{A}_n$, and let $f(\boldsymbol{x}) = A\boldsymbol{x} + \boldsymbol{b}$ be a symmetry of $\mathbb{A}_n$. We say that f is a* symmetry *of $\boldsymbol{V}$ if*

$$\boldsymbol{V}(A\boldsymbol{x} + \boldsymbol{b}) = A\boldsymbol{V}(\boldsymbol{x}). \tag{4.4}$$

Furthermore, we say that f, in the same conditions, is an antisymmetry *of $\mathbb{A}_n$ if*

$$\boldsymbol{V}(A\boldsymbol{x} + \boldsymbol{b}) = -A\boldsymbol{V}(\boldsymbol{x}). \tag{4.5}$$

The picture in Fig. 4.3 shows four examples of symmetric planar vector fields. In the first row, left, we have a field with translational symmetry, i.e., $f(\boldsymbol{x}) = \boldsymbol{x} + \boldsymbol{b}$ with $\boldsymbol{b}$ shown in black (a type of symmetry that, in the case of planar curves, appears only with collections of parallel lines); also in the first row, right, we have rotational symmetry, the black dot being the rotation center. In the second row, left, we provide an example of a vector field with reflectional symmetry, where the reflection axis is plotted in black; at the right we see central symmetry, the black dot being the center of symmetry. One does not need to think of any of the black dots in this picture as the origin; even if the black dot does not coincide with the origin, when $\mathbf{q} = A\mathbf{p} + \boldsymbol{b}$, i.e., when $\mathbf{p}, \mathbf{q}$ are symmetric, $\boldsymbol{V}(\mathbf{q}) = A\boldsymbol{V}(\mathbf{p})$, i.e., the vectors $\boldsymbol{V}(\mathbf{q}), \boldsymbol{V}(\mathbf{p})$ are related by the linear mapping $\boldsymbol{v} \to A\boldsymbol{v}$ associated with the affine mapping $f(\boldsymbol{x}) = A\boldsymbol{x} + \boldsymbol{b}$.

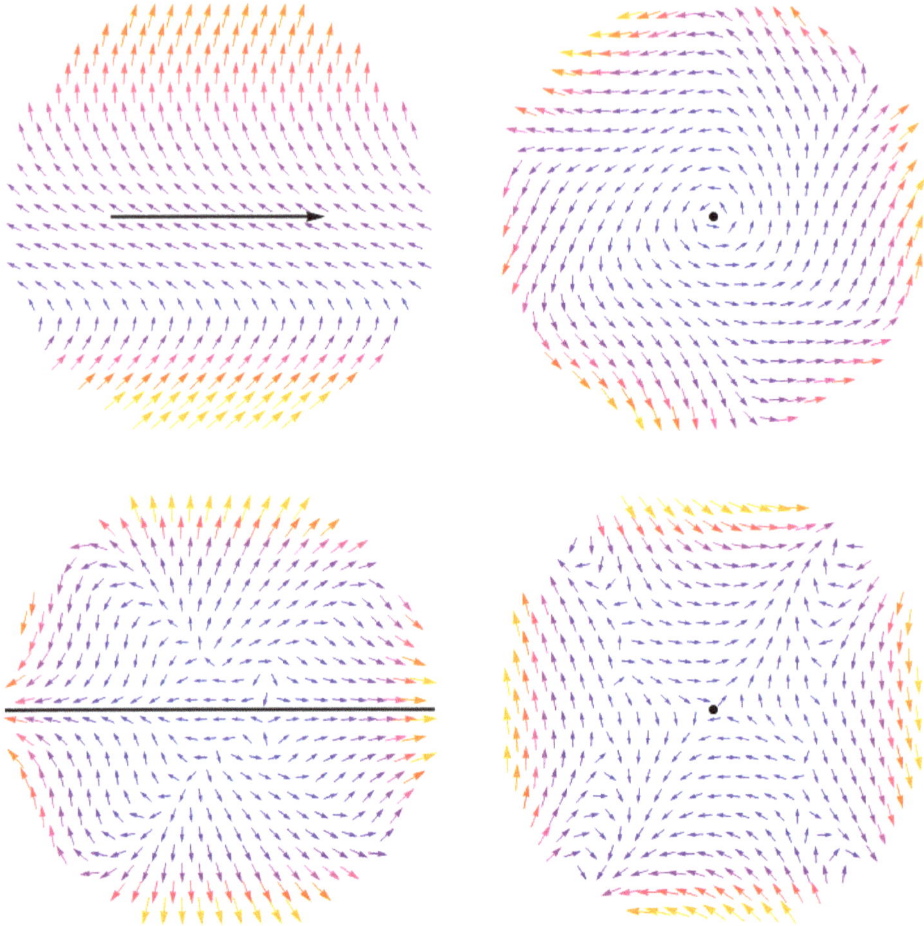

FIGURE 4.3 First row (from left to right): translational symmetry, rotational symmetry; second row (from left to right) reflectional symmetry, central symmetry

From now on we will focus on symmetries of vector fields; antisymmetries will only be invoked at the end of the chapter. We will represent the set of symmetries of a vector field $\boldsymbol{V}$ by $\mathrm{Sym}(\boldsymbol{V})$, which is a group. Also, we will restrict ourselves to the case when $\boldsymbol{V}$ is *planar*, i.e., defined in $\mathbb{A}_2$, so $\boldsymbol{V} = \boldsymbol{V}(\boldsymbol{x}) =$

$(V_1(x, y), V_2(x, y))$, and *algebraic*, implying that the components V_1, V_2 are non-constant rational functions of x, y. In fact, for simplicity we will be assuming that $\boldsymbol{V}$ is *polynomial*; we will consider the case where the V_i are rational functions at the end of the chapter.

To finish this section, let us relate the symmetries of $\boldsymbol{V}$ with those of a certain algebraic curve. Let

$$\|\boldsymbol{V}(\boldsymbol{x})\|^2 = \boldsymbol{V}(\boldsymbol{x}) \cdot \boldsymbol{V}(\boldsymbol{x}) = V_1^2(x, y) + V_2^2(x, y),$$

which corresponds to the pointwise dot product of $\boldsymbol{V}$ with itself, and let $\mathrm{Sym}(\|\boldsymbol{V}\|^2)$ be the symmetry group of the algebraic planar curve defined by $\|\boldsymbol{V}(\boldsymbol{x})\|^2$.

Lemma 67. $\mathrm{Sym}(\boldsymbol{V}) \subset \mathrm{Sym}(\|\boldsymbol{V}\|^2)$

Proof. Let $f(\boldsymbol{x}) = A\boldsymbol{x}+\boldsymbol{b}$ be a symmetry of $\boldsymbol{V}$. Thus, $\boldsymbol{V}(A\boldsymbol{x}+\boldsymbol{b}) = A\boldsymbol{V}(\boldsymbol{x})$. Since A is orthogonal it preserves dot products and norms, and therefore $\|\boldsymbol{V}(A\boldsymbol{x} + \boldsymbol{b})\|^2 = \|\boldsymbol{V}(\boldsymbol{x})\|^2$. □

Since we know how to compute the symmetries of a planar curve, Lemma 67 leads to a method to compute the symmetries of a vector field: we compute $\mathrm{Sym}(\|\boldsymbol{V}\|^2)$, and then, for each $f \in \mathrm{Sym}(\|\boldsymbol{V}\|^2)$, we check the condition in Eq. (4.4). However, $\mathrm{Sym}(\|\boldsymbol{V}\|^2)$ might be much bigger than $\mathrm{Sym}(\boldsymbol{V})$, and even infinite in case that $\|\boldsymbol{V}\|^2$ corresponds to a union of circles with the same center, or to a collection of parallel lines. Because of this, we will develop a different method, presented in Section 4.3, that has the additional advantage of providing closed forms for the sets of fixed points of the symmetries of $\boldsymbol{V}$. This method works in the generic case, but there will be also some special cases that we will analyze separately, and that will eventually lead us to study vector fields with infinitely many symmetries; at that point, Lemma 67 will prove useful.

4.2 A USEFUL TOOL: THE RESULTANT

Here we briefly recall the notion and properties of the *resultant* of two polynomials, that will appear later. We recommend

[34, 45, 89] for further reading on the topic. The resultant of two polynomials has its origin in the problem of checking whether or not two univariate polynomials $g(x), h(x)$, with coefficients in some field $\mathbb{K}$, have a common root in $\overline{\mathbb{K}}$, the algebraic closure of $\mathbb{K}$. Writing

$$\begin{array}{rcl} g(x) & = & a_n x^n + a_{n-1}x^{n-1} + \cdots + a_0, \\ h(x) & = & b_m x^m + b_{m-1}x^{m-1} + \cdots + b_0, \end{array}$$

with $a_n \neq 0$, $b_m \neq 0$, we define the *Sylvester matrix* of g, h, $S(g,h)$ as

$$S(g,h) = \begin{bmatrix} a_n & \cdots & \cdots & \cdots & a_0 & & & \\ & \ddots & & & & \ddots & & \\ & & a_n & \cdots & \cdots & \cdots & a_n \\ b_m & \cdots & \cdots & b_0 & & & \\ & \ddots & & & \ddots & & \\ & & \ddots & & & \ddots & \\ & & & b_m & \cdots & \cdots & b_0 \end{bmatrix}, \tag{4.6}$$

where the m first rows contain the coefficients of $g(x)$, and the n last rows contain the coefficients of $h(x)$; the other entries are zero. Thus, $S(g,h)$ is an $m+n$ square matrix.

Definition 68. *Let $g(x), h(x) \in \mathbb{K}[x]$, with $\mathbb{K}$ a field. The* resultant $\mathrm{Res}(g,h)$ *of $g(x), h(x)$ is the determinant of the Sylvester matrix $S(g,h)$.*

Since the $a_i, b_j \in \mathbb{K}$, $\mathrm{Res}(g,h) \in \mathbb{K}$. Then we have the following result, see [45] for a proof.

Theorem 69. *Two non-constant polynomials $g(x), h(x)$ have a common root if and only if* $\mathrm{Res}(g,h) = 0$.

Additionally, resultants satisfy the following two properties (that can also be taken as definitions of the resultant): calling $\alpha_1, \ldots, \alpha_n \in \overline{\mathbb{K}}$ to the roots of g, and $\beta_1, \ldots, \beta_m \subset \overline{\mathbb{K}}$ to the roots of h, then

$$\mathrm{Res}(g,h) = a_n^m b_m^n \prod_{i=1}^{n} \prod_{j=1}^{m} (\alpha_i - \beta_j),$$

and

$$\text{Res}(g,h) = a_n^m \prod_{j=1}^{n} h(\alpha_i) = b_m^n \prod_{i=1}^{m} g(\beta_j). \tag{4.7}$$

This last property is useful to define the resultant when one of the polynomials g, h is a constant, in which case the Sylvester matrix cannot be constructed; in particular, if one of g, h is the zero constant, then the resultant is zero as well.

Resultants can also be used with multivariate polynomials. For instance, if we have $g(x, y), h(x, y) \in \mathbb{K}[x, y]$, we can define the *resultant with respect to*, say, y, $\text{Res}_y(g, h)$, by seeing $g(x, y), h(x, y)$ as polynomials in the variable y, with coefficients in the ring $\mathbb{K}[x]$. Taking $n = \deg_y(g(x, y))$, $m = \deg_y(h(x, y))$, we can construct a Sylvester matrix for g, h with entries in $\mathbb{K}[x]$, and then take the determinant, which belongs to $\mathbb{K}[x]$; similarly for $\text{Res}_x(g, h)$. Notice that when we take the resultant of g, h with respect to y, the y variable disappears, which makes the resultant a useful tool for *elimination*. In particular, it can be applied to solve a bivariate system like

$$g(x, y) = 0, \; h(x, y) - 0. \tag{4.8}$$

The reason is the following lemma. Here, if $n = \deg_y(g)$, in which case $g(x, y) = a_n(x)y^n + a_{n-1}(x)y^{n-1} + \cdots + a_0(x)$, we say that $a_n(x)$ is the *leading* coefficient of $g(x, y)$ with respect to y; similarly for h.

Lemma 70. *Let $a_n(x), b_m(x)$ be the leading coefficients of $g(x, y), h(x, y)$ with respect to y, and let $x_0 \in \overline{\mathbb{K}}$ be a root of $\text{Res}_y(g, h)$. Then one of the following conditions hold: (1) x_0 is a common root of $a_n(x), b_m(x)$; (2) $f(x_0, y) = 0$ or $g(x_0, y) = 0$; (3) there exists $y_0 \in \overline{\mathbb{K}}$ such that (x_0, y_0) is a solution of the system in Eq.* (4.8).

Proof. If x_0 is not a common root of $a_n(x), b_m(x)$, then using properties of determinants one can check that $\text{Res}_y(f(x, y), g(x, y))(x_0)$ coincides, up to a nonzero constant, with $\text{Res}(g(x_0, y), h(x_0, y))$. Thus, if $\text{Res}_y(g(x, y), h(x, y))$ vanishes at $x = x_0$, then so does $\text{Res}(g(x_0, y), h(x_0, y))$. Then either $g(x_0, y)$ or $h(x_0, y)$ vanishes identically (recall Eq. (4.7)) or, by Theorem 69, the univariate polynomials $g(x_0, y), h(x_0, y)$ have a common root $y_0 \in \overline{\mathbb{K}}$, and (x_0, y_0) is a solution of Eq. (4.8). □

Thus, solving a system like Eq. (4.8) using resultants amounts to computing the roots of, say, $\text{Res}_y(g, h)$, denoted by $x_0, x_1, \ldots, x_\ell$, and computing afterwards the gcd of $g(x_i, y), h(x_i, y)$ for $i = 0, 1, \ldots, \ell$. It can also happen that $\text{Res}_y(g, h)$ (resp. $\text{Res}_x(g, h)$) is identically zero, and this occurs if and only if $g(x, y), h(x, y)$ have a common factor depending on y (resp. x), i.e., if and only if the planar algebraic curves defined by g, h have some common component.

Remark 6. *From the previous discussion, given a system like Eq. (4.8), if none of the resultants* $\text{Res}_x(g, h)$, $\text{Res}_y(g, h)$ *are zero, the planar curves defined by* g, h *intersect at finitely many points. Then we can use, say,* $\text{Res}_y(g, h)$, *to derive a polynomial* $\eta(x)$ *containing the* x*-coordinates of the affine intersection points; in the worst case, to find* $\eta(x)$ *we may need to factor out from* $\text{Res}_y(g, h)$ *some factor(s) of the leading coefficients of* g, h *with respect to* y. *If none of these intersections is affine, then* $\eta(x)$ *is a nonzero constant.*

We can do the same thing for polynomials in a higher number of variables. Given two polynomials $g(x_1, \ldots, x_{r-1}, x_r)$, $h(x_1, \ldots, x_{r-1}, x_r)$, seeing g, h as polynomials in, say, x_r, with coefficients in $\mathbb{K}[x_1, \ldots, x_{r-1}]$, $\text{Res}_{x_r}(g, h) \in \mathbb{K}[x_1, \ldots, x_{r-1}]$. In that case, if $(a_1, \ldots, a_{r-1}, a_r)$ is a common zero of g, h, then $\text{Res}_{x_r}(g, h)$ vanishes at $(a_1, \ldots, a_{r-1})$. Conversely, if $\text{Res}_{x_r}(g, h)$ vanishes at $(a_1, \ldots, a_{r-1})$, then one of the following holds:

1. the leading coefficients of g, h vanish at $(a_1, \ldots, a_{r-1})$.

2. Either $g(a_1, \ldots, a_{r-1}, x_r)$ or $h(a_1, \ldots, a_{r-1}, x_r)$ is identically zero.

3. There exists $a_r \in \overline{\mathbb{K}}$ such that $(a_1, \ldots, a_{r-1}, a_r)$ is a common zero of g, h.

4.3 COMPUTATION OF SYMMETRIES (I): GENERAL CASE

The idea to compute Sym($\boldsymbol{V}$) is to use the notion of *singular point*, introduced in Definition 30 (see Section 2.2 in Chapter 2). However, now we must allow not only for real singular points, but also for complex singular points, which are the solutions in $\mathbb{C}^2$ of

$$V_1(x,y)=0,\ V_2(x,y)=0.$$

Furthermore, we need to consider the following change of variables in $\mathbb{C}^2$,

$$\psi:\mathbb{C}^2\to\mathbb{C}^2,\ (x,y)\to\psi(x,y)=(z,\omega)=(x+\mathbf{i}y,x-\mathbf{i}y),\quad(4.9)$$

which is invertible, with (compare to Eq. (2.8) in Section 2.2, Chapter 2)

$$\psi^{-1}(z,\omega)=(x,y)=\left(\frac{z+\omega}{2},\frac{z-\omega}{2\mathbf{i}}\right).\qquad(4.10)$$

Compared to the strategy in Section 2.2, there is a slight change here. In Section 2.2, we used $(z,\overline{z})$ instead of (z,ω), because we were assuming that $(x,y)\in\mathbb{R}^2$ was a *real* point of $H(x,y)=0$, with $H(x,y)$ a bivariate *harmonic* polynomial; if $x,y\in\mathbb{R}$, obviously the value of the variable ω in Eq. (4.9) coincides with that of $\overline{z}$. The reason for that assumption was the fact that $H(x,y)$ was harmonic, which allowed us to replace the vector field $\boldsymbol{V}$ (see Eq. (2.7)) with a *holomorphic* complex function, therefore not depending on $\overline{z}$. Because of this, the singular points of the vector field in Eq. (2.7) had to be real. However, this is not true for just any vector field, in general.

The crucial, although essentially obvious, idea is the following.

Lemma 71. *Let* $f(\boldsymbol{x})=A\boldsymbol{x}+\boldsymbol{b}$ *be a symmetry of* $\boldsymbol{V}$. *Then* f *maps singular points of* $\boldsymbol{V}$ *onto singular points of* $\boldsymbol{V}$.

Proof. Let $\boldsymbol{x}_0$ satisfy that $\boldsymbol{V}(\boldsymbol{x}_0)=0$. By Eq. (4.4), $V(A\boldsymbol{x}_0+\boldsymbol{b})=0$, so $f(\boldsymbol{x}_0)$ is also a singular point of $\boldsymbol{V}$. □

One is therefore tempted to say that any symmetry of f is also a symmetry of the set $\mathrm{Sing}(\boldsymbol{V})$ of singular points of $\boldsymbol{V}$. There is a subtle observation here, though, because we mentioned, and we will need to use, the fact that $\mathrm{Sing}(\boldsymbol{V})$ is a subset of $\mathbb{C}^2$. So it is just natural to bring up the question of what a symmetry of $\mathbb{C}^2$ is. In Chapter 1, we defined the Euclidean symmetries as the linear mappings $f(\boldsymbol{x}) = A\boldsymbol{x} + \boldsymbol{b}$ preserving the usual dot product in $\mathbb{R}^n$, which were the mappings of this type with A an orthogonal matrix. Thus, "natural" symmetries of $\mathbb{C}^2$ can be defined as the linear mappings preserving the usual dot product in $\mathbb{C}^2$, i.e., the *Hermitian* dot product: given $\mathbf{u}, \mathbf{v} \in \mathbb{C}^2$, $\mathbf{u} = (u_1, u_2)$, $\mathbf{v} = (v_1, v_2)$, the Hermitian dot product of $\mathbf{u}, \mathbf{v}$ is

$$\langle \mathbf{u}, \mathbf{v} \rangle = \mathbf{u} \cdot \overline{\mathbf{v}}^T = u_1\overline{v_1} + u_2\overline{v_2},$$

where $\overline{\mathbf{v}}$ denotes the conjugate of $\mathbf{v}$. The matrices preserving the norm associated with the Hermitian dot product are the *unitary* matrices, which satisfy that $\overline{A}^T A = I$, with $\overline{A}$ the conjugate of A. Thus, the symmetries in $\mathbb{C}^2$ are the mappings $f(\boldsymbol{x}) = A\boldsymbol{x} + \boldsymbol{b}$, with A unitary, $\boldsymbol{b} \in \mathbb{C}^2$. In particular any real orthogonal matrix is unitary, so we finally state the following, as a corollary of Lemma 71.

Corollary 72. *Let* $f(\boldsymbol{x}) = A\boldsymbol{x} + \boldsymbol{b}$ *be a real symmetry of* $\boldsymbol{V}$. *Then* f *is also a symmetry of* $\mathrm{Sing}(\boldsymbol{V})$.

Thus, the idea is to aim for the set $\mathrm{Sym}(\mathrm{Sing}(\boldsymbol{V}))$, and extract $\mathrm{Sym}(\boldsymbol{V})$ from there. However, for reasons that will be clear later, we will be considering not $\boldsymbol{V}$, but the vector field $\widehat{\boldsymbol{V}} = \boldsymbol{V} \circ \psi^{-1}$, with ψ^{-1} in Eq. (4.10), where $\widehat{V}(z, \omega) = (\widehat{V}_1(z, \omega), \widehat{V}_2(z, \omega))$, $\widehat{V}_i = V_i \circ \psi^{-1}$. The mapping ψ in Eq. (4.9) defines a bijection between $\mathrm{Sing}(\boldsymbol{V})$ and the set $\mathrm{Sing}(\widehat{\boldsymbol{V}})$ of singular points of $\widehat{\boldsymbol{V}}$, which is the solution to the system

$$\widehat{V}_1(z, \omega) = 0, \;\; \widehat{V}_2(z, \omega) = 0. \tag{4.11}$$

Additionally, any real symmetry $f(\boldsymbol{x}) = A\boldsymbol{x} + \boldsymbol{b}$ of $\boldsymbol{V}$ gives rise to a mapping $\widehat{f} : \mathbb{C}^2 \to \mathbb{C}^2$, $\widehat{f} = f \circ \psi^{-1}$. In order to explore $\widehat{f}$, one can observe that Eq. (4.10) is nothing else that a linear change of parameters,

$$\begin{bmatrix} z \\ \omega \end{bmatrix} = \begin{bmatrix} 1 & \mathbf{i} \\ 1 & -\mathbf{i} \end{bmatrix} \begin{bmatrix} x \\ y \end{bmatrix}, \quad \begin{bmatrix} x \\ y \end{bmatrix} = \begin{bmatrix} 1/2 & 1/2 \\ -\mathbf{i}/2 & \mathbf{i}/2 \end{bmatrix} \begin{bmatrix} z \\ \omega \end{bmatrix}. \tag{4.12}$$

Taking into account the possible shapes for the matrix A (see Eq. (1.10) in Chapter 1), the mapping $\widehat{f}$ can be as follows: here $\alpha, \beta \in \mathbb{R}$, with $\alpha^2 + \beta^2 = 1$, and the primes are used to distinguish the original point from the image of the point under f or $\widehat{f}$; the mapping $\widehat{f}$ appears at the right (f is at the left):

(i) Translations:

$$\begin{bmatrix} x' \\ y' \end{bmatrix} = \begin{bmatrix} x \\ y \end{bmatrix} + \begin{bmatrix} b_1 \\ b_2 \end{bmatrix} \Rightarrow \begin{bmatrix} z' \\ \omega' \end{bmatrix} = \begin{bmatrix} z \\ \omega \end{bmatrix} + \begin{bmatrix} b_1 + \mathbf{i}b_2 \\ b_1 - \mathbf{i}b_2 \end{bmatrix}. \tag{4.13}$$

(ii) Rotations:

$$\begin{aligned} \begin{bmatrix} x' \\ y' \end{bmatrix} &= \begin{bmatrix} \alpha & -\beta \\ \beta & \alpha \end{bmatrix} \begin{bmatrix} x \\ y \end{bmatrix} + \begin{bmatrix} b_1 \\ b_2 \end{bmatrix} \Rightarrow \begin{bmatrix} z' \\ \omega' \end{bmatrix} \\ &= \begin{bmatrix} \alpha + \mathbf{i}\beta & 0 \\ 0 & \alpha - \mathbf{i}\beta \end{bmatrix} \begin{bmatrix} z \\ \omega \end{bmatrix} + \begin{bmatrix} b_1 + \mathbf{i}b_2 \\ b_1 - \mathbf{i}b_2 \end{bmatrix}. \end{aligned} \tag{4.14}$$

(iii) Reflections:

$$\begin{aligned} \begin{bmatrix} x' \\ y' \end{bmatrix} &= \begin{bmatrix} \alpha & \beta \\ \beta & -\alpha \end{bmatrix} \begin{bmatrix} x \\ y \end{bmatrix} + \begin{bmatrix} b_1 \\ b_2 \end{bmatrix} \Rightarrow \begin{bmatrix} z' \\ \omega' \end{bmatrix} \\ &= \begin{bmatrix} 0 & \alpha + \mathbf{i}\beta \\ \alpha - \mathbf{i}\beta & 0 \end{bmatrix} \begin{bmatrix} z \\ \omega \end{bmatrix} + \begin{bmatrix} b_1 + \mathbf{i}b_2 \\ b_1 - \mathbf{i}b_2 \end{bmatrix}. \end{aligned} \tag{4.15}$$

One can check that in all the cases the matrix appearing in $\widehat{f}$ is a unitary matrix. In fact, we have the following.

Lemma 73. $f \in \text{Sym}(\text{Sing}(\boldsymbol{V})) \Leftrightarrow \widehat{f} \in \text{Sym}(\text{Sing}(\widehat{\boldsymbol{V}}))$.

So we must look for $\text{Sym}(\text{Sing}(\widehat{\boldsymbol{V}}))$, now. In order to do this, we observe that $\text{Sing}(\widehat{\boldsymbol{V}})$ is the solution set of the bivariate system in Eq. (4.11). As we saw in Section 4.2, we can solve this system by using resultants. Let

$$g^\star(z) = \text{Res}_\omega(\widehat{V}_1, \widehat{V}_2). \tag{4.16}$$

If $\widehat{V}_1, \widehat{V}_2$ share a factor, as noted in Section 4.2, $g^\star(z) = 0$. Otherwise, from Remark (6), we can compute, from $g^\star(z)$, a polynomial $g(z)$ whose roots correspond exactly to the z-coordinates of the affine solutions of Eq. (4.11), perhaps after removing from $g^\star(z)$ a spurious factor related to the leading coefficients of $\widehat{V}_1, \widehat{V}_2$ with respect to ω. For now, we will assume that $g(z)$ is not constant, and postpone the special case of $g(z)$ being constant for the next section; in that section we will also include the case when $g^\star(z)$ is identically zero, where we define $g(z)$ to be identically zero as well.

The surprising result is that the symmetries of Sing($\boldsymbol{V}$) are symmetries of the roots of $g(z)$ (see Example 7), identified with points of $\mathbb{R}^2$ via

$$\varphi : \mathbb{C} \mapsto \mathbb{R}^2,\ z = x + \mathbf{i}y \mapsto \varphi(z) = (x, y), \tag{4.17}$$

which is the reason why we worked with the vector field $\widehat{\boldsymbol{V}}$, instead of $\boldsymbol{V}$. This would not be surprising if all the singular points of $\boldsymbol{V}$ were real, but recall that we may also have complex singular points.

In order to justify this, we start by examining in more detail the transformations in Eqs. (4.13), (4.14), and (4.15). Translations, in Eq. (4.13), give rise to mappings

$$\widehat{f}(z, \omega) = (z + B, \omega + \overline{B}) \tag{4.18}$$

with $B = b_1 + \mathbf{i}b_2$. Rotations, in Eq. (4.14), give rise to mappings

$$\widehat{f}(z, \omega) = (\mathbf{u}z + B, \overline{\mathbf{u}}\omega + \overline{B}) \tag{4.19}$$

with $\mathbf{u} = \alpha + i\beta$, $B = b_1 + ib_2$. And reflections, in Eq. (4.15), correspond to

$$\widehat{f}(z, \omega) = (\mathbf{u}\omega + B, \overline{\mathbf{u}}z + \overline{B}). \tag{4.20}$$

Since $\alpha, \beta \in \mathbb{R}$, $\overline{\mathbf{u}}, \overline{B}$ are the conjugates of $\mathbf{u}, B$. Calling

$$\widehat{f}(z) = (\widehat{f}_1(z,\omega), \widehat{f}_2(z,\omega)),$$

we notice that when f is either a translation or a rotation, $\widehat{f}_1(z,\omega) = \widehat{f}_1(z)$ corresponds to the complex form of the mapping f, see Section 3.3 in Chapter 3. This is not true for reflections, though, since z and ω are swapped. This means that if $\mathrm{Sing}(\boldsymbol{V})$ undergoes either a translation or a rotation, and in general an orientation-preserving symmetry (written in complex form as $f(z) = \mathbf{u}z + B$), the roots of $g(z)$, which are the projections onto the z-coordinate of the solutions of Eq. (4.11), undergo the same transformation. Thus we have the following result; here we denote by $\mathrm{Sym}(g)$ the set of symmetries of the roots of $g(z)$ (identified with points of $\mathbb{R}^2$ via Eq. (4.17)).

Lemma 74. *Let f be an orientation-preserving Euclidean symmetry. The images of* $\mathrm{Sing}(\boldsymbol{V})$ *under f correspond pointwise to the images of the roots of $g(z)$ under f (written in complex form). In particular, if $f \in$* $\mathrm{Sym}(\mathrm{Sing}(\boldsymbol{V}))$, *then $f \in$* $\mathrm{Sym}(g)$ *as well.*

A consequence of this result is that $\boldsymbol{V}$ cannot have translational symmetry unless $g(z)$ is a constant.

Corollary 75. *If $g(z)$ is not a constant, then $\boldsymbol{V}$ does not have translational symmetry and is not invariant under any glide reflection.*

Proof. By Lemma 74, any translational symmetry of $\boldsymbol{V}$ is also a translational symmetry of the roots of $g(z)$. However, the roots of a non-constant polynomial cannot have non-trivial translational symmetry: indeed, the only solution for $g(z+\mathbf{u}) = \lambda g(z)$, with $g(z)$ a polynomial, is $\mathbf{u} = 0$ and $\lambda = 1$; if $\mathbf{u} \neq 0$, we would deduce that the roots of $g(z)$ and $g(z+\mathbf{u})$ are the same, which cannot happen because the roots of the latter are shifted with respect to the roots of the former. Since the composition of a glide reflection with itself is a translation and the symmetries of $\boldsymbol{V}$ form a group, the second part of the corollary is a consequence of the first part.
□

For reflections we cannot argue as in Lemma 74, because $\widehat{f_1}(z,\omega) \neq \widehat{f_1}(z)$. However, since we have already proven the result for orientation-preserving symmetries, it is enough to prove the following lemma.

Lemma 76. *Assume that* $f(\boldsymbol{x}) = f(x,y) = (x,-y)$ *is a symmetry of the vector field* $\boldsymbol{V}$. *Then* f *is a symmetry of the set of roots of* $g(z)$.

Proof. Since $f(\boldsymbol{x}) = f(x,y) = (x,-y) \in \mathrm{Sym}(\boldsymbol{V})$, $f \in \mathrm{Sym}(\mathrm{Sing}(\boldsymbol{V}))$ too. Thus, $(x_i,y_i) \in \mathrm{Sing}(\boldsymbol{V})$, $x_i,y_i \in \mathbb{C}$, iff $(x_i,-y_i) \in \mathrm{Sing}(\boldsymbol{V})$, so the elements of $\mathrm{Sing}(\boldsymbol{V})$ can be grouped in pairs of symmetric (complex) points $\{(x_i,y_i),(x_i,-y_i)\}$. Using Eq. (4.9) and since ψ is a bijection between $\mathrm{Sing}(\boldsymbol{V})$ and $\mathrm{Sing}(\widehat{\boldsymbol{V}})$, the points of $\mathrm{Sing}(\widehat{\boldsymbol{V}})$ can be grouped in pairs $\{(x_i+\mathbf{i}y_i, x_i-\mathbf{i}y_i),(x_i-\mathbf{i}y_i, x_i+\mathbf{i}y_i)\}$. Since the roots of $g(z)$ are exactly the first coordinates of the points of $\mathrm{Sing}(\widehat{\boldsymbol{V}})$, we deduce that the roots of $g(z)$ can be grouped in pairs $\{x_i+\mathbf{i}y_i, x_i-\mathbf{i}y_i\}$, so f is a symmetry of the set of roots of $g(z)$, identified with points of $\mathbb{R}^2$ via Eq. (4.17). □

Then they have the following result.

Theorem 77. *Let* f *be a symmetry of* $\mathrm{Sing}(\boldsymbol{V})$. *Then* $f \in \mathrm{Sym}(g)$.

Proof. The result was proven for orientation-preserving symmetries in Lemma 74, so we just need to prove it for reflections. Thus, assume that f is a reflection on the line $\mathcal{L}$. Let h be the orientation-preserving symmetry mapping $\mathcal{L}$ onto the x-axis, and let $\tilde{f} = f \circ h$, so $\tilde{f}$ is a reflection on the x-axis. Also, let $\tilde{\boldsymbol{V}} = \boldsymbol{V} \circ h$. Thus, $\mathrm{Sing}(\tilde{\boldsymbol{V}}) = h(\mathrm{Sing}(\boldsymbol{V}))$. Furthermore, by the first part of Lemma 74, $\tilde{g}(z)$, which is the univariate polynomial corresponding to Eq. (4.16) for $\tilde{\boldsymbol{V}}$, satisfies that $\tilde{g} = g \circ h$, where g corresponds to Eq. (4.16) for $\boldsymbol{V}$, and h is written in complex form. Since $f \in \mathrm{Sym}(\mathrm{Sing}(\boldsymbol{V}))$, $\tilde{f} \in \mathrm{Sym}(\mathrm{Sing}(\tilde{\boldsymbol{V}}))$, and by Lemma 76 we get that $\tilde{f} \in \mathrm{Sym}(\tilde{g})$. Since $\tilde{g} = g \circ h$, the result follows. □

The symmetries of the roots of a complex polynomial $g(z)$ were already addressed in Section 2.2.1, when we studied symmetries of harmonic planar curves. In particular, any rotational symmetry of the roots of $g(z)$ corresponds to a rotation with center at the *barycenter* of the roots of $g(z)$ (see Lemma 33). Furthermore, the following lemma shows that the center of any rotationally symmetric vector field $\boldsymbol{V}$ is found among its singular points.

Lemma 78. *If $\boldsymbol{V}$ has rotational symmetry with center of rotation $\mathbf{c}$, then $\mathbf{c} \in \mathrm{Sing}(\boldsymbol{V})$.*

Proof. Let $f(\boldsymbol{x}) = A\boldsymbol{x} + \boldsymbol{b}$ define a rotational symmetry of $\boldsymbol{V}$, and let $\mathbf{c}$ be the rotation center, which is a fixed point of f. Since $f(\mathbf{c}) = A\mathbf{c} + \boldsymbol{b} = \mathbf{c}$, from Eq. (4.4) we deduce that $(A - I)\boldsymbol{V}(\mathbf{c}) = \mathbf{0}$. If the rotation angle is π, in which case f corresponds to a central symmetry, the only eigenvalue of A is -1; otherwise, the eigenvalues of A are not real. Thus, in particular $\lambda = 1$ is not an eigenvalue of A, so $\det(A - I) \neq 0$ and $\boldsymbol{V}(\mathbf{c}) = \mathbf{0}$. Hence, $\mathbf{c} \in \mathrm{Sing}(\boldsymbol{V})$. □

Writing $g(z)$ as in Eq. (2.9), we can efficiently give the coordinates of the barycenter by using Cardano-Vieta relationships; the complex form of the barycenter appears in Eq. (2.11). Additionally, after computing the barycenter, we can always perform a translation so that the barycenter is taken to the origin. In the rest of the section (and also in some parts of the following section), we will assume that this has been done. Now to compute the rotation angle θ, we will assume that $g(z)$ has at least two different roots, i.e., that after performing the translation taking the barycenter to the origin we have $g(z) \neq z^m$; the special case $g(z) = z^m$ (up to multiplication by constants) will also be considered in the next section.

Assuming that $g(z)$ is not a constant and $g(z) \neq z^m$, the rotational symmetries of $g(z)$ are the symmetries of a finite set of points, so $\theta = \frac{2\pi}{k}$, with $k \in \mathbb{N}$. As we saw in Section 2.3, rotations around the origin correspond to complex mappings $z \mapsto e^{\mathbf{i}\theta} z$. Writing $g(z) = z^{m_1} \sum_{j=0}^{m_2} \beta_j z^j$, where $m_1 + m_2 = m$, $m_2 \geq 1$, the condition that both $z, e^{\frac{2\pi j}{k}\mathbf{i}}$ are roots of $g(z)$ is equivalent to

$$z^{m_1}\sum_{j=0}^{m_2}\beta_j\left(1-e^{\frac{2\pi j}{k}\mathbf{i}}\right)z^j=0.$$

But this implies that the potential values of k are the common divisors of the js whose coefficients β_j are nonzero. We summarize the results concerning rotations in the following proposition.

Proposition 79. *Let $g(z)=\sum_{j=0}^{m}\alpha_j z^j$. If $\mathbf{V}$ has rotational symmetry, the center of rotation, in complex form, is*

$$\mathbf{c}=-\frac{\alpha_{m-1}}{m\alpha_m}. \tag{4.21}$$

Let $\tilde{g}(z)=g(z-\mathbf{c})=z^{m_1}\sum_{j=0}^{m_2}\beta_j z^j$. Then the potential rotation angles are $\theta=\frac{2\pi}{k}$, where k is a common divisor of the js whose coefficients β_j are nonzero.

For reflections, whenever $g(z)\neq z^m$, the reflection axis can be computed using Eq. (2.12). The ideas in this section are illustrated in the following example, taken from [14].

Example 7. *Let $\mathbf{V}(x,y)=(-3x^3-3xy^2-9x^2-6xy-3y^2-6y+6,-3x^2y-3y^3-3x^2-6xy+15y^2-6x-24y+6)$, plotted in Fig. 4.4, left. First, we compute*

$$g(z)=\tfrac{81}{2}\mathbf{i}\,z^7+\left(\tfrac{567}{2}+\tfrac{567}{2}\mathbf{i}\right)z^6+1701z^5+(2511-2835\mathbf{i})z^4$$
$$-(1296+4374\mathbf{i})z^3-(3402-486\mathbf{i})z^2+(324+2592\mathbf{i})z+(972+324\mathbf{i}).$$

The barycenter of the roots of $g(z)$ (and thus the center of a potential rotation) is

$$b_\sigma=-\frac{\frac{567}{2}+\frac{567}{2}\mathbf{i}}{7\cdot\frac{81}{2}\mathbf{i}}=-1+\mathbf{i}.$$

Hence, we apply the translation $z\mapsto z+(1-\mathbf{i})$ and arrive at the new polynomial

$$\tilde{g}(z)=\frac{81}{2}\mathbf{i}z^7-324z^4=z^4\left(\frac{81}{2}\mathbf{i}z^3-324\right).$$

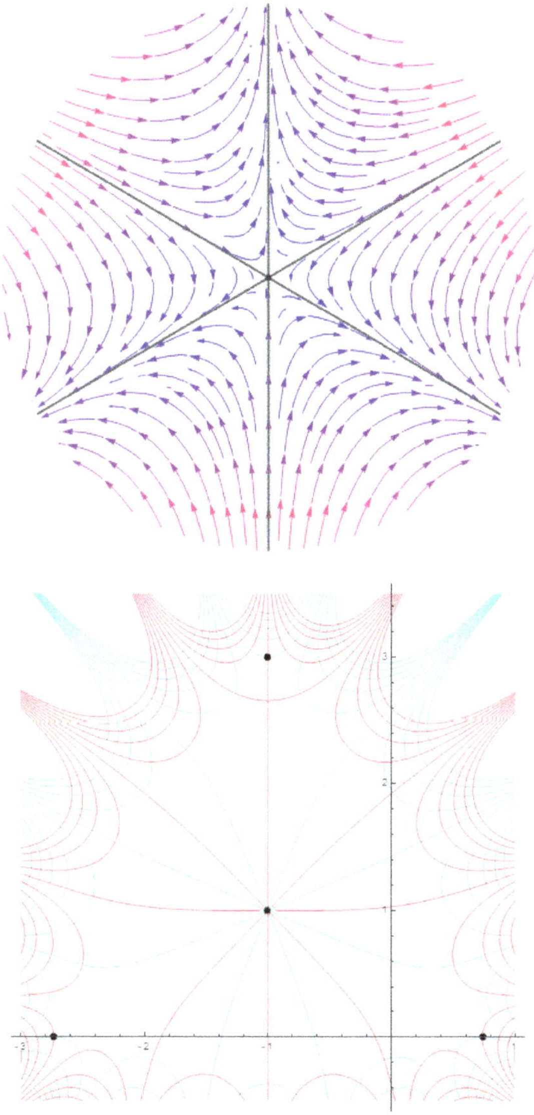

FIGURE 4.4 Left: The vector field from Example 7 with its reflection lines highlighted. Right: Symmetrically distributed roots of the associated polynomial $g(z)$ with the same group of symmetries (green and red curves visualize the contour lines of real and imaginary part of $g(z)$).

From Proposition 79 we obtain $k = 3$. *It can be verified by direct computation that* $\boldsymbol{V}(x, y)$ *has the rotational symmetry of the equilateral triangle with the center* $\mathbf{c} = (-1, 1)$.

Next using Eq. (2.12), *where* $m = 7$ *and* $\ell = 4$, *we get*

$$3\theta = \arg(-8\mathbf{i}) \bmod 2\pi,$$

which yields three angles $\frac{\pi}{2}, \frac{7\pi}{6}, \frac{11\pi}{6}$ *and thus three lines through the point* $[-1, 1]$,

$$x + 1 = 0, \quad x - \sqrt{3}y + 1 + \sqrt{3} = 0, \quad x + \sqrt{3}y + 1 - \sqrt{3} = 0.$$

Again, it can be verified by direct computation that all three reflections given by the above lines are symmetries of $\boldsymbol{V}(x, y)$.

4.4 COMPUTATION OF SYMMETRIES (II): SPECIAL CASES

Here we will be considering the cases left out in Section 4.3, namely the cases when the polynomial $g(z)$ has just one (possibly multiple) root, or is a constant; when $g(z)$ is not constant, we will assume that a translation has been performed so that the only root of $g(z)$ is zero. Thus, we distinguish three possibilities: (a) $g(z) = z^m$ (up to multiplication by constants), $m \neq 0$; (b) $g(z) = c \neq 0$, i.e., the case where $g(z)$ is a nonzero constant; (c) $g(z) = 0$. Among these cases we will find a situation that did not come up in Section 4.3, namely the existence of infinitely many symmetries, including the possibility of translational symmetry.

4.4.1 Case (a): $g(z) = z^m$, $m \neq 0$

Since $\mathrm{Sing}(\boldsymbol{V}) = \{(0, 0)\}$, by Lemma 71 we deduce that the origin must be a fixed point of any symmetry of $\boldsymbol{V}$. Thus, the symmetries of $\boldsymbol{V}$, in this case, have the form $f(\boldsymbol{x}) = A\boldsymbol{x}$. Since in this case we are looking for symmetries fixing the origin, we will introduce two auxiliary scalar fields that will be helpful. First, we denote the identity vector field, also called the *Euler vector field*, by $\boldsymbol{x} = (x, y)$, and we consider

$$\boldsymbol{V}(\boldsymbol{x})\cdot\boldsymbol{x} = xV_1(x, y) + yV_2(x, y), \quad \boldsymbol{V}(\boldsymbol{x})\times\boldsymbol{x} = (yV_1(x, y) - xV_2(x, y))\mathbf{k},$$

where $\mathbf{k}$ is a unit vector normal to the x, y coordinate axes. Then let us define

$$\begin{aligned} X_1(\boldsymbol{x}) &= xV_1(x,y) + yV_2(x,y), \\ X_2(\boldsymbol{x}) &= yV_1(x,y) - xV_2(x,y). \end{aligned} \tag{4.22}$$

Since $\boldsymbol{V}$ is a polynomial vector field, X_1, X_2 are polynomials, therefore defining two algebraic curves.

A first result on the connections of these polynomials with the symmetries of $\boldsymbol{V}$ is the following proposition, that shows how to quickly compute the reflection axes for $\boldsymbol{V}$, if any.

Proposition 80. *Let $\mathcal{L}$ be a reflectional axis through the origin of a polynomial planar vector field $\boldsymbol{V}$. Then $\mathcal{L}$ is a linear factor of X_2.*

Proof. Let $f(\boldsymbol{x}) = A\boldsymbol{x}$ be the reflection whose set of fixed points is $\mathcal{L}$. Since f is a reflection of $\boldsymbol{V}$, $\boldsymbol{V}(A\boldsymbol{x}) = A\boldsymbol{V}(\boldsymbol{x})$. Next every $\mathbf{p} \in \mathcal{L}$ satisfies that $A\mathbf{p} = \mathbf{p}$. Thus, the $\mathbf{p}$'s, seen as vectors of $\mathbb{R}^2$, are eigenvectors of the matrix A associated with the eigenvalue $\lambda = 1$. Substituting $A\mathbf{p} = \mathbf{p}$ into $\boldsymbol{V}(A\boldsymbol{x}) = A\boldsymbol{V}(\boldsymbol{x})$ we get that $(A - I)\boldsymbol{V}(\mathbf{p}) = \mathbf{0}$, so the vectors $\boldsymbol{V}(\mathbf{p})$ with $\mathbf{p} \in \mathcal{L}$ are also eigenvectors associated with the eigenvalue $\lambda = 1$. Since the eigenspace associated with $\lambda = 1$ has dimension one (this eigenspace consists of the vectors parallel to $\mathcal{L}$), then $\mathbf{p}$ and $\boldsymbol{V}(\mathbf{p})$ are parallel, so $\boldsymbol{V}(\mathbf{p}) \times \mathbf{p} = \mathbf{0}$. Since $\boldsymbol{V}(\boldsymbol{x}) \times \boldsymbol{x}$ vanishes at all the points of $\mathcal{L}$, the algebraic curve defined by X_2 and $\mathcal{L}$ must have a common factor. Since $\mathcal{L}$ has degree one, the result follows. □

Remark 7. *Proposition 80 can also be used in the case of Section 4.3, whenever we have performed a translation taking the barycenter of the roots of $g(z)$ to the origin. For instance, in Example 7, after translating the vector field $\boldsymbol{V}(x, y)$ so that the barycenter of the roots is at the origin, we obtain, for the translated vector field $\boldsymbol{V}_0(x, y)$, the polynomial $X_2(x, y) = 6x^3 - 18xy^2$. This can be factorized as $6x(x - \sqrt{3}y)(x + \sqrt{3}y)$. The backward translation yields the same axes computed in Example 7.*

But the connection between the symmetries of $\boldsymbol{V}$ and the symmetries of the curves defined by X_1, X_2 is deeper. Before we address this connection we need some previous lemmas. To formulate these results, we denote the set of symmetries of $\boldsymbol{V}$ fixing the origin by $\mathrm{Sym}_0(\boldsymbol{V})$; also, we denote by $\mathrm{Sym}_0(X_1)$, $\mathrm{Sym}_0(X_2)$ the symmetry groups of the algebraic curves defined by X_1, X_2 fixing the origin. It is clear though that in the case addressed in this section, $\mathrm{Sym}_0(\boldsymbol{V}) = \mathrm{Sym}(\boldsymbol{V})$.

Lemma 81. *Let $f \in \mathrm{Sym}_0(\boldsymbol{V})$. Then $X_1 \circ f = X_1$; in particular, $f \in \mathrm{Sym}_0(X_1)$.*

Proof. Let $f(\boldsymbol{x}) = A\boldsymbol{x}$. Since $\boldsymbol{V}(A\boldsymbol{x}) = A\boldsymbol{V}(\boldsymbol{x})$ and A is orthogonal,

$$(X_1 \circ f)(\boldsymbol{x}) = \boldsymbol{V}(A\boldsymbol{x}) \cdot A\boldsymbol{x} = \boldsymbol{V}(\boldsymbol{x}) \cdot \boldsymbol{x} = X_1(\boldsymbol{x}).$$

□

Lemma 82. *Let $f \in \mathrm{Sym}_0(\mathbf{V})$. Then $X_2 \circ f = \det(A) \cdot X_2$; in particular, $f \in \mathrm{Sym}_0(X_2)$.*

Proof. Let $f(\boldsymbol{x}) = A\boldsymbol{x}$. Since $\boldsymbol{V}(A\boldsymbol{x}) = A\boldsymbol{V}(\boldsymbol{x})$, using Eq. (3.9) in Chapter 3, we get that

$$\boldsymbol{V}(A\boldsymbol{x}) \times (A\boldsymbol{x}) = (A\boldsymbol{V}(\boldsymbol{x})) \times (A\boldsymbol{x}) = \det(A) \cdot (\boldsymbol{V}(\boldsymbol{x}) \times \boldsymbol{x}).$$

Since $\boldsymbol{V}(A\boldsymbol{x}) \times (A\boldsymbol{x}) = (X_2 \circ \phi)(\boldsymbol{x})\mathbf{k}$ and $\boldsymbol{V}(\boldsymbol{x}) \times \boldsymbol{x} = X_2(\boldsymbol{x})\mathbf{k}$, the result follows. □

Corollary 83. $\mathrm{Sym}_0(\boldsymbol{V}) \subset \mathrm{Sym}_0(X_1) \cap \mathrm{Sym}_0(X_2)$.

In particular, whenever $\mathrm{Sym}_0(X_1)$ or $\mathrm{Sym}_0(X_2)$ is finite, $\mathrm{Sym}_0(\mathbf{V})$ is finite as well, and the symmetries of $\boldsymbol{V}$ with a fixed point can be computed from those of X_1, X_2. We can be even more precise about the connection between $\mathrm{Sym}_0(\boldsymbol{V})$, $\mathrm{Sym}_0(X_1)$ and $\mathrm{Sym}_0(X_2)$.

Theorem 84. *Let $f(\boldsymbol{x}) = A\boldsymbol{x}$ be a Euclidean symmetry. Then f is a symmetry of $\mathbf{V}$ if and only if f is a symmetry of both X_1, X_2, satisfying that $X_1 \circ f = X_1$, $X_2 \circ f = \det(A) \cdot X_2$.*

Proof. The implication ($\Rightarrow$) follows from Lemma 81 and Lemma 82, so let us focus on the implication ($\Leftarrow$). Let $f \in \mathrm{Sym}_0(X_1) \cap$

$\mathrm{Sym}_0(X_2)$, $f(\boldsymbol{x}) = A\boldsymbol{x}$, satisfy that $X_1 \circ f = X_1$, $X_2 \circ f = \det(A) \cdot X_2$. Additionally, let $\tilde{V}_1 = V_1 \circ f$, $\tilde{V}_2 = V_2 \circ f$. We prove the implication for rotations; similarly for reflections. Thus, assume that $f(\boldsymbol{x}) = (\alpha x - \beta y, \beta x + \alpha y)$, $\alpha^2 + \beta^2 = 1$. Then $X_1 \circ f = X_1$ implies that

$$(\alpha x - \beta y)\tilde{V}_1 + (\beta x + \alpha y)\tilde{V}_2 = xV_1 + yV_2. \tag{4.23}$$

Furthermore, since f is a rotation we have that $\det(A) = 1$; then $X_2 \circ \phi = X_2$ implies that

$$(\beta x + \alpha y)\tilde{V}_1 - (\alpha x - \beta y)\tilde{V}_2 = yV_1 - xV_2. \tag{4.24}$$

Seeing Eqs. (4.23) and (4.24) as a linear system in $\tilde{V}_1, \tilde{V}_2$ and applying Cramer's rule, we get that

$$\tilde{V}_1 = \alpha V_1 - \beta V_2,\ \tilde{V}_2 = \beta V_1 + \alpha V_2.$$

But this implies that $\boldsymbol{V} \circ f = f \circ \boldsymbol{V}$, i.e., that f is a rotational symmetry of $\boldsymbol{V}$. For reflections, the proof is analogous, taking into account that $\det(A) = -1$. □

This theorem opens up the possibility of finding vector fields with infinitely many symmetries: as a trivial example, for the Euler field $\boldsymbol{V} = (x, y)$ we have $X_1 = x^2 + y^2$ and $X_2 = 0$; since any rotation around the origin leaves X_1 invariant, from Theorem 84 we deduce that $\boldsymbol{V}$ has infinitely many symmetries. We wonder if we can give a more complete description of vector fields within this case and with that property, i.e., $\mathrm{Sym}(\boldsymbol{V})$ being infinite. To do that, we first need this result.

Lemma 85. *If $\boldsymbol{V}$ has infinitely many symmetries fixing the origin, then X_1, X_2 are products of (possibly imaginary) circles centered at the origin.*

Proof. By Corollary 83, if $\boldsymbol{V}$ has infinitely many symmetries fixing the origin, then so do X_1, X_2. By Theorem 26 in Chapter 2, the only algebraic curves with infinitely many symmetries are unions of parallel lines and unions of circles with a common center. However, the first possibility cannot occur because the number of symmetries of a union of parallel lines that fix the origin is finite (although there are infinitely many translations leaving such a curve invariant). □

Thus we can provide the following characterization for vector fields within this case, with infinitely many symmetries fixing the origin.

Theorem 86. *Let $\boldsymbol{V}(\boldsymbol{x}) = (V_1(\boldsymbol{x}), V_2(\boldsymbol{x}))$ be a planar, polynomial, real vector field with $g(z) = z^m$, $m \neq 0$. Then $\mathbf{V}$ has infinitely many symmetries fixing the origin if and only if*

$$\begin{aligned} V_1(x,y) &= xP_1(x^2+y^2) + yP_2(x^2+y^2), \\ V_2(x,y) &= yP_1(x^2+y^2) - xP_2(x^2+y^2), \end{aligned} \tag{4.25}$$

where P_1, P_2 are univariate polynomials (possibly constant).

Proof. ($\Rightarrow$) From the definition of X_1, X_2 in Eq. (4.22) and Lemma 85, we get that

$$\begin{aligned} xV_1 + yV_2 &= \tilde{P}_1(x^2+y^2), \\ yV_1 - xV_2 &= \tilde{P}_2(x^2+y^2), \end{aligned} \tag{4.26}$$

with $\tilde{P}_1, \tilde{P}_2$ univariate polynomials. If we substitute $x = 0, y = 0$ into Eq. (4.26), we get that $\tilde{P}_1(0) = \tilde{P}_2(0) = 0$, implying that x^2+y^2 is a factor of both $\tilde{P}_1(x^2+y^2)$ and $\tilde{P}_2(x^2+y^2)$. Now solving Eq. (4.26) for V_1, V_2, we get

$$V_1 = \frac{x\tilde{P}_1(x^2+y^2) + y\tilde{P}_2(x^2+y^2)}{x^2+y^2}, \quad V_2 = \frac{y\tilde{P}_1(x^2+y^2) - x\tilde{P}_2(x^2+y^2)}{x^2+y^2}.$$

Since x^2+y^2 is a factor of both $\tilde{P}_1, \tilde{P}_2$,

$$P_1 = \frac{\tilde{P}_1}{x^2+y^2}, \quad P_2 = \frac{\tilde{P}_2}{x^2+y^2}$$

are polynomials, and the implication follows.

($\Leftarrow$) Let us write $\|\boldsymbol{x}\|^2 = x^2+y^2$. By hypothesis,

$$\boldsymbol{V}(\boldsymbol{x}) = \boldsymbol{V}(x,y) = \begin{bmatrix} x & y \\ y & -x \end{bmatrix} \begin{bmatrix} P_1(\|\boldsymbol{x}\|^2) \\ P_2(\|\boldsymbol{x}\|^2) \end{bmatrix}.$$

Now let $f(\boldsymbol{x}) = A\boldsymbol{x}$, with A any rotation matrix. Taking into account that f leaves $\|\boldsymbol{x}\|^2$ invariant and using the above expression, one can check that $\boldsymbol{V}(A\boldsymbol{x}) = A\boldsymbol{V}(\boldsymbol{x})$. □

Remark 8. *Notice that the vector fields in Eq.* (4.25) *are invariant under* all *the rotations around the origin.*

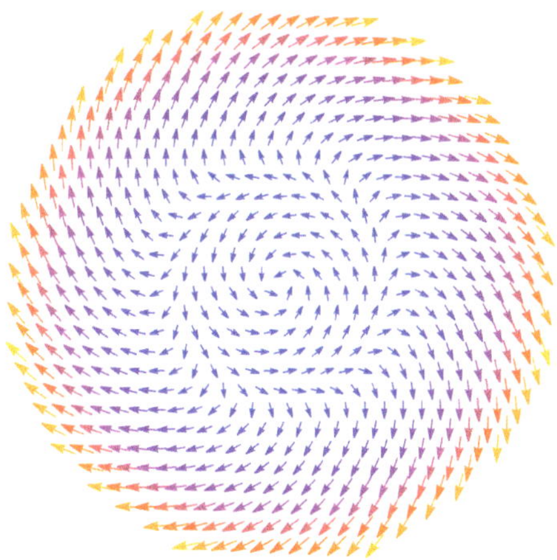

FIGURE 4.5 The vector field with infinitely many symmetries from Example 8.

Example 8. *Let* $\mathbf{V}(x, y) = (x^5 + 2x^4y + 2x^3y^2 + 4x^2y^3 + xy^4 + 2y^5 + x^3 - 2x^2y + xy^2 - 2y^3 - 2x - 4y, -2x^5 + x^4y - 4x^3y^2 + 2x^2y^3 - 2xy^4 + y^5 + 2x^3 + x^2y + 2xy^2 + y^3 + 4x - 2y)$, *see Fig. 4.5. The polynomial* $g(z)$ *is*

$$g(z) = -160\mathbf{i}z^9.$$

Since the only root of $g(z)$ *is already at the origin, no translation is needed. One can check that the components of* $\mathbf{V}(x, y)$ *can be written in the form of Eq.* (4.25) *with*

$$P_1(\xi) = \xi^2 + \xi - 2, \quad P_2(\xi) = 2\xi^2 - 2\xi - 4,$$

so from Theorem 86, $\mathbf{V}(x, y)$ *has infinitely many symmetries, and in particular is invariant under any rotation around the origin. Additionally, the polynomials* X_1, X_2 *are*

$$\begin{aligned} X_1(x,y) &= (x^2+y^2-1)(x^2+y^2)(x^2+y^2+2), \\ X_2(x,y) &= 2(x^2+y^2-2)(x^2+y^2)(x^2+y^2+1), \end{aligned}$$

which factor as products of circles centered at the origin.

4.4.2 Case (b): $g(z) = c$, $c \neq 0$

In this case $\mathrm{Sing}(\boldsymbol{V}) = \emptyset$, so because of Lemma 78 we cannot have rotational symmetry. However, this is the first case when we can have translational symmetry. Let us analyze this type of symmetry: if $\boldsymbol{V}$ is invariant under the translation by a vector, say, $\boldsymbol{b} = (b, 0)$, then by Eq. (4.4) we have that

$$\boldsymbol{V}(\boldsymbol{x}+\boldsymbol{b}) = \boldsymbol{V}(x+b, y) = \boldsymbol{V}(x,y),$$

so writing $\boldsymbol{V}(x,y) = (V_1(x,y), V_2(x,y))$, we get that $V_i(x+b,y) = V_i(x,y)$ for $i = 1,2$; in particular, $\boldsymbol{V}$ is constant over every line parallel to the x-axis, i.e., over every line parallel to $\boldsymbol{b}$. However, the only possibility for $V_i(x+b,y) = V_i(x,y)$ to happen is that V_1, V_2 do not depend on x, so $\boldsymbol{V}(x,y) = \boldsymbol{V}(y) = (V_1(y), V_2(y))$. Since for any translation by a vector $\boldsymbol{b} = (b_1, b_2)$ we can always consider an orthogonal transformation mapping (b_1, b_2) onto $(b, 0)$ for an appropriate b, we deduce that if $\boldsymbol{V}$ is invariant under the translation by $\boldsymbol{b} = (b_1, b_2)$, then

$$\boldsymbol{V}(x,y) = (V_1(x,y), V_2(x,y)) = (p_1(-b_2x+b_1y), p_2(-b_2x+b_1y)), \tag{4.27}$$

where the p_i are univariate polynomials. This structure can be discovered by observing that if $V_i(x,y) = p_i(-b_2x+b_1y)$, then

$$\nabla V_i(x,y) = p_i'(-b_2x+b_1y)(-b_2, b_1), \tag{4.28}$$

where p_i' is the derivative of the univariate function f_i. Thus, after looking for the common factor of both components of the ∇V_i, the structure of Eq. (4.28) can be recognized, and $\boldsymbol{b}$ can be computed. A field possessing a translational symmetry can also be invariant under a reflection on an axis parallel or orthogonal to $\boldsymbol{b}$. Such reflections can be easily determined from p_1, p_2. Notice also that if a vector field is invariant under translation by two different vectors, then it is also invariant under all their linear combinations, so the only vector fields invariant under translations in two different directions are constant fields.

The remaining case is the computation of the symmetries of fields $\boldsymbol{V}$ without any translational symmetry. Such fields have at most one reflection symmetry: indeed, since symmetries have the structure of a group, if $\boldsymbol{V}$ has two reflectional symmetries, then $\boldsymbol{V}$ must also have the symmetry corresponding to their composition, which is either a translation, when the reflection axes are parallel, or a rotation, when the axes intersect. However, we are addressing now the case when no translations occur, and we know that rotations cannot appear. Because of the following lemma we can find the only reflection axis, if any, among the reflection axes of the curve defined by the polynomial $\|\boldsymbol{V}\|^2$ (recall Lemma 67).

Lemma 87. *Let* $\boldsymbol{V}$ *be a vector field. Then* $\mathrm{Sym}(\boldsymbol{V})$ *contains a translation* τ *if and only if* $\tau \in \mathrm{Sym}(\|\boldsymbol{V}\|^2)$.

Proof. The implication $(\Rightarrow)$ is a consequence of Lemma 67, so let us see $(\Leftarrow)$. Thus, let $\tau \in \mathrm{Sym}(\|\boldsymbol{V}\|^2)$. Assume for the sake of simplicity that the direction of the translation is $(1, 0)$. Then $\|\boldsymbol{V}\|^2 = p(y)$ does not depend on x. However, since $\|\boldsymbol{V}\|^2 = V_1^2(x, y)+V_2^2(x, y)$ there is no cancellation in the sum of the squares of V_1, V_2, and therefore both V_1 and V_2 depend only on y. This means that $\tau \in \mathrm{Sym}(\boldsymbol{V})$. □

By the previous lemma, if $\boldsymbol{V}$ is not invariant under a translation, then $\mathrm{Sym}(\|\boldsymbol{V}\|^2)$ is either finite, in which case it can be used to find a tentative reflection axis of $\boldsymbol{V}$, or it contains the full rotation group as a subgroup. In the latter case $\|\boldsymbol{V}\|^2$ is a product

$$\|\boldsymbol{V}\|^2 = \prod_{i=1}^{k} \left((x - c_1)^2 + (y - c_2)^2 - \rho_i\right), \tag{4.29}$$

where (c_1, c_2) is the center of rotational symmetry. After translating the center to the origin we can apply Proposition 80: indeed, let $\boldsymbol{V}_0$ be the translated field. Then $\boldsymbol{V}_0(\boldsymbol{x}) \times \boldsymbol{x}$ is not identically zero, since otherwise this would imply that $\boldsymbol{V}_0(\boldsymbol{x})$ is a multiple of the Euler vector field, in which case $(0,0)$ is a singular point; but this cannot happen because $\text{Sing}(\boldsymbol{V}) = \emptyset$. Hence the tentative axis can be found as a linear factor of $\boldsymbol{V}_0(\boldsymbol{x}) \times \boldsymbol{x}$.

4.4.3 Case (c): $g(z) = 0$

From Section 4.2, $g(z) = 0$ implies that the components $\widehat{V}_1, \widehat{V}_2$ of $\widehat{\boldsymbol{V}}$, and therefore the components V_1, V_2 of $\boldsymbol{V}$, have a nontrivial common factor, i.e., $\boldsymbol{V} = h\boldsymbol{W}$, where $\boldsymbol{W} = (W_1, W_2)$ with W_1, W_2 being relatively prime. Using Eq. (4.4), any symmetry $f(\boldsymbol{x}) = A\boldsymbol{x} + \boldsymbol{b}$ of $\boldsymbol{V}$ satisfies that

$$h(A\boldsymbol{x} + \boldsymbol{b})\boldsymbol{W}(A\boldsymbol{x} + \boldsymbol{b}) = h(\boldsymbol{x})A\boldsymbol{W}(\boldsymbol{x}).$$

Since the components of $\boldsymbol{W}$ are relatively prime, $h(\boldsymbol{x})$ must divide $h(A\boldsymbol{x} + \boldsymbol{b})$. Since $h(\boldsymbol{x}), h(A\boldsymbol{x} + \boldsymbol{b})$ are polynomials of the same degree we get that $h(A\boldsymbol{x} + \boldsymbol{b}) = \lambda h(\boldsymbol{x})$, $\lambda \in \mathbb{R}$, and therefore $\lambda\boldsymbol{W}(A\boldsymbol{x} + \boldsymbol{b}) = A\boldsymbol{W}(\boldsymbol{x})$.

Lemma 88. *Let h be a bivariate polynomial, and let $f(\boldsymbol{x}) = A\boldsymbol{x}+\boldsymbol{b}$ be a Euclidean symmetry fulfilling that $h(A\boldsymbol{x}+\boldsymbol{b}) = \lambda h(\boldsymbol{x})$, $\lambda \in \mathbb{R}$. Then $\lambda = \pm 1$.*

Proof. If there exists $n \in \mathbb{N}$ such that $f^{(n)}$ is the identity, then the result holds (see the proof of Corollary 27 in Chapter 2). So we just need to prove the result for translations, and for rotations where the rotation angle is $\theta \neq 2\pi/k$, with $k \in \mathbb{N}$. For translations, we can assume without loss of generality that $f(\boldsymbol{x}) = \boldsymbol{x} + \boldsymbol{b}$ with $\boldsymbol{b} = (b, 0)$, $b \neq 0$. In that case, $h = h(y)$ (see Section 4.4.2) and $\lambda = 1$. For rotations, from Corollary 25 we just need to consider the case when h corresponds to a union of concentric circles, whose center we can assume to be the origin. Thus,

$$h(x, y) = \prod_{i=1}^{k} \left(x^2 + y^2 - \rho_i\right).$$

Since any rotation around the origin leaves the form $x^2 + y^2$ invariant, we conclude that $\lambda = 1$ in that case. □

As a consequence, we get $h(A\boldsymbol{x}+\boldsymbol{b}) = \pm h(\boldsymbol{x})$ and $\boldsymbol{W}(A\boldsymbol{x}+\boldsymbol{b}) = \pm A\boldsymbol{W}(\boldsymbol{x})$, where the sign in $\pm$ must be the same in both equalities. The first equality implies that $f(\boldsymbol{x}) = A\boldsymbol{x}+\boldsymbol{b}$ is a symmetry of the algebraic curve defined by h, i.e., $f \in \mathrm{Sym}(h)$; we will write $\mathrm{Sym}(h) = \mathrm{Sym}(h^+) \cup \mathrm{Sym}(h^-)$, where $\mathrm{Sym}(h^+)$ (resp. $\mathrm{Sym}(h^-)$) is the collection of symmetries of h where the sign in $h(A\boldsymbol{x}+\boldsymbol{b}) = \pm h(\boldsymbol{x})$ is $+$ (resp. $-$). The second equality, $\boldsymbol{W}(A\boldsymbol{x}+\boldsymbol{b}) = \pm A\boldsymbol{W}(\boldsymbol{x})$, implies that $f(\boldsymbol{x}) = A\boldsymbol{x}+\boldsymbol{b}$ is a symmetry of $\boldsymbol{W}$ when the sign in $h(A\boldsymbol{x}+\boldsymbol{b}) = \pm h(\boldsymbol{x})$ is $+$, and an antisymmetry (see Definition 66 and Eq. (4.5)) of $\boldsymbol{W}$ when the sign is $-$. Thus, we get the following result, where we denote by $\mathrm{Asym}(\boldsymbol{W})$ the set of antisymmetries of $\boldsymbol{W}$.

Proposition 89. *Assume that $g(z)$ is identically zero. Then*

$$\mathrm{Sym}(\boldsymbol{V}) = [\mathrm{Sym}(h^+) \cap \mathrm{Sym}(\boldsymbol{W})] \cup [\mathrm{Sym}(h^-) \cap \mathrm{Asym}(\boldsymbol{W})].$$

The antisymmetries of $\boldsymbol{W}$ can be computed with the same strategy of symmetries, so we conclude that the case $g(z) = 0$ essentially reduces to the previous ones.

4.5 RATIONAL VECTOR FIELDS

Finally, let us consider *rational* vector fields, so that the components $V_i(x,y)$ of $\boldsymbol{V}(x,y) = (V_1(x,y), V_2(x,y))$ are rational functions. After computing the common denominator of the V_i, we can write

$$\boldsymbol{V} = \frac{1}{h}\boldsymbol{W},$$

where h is a bivariate polynomial, and $\boldsymbol{W}$ is a polynomial vector field. From Eq. (4.4), we get that the symmetries $f(\boldsymbol{x}) = A\boldsymbol{x}+\boldsymbol{b}$ of $\boldsymbol{V}$ fulfill

$$\frac{1}{h(A\boldsymbol{x}+\boldsymbol{b})}\boldsymbol{W}(A\boldsymbol{x}+\boldsymbol{b}) = \frac{1}{h(\boldsymbol{x})}A\boldsymbol{W}(\boldsymbol{x}).$$

Thus, $h(A\boldsymbol{x}+\boldsymbol{b}) = \lambda h(\boldsymbol{x})$, and arguing as in Section 4.4.3 we deduce that $\lambda = \pm 1$ and Proposition 89 applies also here. So the rational case can be essentially reduced to the polynomial case.

4.6 FINAL REMARKS AND OPEN PROBLEMS

Symmetries and antisymmetries of vector fields have also been studied in the context of autonomous Dynamical Systems; some references in this sense are [66, 67, 78, 81, 82, 108]. One of the main motivations has to do with the notion of *limit cycle*: a limit cycle is a closed curve such that there exists at least one solution of the system that approaches it asymptotically; see Fig. 4.6, where the limit cycle of a well-known dynamical system, the *Van Der Pol Oscillator* in 2D, is reproduced (figure taken from `https://en.wikipedia.org`). In particular, limit cycles correspond to periodic solutions of the dynamical system, enclose at least a singular point of the system, and define regions where certain solutions of the system must be confined. A still open problem, posed by David Hilbert within the 16th problem of his celebrated list, is to determine an upper bound on the number of limit cycles that a polynomial dynamical system can have. In this sense, vector fields exhibiting rotational symmetry lead to examples of systems with several limit cycles, which helps to explore this question [66].

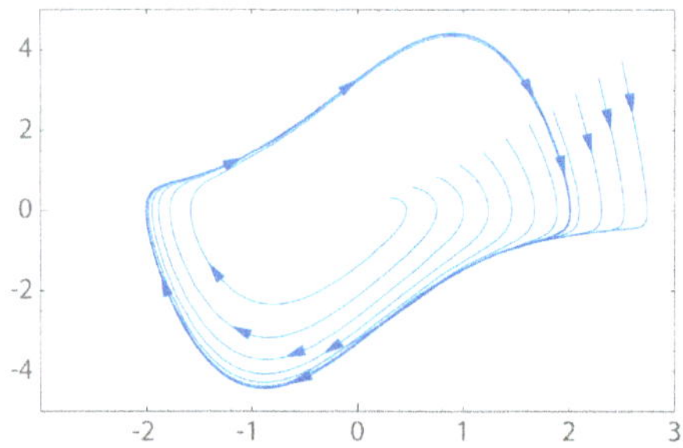

FIGURE 4.6 Limit cycle of the Van der Pol oscillator

Another reason for the interest of the problem in Dynamical Systems is what in this context is called *time-reversibility*, which corresponds to what we called *antisymmetry*. If a vector field has an antisymmetry that fixes a singular point, then one can guarantee the existence of a local analytic integral of the system around the point [67, 81, 82, 108].

Although in our case we considered the problem from a purely geometric point of view, Dynamical Systems seem to be an interesting source of potential problems regarding this question. Also, here we explored the problem for planar fields; higher dimensions are open. Thus, we can suggest the following open questions.

Open Problem 11. *Rotationally symmetric planar fields, and limit cycles.* In Section 2.3 of Chapter 2, we provide a description for all the planar algebraic curves exhibiting rotational symmetry around a point. One can consider the analogous problem for vector fields, i.e., generating planar vector fields with rotational symmetry. One can also wonder under what conditions the corresponding dynamical systems can have limit cycles, starting by the simpler question about when they have circles as limit cycles.

Open Problem 12. *Affine mappings transforming a vector field into an antisymmetric field.* It can happen that a vector field with singular points is not antisymmetric, but admits an affine mapping, fixing some singular point, transforming the vector field into an antisymmetric vector field. The dynamical system corresponding to the latter can be guaranteed to have a local analytic integral around the fixed singular point, which serves as a guarantee for the original dynamical system around the fixed singular point, as well. Thus, we wonder if, given a vector field, we can find a criterion for the existence of affine mappings with these properties, and how to compute them.

Open Problem 13. *Special planar fields.* Can we say more, or can we do better, if the vector field is $\boldsymbol{V}(x,y) = (U_x, U_y)$ or $\boldsymbol{V}(x,y) = (-U_y, U_x)$, for $U = U(x,y)$ a polynomial? The first case implies that $\boldsymbol{V}$ is the gradient of U, and the second case is related to *Hamiltonian* systems.

Open Problem 14. *Three-dimensional vector fields.* The problem for a higher number of dimensions is certainly open. This analysis can benefit from the results obtained so far for implicit algebraic surfaces (see Chapter 6). Also, a possibly simpler problem would be to analyze the symmetries of vector fields on the sphere.

CHAPTER 5

Space algebraic curves

In this chapter we begin the study of symmetries of algebraic varieties in $\mathbb{R}^3$. We will study two types of algebraic varieties: space curves, i.e., algebraic varieties in $\mathbb{R}^3$ of dimension one, in this chapter, and surfaces, in the remaining chapters, which are algebraic varieties in $\mathbb{R}^3$ of dimension two. As in the case of planar algebraic curves, space algebraic curves come in two different flavors, implicit and rational. For implicit space curves, we will provide some ideas, but we will postpone the algorithms for the next chapter, where algebraic surfaces will come into the picture to provide some help; still, we advance that the completely general case, for implicit space curves, is, at least partially, still open. For rational curves the situation, however, is much better.

5.1 REPRESENTATIONS OF SPACE ALGEBRAIC CURVES

We had two possible representations for planar algebraic curves, that we called implicit and rational; the second one was not always available, and the curves admitting such a representation were called rational planar curves. We expect something similar for space algebraic curves, although in this case we will discover that the question is more complex.

For space algebraic curves the affine and the projective setting will show some differences, and we need to address both. We begin with affine space curves. Thus, let us consider an algebraic variety

DOI: 10.1201/9781003648710-5

$\mathcal{V} \subset \mathbb{K}[x, y, z]$, with $\mathbb{K}$ a field, in our case $\mathbb{K} = \mathbb{R}$ or $\mathbb{K} = \mathbb{C}$,

$$\mathcal{V} = V(F_1(x, y, z), \ldots, F_m(x, y, z)). \tag{5.1}$$

The nature of $\mathcal{V}$ depends on its *dimension*; for a formal treatment of the notion of dimension in the context of Algebraic Geometry, the interested reader can check, for instance, [45]. We have three possibilities: the dimension of $\mathcal{V}$ could be 0, in which case we have finitely many points; it could be 1, in which case we say that we have a *space algebraic curve*, that we will denote by $\mathcal{C}$; or it could be 2, in which case we have an *algebraic surface*. Here we focus on space curves. For instance, consider the set $\mathcal{C}_1$ of common zeroes of

$$F_1(x, y, z) = y - x^2, \;\; F_2(x, y, z) = z - x^3. \tag{5.2}$$

The set $\mathcal{C}_1$, which is a well-known space curve called the *twisted cubic*, is the intersection of two algebraic surfaces, $y - x^2 = 0$ and $z - x^3 = 0$. The polynomials $F_1(x, y, z)$, $F_2(x, y, z)$ are relatively prime, i.e., $\gcd(F_1, F_2) = 1$, so intuitively $\mathcal{C}_1$ must be a space curve; this curve is plotted in Fig. 5.1, left. It is just natural to call

$$F_1(x, y, z) = 0, \;\; F_2(x, y, z) = 0$$

an *implicit* representation of $\mathcal{C}_1$. However, $\mathcal{C}_1$ can also be seen as the set of common zeroes of

$$F_1(x, y, z) = y - x^2, \;\; F_3(x, y, z) = xy - z, \tag{5.3}$$

where $F_3(x, y, z) = xF_1(x, y, z) - F_2(x, y, z)$, so

$$F_1(x, y, z) = 0, \;\; F_3(x, y, z) = 0$$

would be another implicit representation of $\mathcal{C}_1$, not consisting of the same polynomials. And we might generate many more, by considering linear combinations of the $F_i(x, y, z)$, with coefficients depending on x, y, z. This is a phenomenon that did not arise with planar algebraic curves, where the implicit equation was defined up to a constant. Additionally, there is no reason why we should

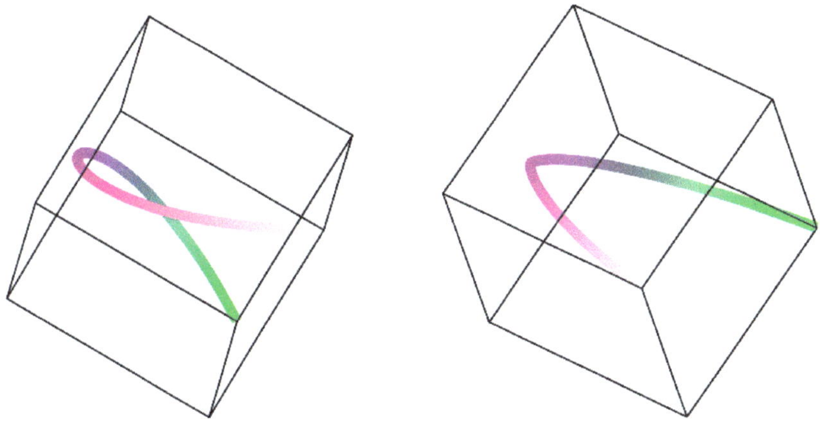

FIGURE 5.1 The curve in Eq. (5.1) (left); the curve in Eq. (5.4) (right)

only have two polynomials. Consider the variety $\mathcal{C}_2$ defined as the set of common zeroes of

$$G_1(x,y,z) = y^2 - xz, \;\; G_2(x,y,z) = x^2 - yz, \;\; G_3(x,y,z) = x^2y - z^2, \tag{5.4}$$

plotted in Fig. 5.1, right. In this case we have three polynomials, and one can check that if one of them is removed one would be adding points, so the resulting variety would not coincide with $\mathcal{C}_2$: for instance, if we remove $G_3(x,y,z)$, we are left with

$$y^2 - xz = 0, \;\; x^2 - yz = 0,$$

which, in particular, contains the z-axis as a component. However, the polynomial $G_3(x,y,z) = 0$ does not vanish at the points $(0,0,z)$, so the z-axis is not a part of $\mathcal{C}_2$. Thus, in this case three polynomials are needed to implicitly define $\mathcal{C}_2$.

All these considerations justify the introduction of a notion which was not so necessary in the case of plane curves, but that lies at the core of Algebraic Geometry, namely the notion of *ideal of a variety*. Although it can be defined with full generality, here we will limit ourselves to space algebraic curves, because we will not really need it for surfaces. To define an ideal we need a ring first, which will be the ring of (maybe complex) polynomials in the variables x, y, z, i.e., $\mathbb{K}[x, y, z]$. Then we have the following definition.

Definition 90. *Let $\mathcal{C} \subset \mathbb{K}^3$ be a space algebraic curve. The* ideal *of $\mathcal{C}$, $\mathbf{I}(\mathcal{C})$, is the subset $\mathbf{I}(\mathcal{C}) \subset \mathbb{K}[x, y, z]$ consisting of all the polynomials $F(x, y, z) \in \mathbb{K}[x, y, z]$ such that $F(\mathbf{p}) = 0$ for all $\mathbf{p} \in \mathcal{C}$.*

Since $F(x, y, z) \in \mathbf{I}(\mathcal{C})$ implies that $G(x, y, z) \cdot F(x, y, z) \in \mathbf{I}(\mathcal{C})$ for any $G \in \mathbb{K}[x, y, z]$, $\mathbf{I}(\mathcal{C})$ certainly has the structure of an ideal; we say that $\mathbf{I}(\mathcal{C})$ is a *polynomial ideal*, because it is an ideal of a polynomial ring, $\mathbb{K}[x, y, z]$ in this case. Now by the celebrated Hilbert Basis Theorem (see e.g. Chapter 2 of [45]), every polynomial ring is finitely generated, i.e., there exist $F_1, \ldots, F_m \in \mathbf{I}(\mathcal{C})$ such that any other $F \in \mathbf{I}(\mathcal{C})$ can be written as

$$F = \alpha_1 F_1 + \cdots + \alpha_m F_m,$$

where $\alpha_i \in \mathbb{K}[x, y, z]$ for $i = 1, \ldots, m$; we say that $F_1, \ldots, F_m$ form a *basis* of $\mathbf{I}(\mathcal{C})$, and we write

$$\mathbf{I}(\mathcal{C}) = \langle F_1, \ldots, F_m \rangle.$$

This explains why the implicit description of an algebraic space curve is not as simple as in the planar case: coming back to the initial examples with $\mathcal{C}_1, \mathcal{C}_2$, we have

$$\mathbf{I}(\mathcal{C}_1) = \langle F_1, F_2 \rangle = \langle F_1, F_3 \rangle, \ \mathbf{I}(\mathcal{C}_2) = \langle G_1, G_2, G_3 \rangle,$$

while for planar algebraic curves (as it will also happen with algebraic surfaces), the corresponding ideals are defined by just one polynomial. In particular, there is not a unique basis for a space algebraic curve, as $\mathcal{C}_1$ shows, although any basis implicitly defines the same geometric object. Furthermore, the minimal number of polynomials in the basis needs not be two, as $\mathcal{C}_2$ shows: for $\mathcal{C}_2$, three is the minimal number.

Thus, given an algebraic variety, in our case a space algebraic curve $\mathcal{C} \subset \mathbb{K}^3$, we can define an algebraic object, the ideal $\mathbf{I}(\mathcal{C})$, associated with it. We can also go the other way around: given a polynomial ideal $\mathbf{I} \subset \mathbb{K}[x, y, z]$ we can define the *variety associated with the ideal* $\mathbf{I}$, $V(\mathbf{I})$, as the set of points of $\mathbb{K}^3$ where *all* the polynomials in $\mathbf{I}$ vanish. In Commutative Algebra (see for instance Chapter 5 of [45]) one can prove that for an algebraic variety $\mathcal{V}$,

$$V(\mathbf{I}(\mathcal{V})) = \mathcal{V},$$

and (which is called the *Nullstellensazt*, although in this case we must require $\mathbb{K}$ to be algebraically closed)

$$\mathbf{I}(V(\mathbf{I})) = \sqrt{\mathbf{I}},$$

where $\sqrt{\mathbf{I}}$ is the *radical* of the ideal $\mathbf{I}$, meaning the set of all the polynomials $F \in \mathbb{K}[x, y, z]$ such that there exists $k \in \mathbb{N}$ with the property that $F^k \in \mathbf{I}$.

Hence, space algebraic curves are identified with polynomial ideals of $\mathbb{K}[x, y, z]$ defining geometric objects of dimension 1. When we move to objects of dimension 2, i.e., algebraic surfaces, we have a situation analogous to planar curves in the sense that one polynomial suffices, so we can, again, speak about the implicit equation of the surface, defined by one polynomial (up to constants), without any ambiguity.

So far we have considered the affine setting. If we move now to $\mathbb{P}^3(\mathbb{K})$ the situation requires some care. As we briefly explained at the beginning of Chapter 2, moving to $\mathbb{P}^3(\mathbb{K})$ implies to consider the projective closure $\widehat{\mathcal{C}}$ of $\mathcal{C}$, where $\mathcal{C} = V(F_1, \ldots, F_m)$, $m \geq 2$. One might think that $\widehat{\mathcal{C}} = V(\widehat{F}_1, \ldots, \widehat{F}_m)$, where $\widehat{F}_i$ is the result of homogenizing $F_i(x, y, z)$ with respect to some homogenization variable, ω, but this is not necessarily true and a good counterexample is, precisely, the twisted cubic $\mathcal{C}_1$ in Eq. (5.1). Indeed, recall that space algebraic curves are essentially defined by the ideal of the curve. For projective curves $\widehat{\mathcal{C}}$, $\mathbf{I}(\widehat{\mathcal{C}})$ is now a *homogeneous* ideal, meaning the ideal of the ring of homogeneous polynomials in x, y, z, ω, consisting of all the homogeneous polynomials vanishing at the (projective) points of $\widehat{\mathcal{C}}$. In the case of the twisted cubic $\mathcal{C}_1$, defined by $F_1(x, y, z) = y - x^2$, $F_2(x, y, z) = z - x^3$, we can see that the affine points of $\mathcal{C}_1$ can be described by means of the parametrization (t, t^2, t^3); thus, $\widehat{\mathcal{C}}_1$ can be parametrized as

$$[ts^2 : t^2 s : t^3 : s^3], \tag{5.5}$$

so that $[0 : 0 : 1 : 0]$ is the point at infinity of $\widehat{\mathcal{C}}_1$. Now if we homogenize the polynomials F_1, F_2, giving rise to $\widehat{F}_1(x, y, z, \omega) = y\omega - x^2$, $\widehat{F}_2(x, y, z, \omega) = z\omega^2 - x^3$, one might expect that $\mathbf{I}(\widehat{\mathcal{C}}_1) = \langle \widehat{F}_1, \widehat{F}_2 \rangle$. However, if we take the polynomial $F_3(x, y, z) = F_2(x, y, z) - xF_1(x, y, z) = z - xy$, which clearly belongs to $\mathbf{I}(\mathcal{C}_1)$, and we homogenize it to get $\widehat{F}_3(x, y, z, \omega) = z\omega - xy$, we observe that $\widehat{F}_3$ vanishes at all the points described by Eq. (5.5), and, however, $\widehat{F}_3 \notin \langle \widehat{F}_1, \widehat{F}_2 \rangle$, which requires an explanation. Indeed, any polynomial in $\langle \widehat{F}_1, \widehat{F}_2 \rangle$ looks like $\alpha_1 \widehat{F}_1 + \alpha_2 \widehat{F}_2$, where α_1, α_2 are homogeneous polynomials. However, the degree of a product of homogeneous polynomials is the sum of their degrees, and the degree of a sum of two homogeneous polynomials is, whenever the sum is not identically zero, equal to the maximum of the degrees of the polynomials. Since $\widehat{F}_2$ has degree 3, the only polynomials of degree 2 in $\langle \widehat{F}_1, \widehat{F}_2 \rangle$ are the polynomials $k\widehat{F}_1$, with k a constant, but obviously $\widehat{F}_3$ is not a multiple of $\widehat{F}_1$. Thus, $\mathbf{I}(\widehat{\mathcal{C}}_1) \neq \langle \widehat{F}_1, \widehat{F}_2 \rangle$. In fact, for the twisted curve we need one more polynomial; for instance, one can check (see Chapter 8, Section 4 of [45]) that

$$\mathbf{I}(\widehat{\mathcal{C}}_1) = \langle y\omega - x^2, xy - \omega z, xz - y^2 \rangle.$$

Therefore, in the case of space algebraic curves, the number of generators needs not be the same for the ideals of $\mathcal{C}$ and $\widehat{\mathcal{C}}$: for $\widehat{\mathcal{C}}$, we may need more. If $\mathbf{I}(\widehat{\mathcal{C}})$ consists of just two generators, we say that $\widehat{\mathcal{C}}$ is a *complete intersection*; the notion of a complete intersection is much wider, i.e., it applies to algebraic varieties of higher dimension in any number of variables, and can be checked, for instance, in Chapter 11 of [68]. It can also be applied to affine curves, but notice the subtlety here: if $\mathcal{C}$ is a complete intersection, $\widehat{\mathcal{C}}$ might not be a complete intersection, and an example is that of the twisted cubic. However, if $\widehat{\mathcal{C}}$ is a complete intersection, then $\mathcal{C}$ is also a complete intersection. We will need these notions later and also in the next chapter, when we come back to space curves with the help of algebraic surfaces.

5.2 SOME FEATURES OF ALGEBRAIC SPACE CURVES

Recalling the notion of irreducibility introduced in Chapter 2, every space algebraic curve $\mathcal{C} \subset \mathbb{K}^3$ can be written as

$$\mathcal{C} = \mathcal{V}_1 \cup \cdots \cup \mathcal{V}_k$$

where $\mathcal{V}_1, \ldots, \mathcal{V}_k$, $k \geq 2$, are algebraic varieties called the *irreducible components* of $\mathcal{C}$; if $k = 1$, then we say that $\mathcal{C}$ is irreducible. Interestingly, the $\mathcal{V}_i$ do not need to be, all of them, space algebraic curves: a reducible space algebraic curve might have 0-dimensional components as well. Irreducibility can also be defined in terms of ideals: it can be proven [45] that $\mathcal{C}$ is irreducible if and only if $\mathbf{I}(\mathcal{C})$ is a *prime* ideal, meaning that whenever $F, G \in \mathbb{K}[x, y, z]$ and $F \cdot G \in \mathbf{I}$, either $F \in \mathbf{I}$ or $G \in \mathbf{I}$. This condition can be used to give an algorithm for checking whether or not an algebraic variety, in particular a space algebraic curve, is irreducible, or for decomposing it into irreducible components. Moreover, for future reference, notice that if two space algebraic curves intersect at infinitely many points, then they must have a common component; this is essentially a consequence of Bezout's Theorem. All these ideas generalize to the projective setting as well.

The notion of *degree* also applies to space algebraic curves. In the planar case, the degree could be directly read from the implicit equation of the curve. However, this is not so simple for space algebraic curves. But we can switch to the geometric interpretation of the degree, which in the case of planar algebraic curves coincides with the number of (possibly complex) intersection points between an algebraic curve and a generic line. This idea generalizes to space algebraic curves so that the degree of a space algebraic curve $\mathcal{C}$, $\deg(\mathcal{C})$, is the number of intersection points between $\mathcal{C}$ and a generic plane.

Since planar algebraic curves are easier to manage compared to space algebraic curves, it is useful to relate a given space algebraic curve to some planar algebraic curve(s), so that we can derive information about the space curve from the planar curve(s). Typically, we do this by projecting the space curve onto some plane, usually one, or several, of the coordinate planes. If $\mathcal{C}$ is a space algebraic curve, then, for instance, the *projection* of $\mathcal{C}$ onto the xy-plane is

$$\Pi_{xy}(\mathcal{C}) = \{(x, y) \in \mathbb{K}^2 | \exists z \in \mathbb{K}, (x, y, z) \in \mathcal{C}\}.$$

If $\mathcal{C} = V(F_1, \ldots, F_m)$, computing $\Pi_{xy}(\mathcal{C})$ amounts to eliminating the z-variable in the polynomial system defined by the F_i's. If $m = 2$ (otherwise the situation gets more messy and Elimination Theory, see [45], comes into the picture), $\Pi_{xy}(\mathcal{C})$ is, essentially, defined by the resultant $\mathrm{Res}_z(F_1, F_2)$, that we introduced in Section 4.2 of Chapter 4. We say that $\Pi_{xy}(\mathcal{C})$ is "essentially" defined by $\mathrm{Res}_z(F_1, F_2)$ because we need to ensure two things: (i) for every point $(x_0, y_0, z_0) \in \mathcal{C}$, (x_0, y_0) must be a point of the planar algebraic curve defined by $\mathrm{Res}_z(F_1, F_2)$; (ii) for every point (x_0, y_0) of the curve defined by $\mathrm{Res}_z(F_1, F_2)$, there exists $z_0 \in \mathbb{K}$ (there might be several) such that $(x_0, y_0, z_0) \in \mathcal{C}$. While (i) is ensured by the properties of the resultant that we mentioned at the end of Section 4.2, (ii) might fail, i.e., we might have spurious components in the resultant. Nevertheless, if the leading coefficient of either F_1 or F_2 with respect to z is a nonzero constant, which is something that we can always achieve by applying to $\mathcal{C}$ a random orthogonal transformation, then we can certainly guarantee that $\Pi_{xy}(\mathcal{C})$ coincides with the planar algebraic curve defined by $\mathrm{Res}_z(F_1, F_2)$.

There is, additionally, a good property that we might request when projecting, which is birationality. Let Π_{xy} denote the projection onto the, say, xy-plane (analogously for other planes),

$$\Pi_{xy} : \mathbb{K}^3 \to \mathbb{K}^2,\ \Pi_{xy}(x, y, z) = (x, y),$$

which is a rational mapping (since their components are rational components of x, y, z). We say that Π_{xy} is *birational* for $\mathcal{C}$ if the restriction $\Pi_{xy}|_{\mathcal{C}}$ is a birational mapping, meaning that $\Pi_{xy}|_{\mathcal{C}}^{-1}$ exists; in fact, if $\Pi_{xy}|_{\mathcal{C}}^{-1}$ exists, then it must be rational. Geometrically, this implies that $\mathcal{C}$ does not have either a component orthogonal to the xy-plane, so that infinitely many points are projected

FIGURE 5.2 Birational (left, middle) and non-birational (right) projections; the space curve appears in red, and the projection of the space curve appears in blue

onto the same planar point, or two different branches whose projections onto the xy-plane overlap. There is no problem with the existence of finitely many points of $\mathcal{C}$ with overlapping projections, though. An algorithm to check whether or not a projection of $\mathcal{C}$ is birational can be found in [49]; furthermore, every space algebraic curve $\mathcal{C} \subset \mathbb{K}^3$ can be birationally projected [132] onto a planar algebraic curve, we just need to choose the right plane to project onto. In Fig. 5.2, we have examples of birational (left, middle) and non-birational (right) projections: of course, birationally is, like properness, a property that holds over the complex numbers, so we should not have overlapping complex branches either, which obviously we cannot capture in the picture; observe also that in Fig. 5.2, middle, the projection has a self-intersection coming from two different points of $\mathcal{C}$ projecting onto the same planar point.

Let us move now to rationality questions. First we have the following definition.

Definition 91. *Let $\mathcal{C} \subset \mathbb{K}^3$ be a space algebraic curve. We say that $\mathcal{C}$ is* rational *if $\mathcal{C}$ admits a parametrization $\boldsymbol{p}(t) = (p_1(t), p_2(t), p_3(t))$ where for $i = 1, 2, 3$, $p_i(t)$ is a rational function; the $p_i(t)$ are called the* components *of $\boldsymbol{p}(t)$.*

As it happens with planar curves, not every space algebraic curve is rational. To decide whether or not an algebraic space curve is rational, we can reduce the problem to planar algebraic curves by computing a birational projection of our space curve: if this projection is rational, the original space curve is also rational, and we can compute a parametrization by first parametrizing the projection, and then lifting this parametrization with the help of the inverse of the projection. For a method to compute this inverse, see also [49].

The two curves $\mathcal{C}_1$ and $\mathcal{C}_2$ in Eqs. (5.1) and (5.3) are rational: a parametrization $\boldsymbol{p}_1(t)$ for $\mathcal{C}_1$ is $\boldsymbol{p}_1(t) = (t, t^2, t^3)$, and a parametrization for $\mathcal{C}_2$ is $\boldsymbol{p}_2(t) = (t^3, t^4, t^5)$, so in these cases the parametrizations are not only rational, but polynomial. Furthermore, every rational space curve is irreducible. The notion of properness for space rational curves is analogous to that of planar curves, and whenever $\boldsymbol{p}(t)$ is proper, the degree of the curve coincides with the degree of $\boldsymbol{p}(t)$, i.e., the maximum of the degrees in the numerators and denominators of $\boldsymbol{p}(t)$.

5.3 GENERAL PROPERTIES

Let us consider now general properties of the symmetries of a *real* space algebraic curve $\mathcal{C} \subset \mathbb{R}^3$: *real* implies that $\mathcal{C}$ has infinitely many real points. We denote by $\text{Sym}(\mathcal{C})$, as in the planar case, the group of symmetries of $\mathcal{C}$; recall that since we are now working in $\mathbb{R}^3$, the potential elements of $\text{Sym}(\mathcal{C})$ are the transformations of Table 1.4 in Chapter 1. We will be assuming that $\mathcal{C}$ is not a planar curve, and by simplicity we will also assume that $\mathcal{C}$ is irreducible; in particular, $\mathcal{C}$ is not either a line or a circle. The final goal is to prove that for irreducible space curves, not lines or circles, $\text{Sym}(\mathcal{C})$ is finite.

We begin with a series of results.

Lemma 92. *Let $\mathcal{C} \subset \mathbb{R}^3$, not contained in any plane, and let $f \in \text{Sym}(\mathcal{C})$ be either a rotation about a line $\mathcal{L}$, by an angle $\theta \in [0, 2\pi)$, or a rotoreflection about a line $\mathcal{L}$, by an angle $\theta \in [0, 2\pi)$, and with reflection plane Π. Then $\theta = 2\pi/k$, with $k \in \mathbb{N}$, $k \leq \deg(\mathcal{C})$.*

Proof. We address rotations first. Let $\mathcal{D}$ be the projection of $\mathcal{C}$ onto a plane Π normal to the line $\mathcal{L}$; notice that since by hypothesis

$\mathcal{C}$ is not planar, $\mathcal{D}$ is not a line. Then $\mathcal{D}$ is an algebraic curve invariant under the rotation about the point $\mathbf{c} = \Pi \cap \mathcal{L}$, by the angle θ. If $\mathcal{D}$ is not a circle, then the result follows from Lemma 24 in Chapter 2. So assume that $\mathcal{D}$ is a circle, in which case $\mathcal{C}$ is contained in a cylinder of axis $\mathcal{L}$. Let Π' be a plane orthogonal to $\mathcal{L}$ such that $\Pi' \cap \mathcal{C} \neq \emptyset$. Then $\Pi' \cap \mathcal{C} = \{\mathbf{p}_1, \ldots, \mathbf{p}_m\}$, and for $i \in \{1, \ldots, m\}$ there exists $j \neq i$, $j \in \{1, \ldots, m\}$ with $f(\mathbf{p}_i) = \mathbf{p}_j$. Since $f^{(m)}(\mathbf{p}_i) = \mathbf{p}_i$, where $f^{(m)} = f \circ \overset{m}{\cdots} \circ f$, for all i, $m\theta$ must be a multiple of 2π. Then one can argue as in statement (ii) of Lemma 24 to conclude that $\theta = 2\pi/k$, with $k \in \mathbb{N}$. Since $m \leq \deg(\mathcal{C})$, $k \leq \deg(\mathcal{C})$.

Now let f be a rotoreflection satisfying that $f(\mathcal{C}) = \mathcal{C}$. Then f^2 is a rotation about the line $\mathcal{L}$, by an angle 2θ. Applying the first part we have $2\theta = 2\pi/k$ for some $k \in \mathbb{N}$, and therefore $\theta = 2\pi/(2k)$. □

Lemma 93. *Let $\mathcal{C} \subset \mathbb{R}^3$, irreducible and not contained in any plane, and let $f \in \mathrm{Sym}(\mathcal{C})$. Then f is not either a translation, or a glide reflection, or a screw motion.*

Proof. Let $\mathbf{p} \in \mathcal{C}$. The first part of the proof is very similar to that of Lemma 21 in Chapter 2. Assume first that f is a translation by the vector $\boldsymbol{b}$. Since $f(\mathcal{C}) = \mathcal{C}$, $\mathbf{p}_n = f^{(n)}(\mathbf{p})$, for $n \in \mathbb{N}$, is an infinite sequence of different points of $\mathcal{C}$ lying on a same line L through $\mathbf{p}$. Since $\mathcal{C}$ and L share infinitely many points, they must have a common component; but L is irreducible, so $L \subset \mathcal{C}$, in which case $\mathcal{C}$ is either reducible or a line; however, both possibilities are forbidden by hypothesis.

Next assume that f is a glide reflection, where $\mathbf{v}$ is the vector defining the translation part. Then the sequence $f^{(2n)}(\mathbf{p})$ is also a sequence of different points of $\mathcal{C}$ belonging both to $\mathcal{C}$ and the line L through $\mathbf{p}$ parallel to $\mathbf{v}$, so we can argue as before.

Finally, let f be a screw motion, and let $\mathcal{L}$ be its axis, θ the rotation angle, and $\mathbf{v}$ the translation vector. Assume, without loss of generality, that $\mathcal{L}$ is the z-axis. Then $f^{(n)}(\mathbf{p})$, for $n \in \mathbb{N}$, yields an infinite sequence of points whose z-coordinates are not bounded, so $\mathcal{C}$ is not bounded. Let Π be a plane normal to $\mathcal{L}$, and let $\mathcal{D}$ be the projection of $\mathcal{C}$ onto Π, which must exhibit rotational symmetry, of angle θ. If $\mathcal{D}$ is not a circle, then $\theta = 2\pi/k$ for some $k \in \mathbb{N}$. Thus, $f^{(kn)}(\mathbf{p})$ is an infinite sequence of points of $\mathcal{C}$, lying on a same line L parallel to $\mathbf{v}$, and we can argue as before to get a contradiction. If $\mathcal{D}$ is a circle, then $\mathcal{C}$ is contained in a cylinder; in fact, $x^2 + y^2 = R^2$ for some $R > 0$, since we assumed that $\mathcal{L}$ was the z-axis, and we assume $\mathcal{C}$ to be real. Since the only real point at infinity of $x^2 + y^2 = R^2$ is $[0 : 0 : 1 : 0]$, this point must also be the only point at infinity of $\mathcal{C}$. But then $\mathcal{C}$ twists infinitely around $\mathcal{L}$ while lying on $x^2 + y^2 = R^2$. Thus, taking a random plane containing $\mathcal{L}$, we get infinitely many intersections of $\mathcal{C}$ with this plane, which is a contradiction because $\mathcal{C}$, by hypothesis, is not planar. □

Lemma 94. *Let $\mathcal{C} \subset \mathbb{R}^3$ be an irreducible space curve, invariant under two rotations f_1, f_2 with axes $\mathcal{L}_1 \neq \mathcal{L}_2$. Then $\mathcal{L}_1, \mathcal{L}_2$ cannot be parallel.*

Proof. Suppose that $\mathcal{L}_1, \mathcal{L}_2$ are parallel. Let Π be a plane normal to $\mathcal{L}_1, \mathcal{L}_2$ that intersects $\mathcal{C}$ in at least one point. The set $\mathcal{C} \cap \Pi$ is invariant under both rotations. But if a set of planar points exhibits rotation symmetry, then the rotation center must be the barycenter of the points, implying that $\mathcal{L}_1 = \mathcal{L}_2$. □

The following result makes use of Lemma 18 in Chapter 1.

Proposition 95. *Let $\mathcal{C} \subset \mathbb{R}^3$ be an irreducible space curve, not contained in a plane. Then $\mathcal{C}$ is invariant under at most:*

(i) one central symmetry;

(ii) finitely many rotational symmetries, whose axes are all concurrent;

(iii) finitely many reflections, whose planes share a point;

(iv) finitely many rotoreflections, whose axes are all concurrent.

Proof. (i) If $\mathcal{C}$ is invariant under two central symmetries with symmetry centers $\mathbf{c}_1$ and $\mathbf{c}_2$, then $\mathcal{C}$ is invariant under their composition, which by statement (1) of Lemma 18 is a translation by a vector parallel to the vector connecting $\mathbf{c}_1, \mathbf{c}_2$. But by Lemma 93 this cannot happen, implying that $\mathbf{c}_1 = \mathbf{c}_2$.

(ii) By Lemma 18, the composition of two rotations with axes $\mathcal{L}_1, \mathcal{L}_2$ is: (a) when the axes are parallel, a rotation about an axis parallel to $\mathcal{L}_1, \mathcal{L}_2$; (b) when the axes intersect, a rotation about an axis passing through $\mathcal{L}_1 \cap \mathcal{L}_2$; (c) when the axes are skew, a screw motion. Cases (a) and (c) are discarded by Lemma 94 and Lemma 93, so we are left with case (b).

A first possibility for case (b) is that we have rotation axes that intersect pairwise, but that are not concurrent. Assume then that $\mathcal{C}$ is invariant under three rotations about different axes $\mathcal{L}_1$, $\mathcal{L}_2$, $\mathcal{L}_3$, where $\mathcal{L}_1 \cap \mathcal{L}_2 \neq \emptyset$ but $\mathcal{L}_1 \cap \mathcal{L}_2 \cap \mathcal{L}_3 = \emptyset$. Then, the composition of the rotations about $\mathcal{L}_1, \mathcal{L}_2$ yields a rotation with axis $\mathcal{L}_4$ transversally intersecting the plane defined by $\mathcal{L}_1$ and $\mathcal{L}_2$ in a point away from the third axis $\mathcal{L}_3$. But then the axes $\mathcal{L}_4$ and $\mathcal{L}_3$ are skew, which corresponds to case (c) and therefore cannot happen. Thus, all the rotation axes of $\mathcal{C}$ must be concurrent.

Let us see now that the number of rotational symmetries of $\mathcal{C}$ must be finite. Suppose that we have an infinite number of axes of rotation $\{\mathcal{L}_i\}_{i\in I}$ meeting in a point P. The axes $\{\mathcal{L}_i\}_{i\in I}$ meet at infinitely many distinct angles. The composition of two rotations with concurrent axes is, by statement (4) of Lemma 18, another rotation, about an axis perpendicular to the concurrent rotation axes. Additionally, if the rotations have rotation angles α, β and their axes meet at an angle ϕ, then the composition is a rotation by an angle γ, satisfying Eq. (1.16) (see Lemma 18). Since there are finitely many α, β by Lemma 92 but infinitely many angles ϕ, we get infinitely many angles γ as well. But by Lemma 92 there are finitely many values for the rotation angles, so we conclude that $\mathcal{C}$ has at most finitely many rotation symmetries, whose axes are concurrent.

(iii) The composition of two reflections on planes Π_1, Π_2 is, by statement (5) of Lemma 18: (a) if the planes are parallel, a translation; (b) if the planes intersect, a rotation about $\Pi_1 \cap \Pi_2$ by an angle twice the angle between Π_1, Π_2. Case (a) can be discarded because of Lemma 93, so any two reflection planes must have a line in common. In particular, if $\mathcal{C}$ has just two reflections, the statement follows.

Suppose $\mathcal{C}$ has three reflections f_1, f_2, f_3 with corresponding planes Π_1, Π_2, Π_3. If these three planes intersect in a point at infinity, then any two pairs, say (Π_1, Π_2) and (Π_1, Π_3), intersect in parallel affine lines $\mathcal{L} = \Pi_1 \cap \Pi_2$ and $\mathcal{L}' = \Pi_1 \cap \Pi_3$. Since the composition of two reflections about intersecting planes is a rotation, we get that $\mathcal{C}$ is invariant under two rotations with parallel axes, which cannot happen by Lemma 94. It follows that the three planes intersect in an affine point $P := \Pi_1 \cap \Pi_2 \cap \Pi_3$.

Suppose there is a fourth reflection f_4 with plane Π_4 not containing P. Then Π_4 intersects Π_3 in a certain line $\mathcal{L}''$. If $\mathcal{L}''$ does not pass through P, then $\mathcal{C}$ has rotational symmetries about the two skew axes $\mathcal{L}$ and $\mathcal{L}''$. But then $\mathcal{C}$ would be invariant under their composition, which is a screw motion, and this cannot happen. So Π_4 also contains P and (iii) holds.

(iv) Let f be a rotoreflection about an axis $\mathcal{L}$, leaving $\mathcal{C}$ invariant. Then f^2 is a rotation about the same axis, and by statement (ii) we know that the number of rotation axes and the number of rotation angles are finite. Thus, if there are infinitely many rotoreflections leaving $\mathcal{C}$ invariant, then there must be infinitely many of them sharing the same axis of rotation. Now one can easily check that the composition of two rotoreflections with the same axis is a rotation. Since, again by statement (ii), the number of rotations is finite, the statement follows. □

Finally, the previous results provide the following theorem.

Theorem 96. *Let $\mathcal{C} \subset \mathbb{R}^3$ be an irreducible, real space algebraic curve, not contained in a plane. Then* $\mathrm{Sym}(\mathcal{C})$ *is finite.*

5.4 IMPLICIT ALGEBRAIC SPACE CURVES (I)

In this section, we will just make some observations about symmetries of implicit space curves, but we will postpone the algorithms for the next chapter, once we know how to find the symmetries of algebraic surfaces. A first observation about implicit space algebraic curves is the fact that the brute-force approach, in this case, is not easy. In the case of planar algebraic curves or planar vector fields this is doable, although increasingly worse as the degree of the curve or of the components of the vector field grows larger: indeed, we just pick a generic planar symmetry $f(\boldsymbol{x}) = A\boldsymbol{x}+\boldsymbol{b}$, and request that Eq. (2.2) or Eq. (4.4) holds. This leads to a polynomial system in the entries of A (that can be written in terms of just two parameters) and the components of the vector $\boldsymbol{b}$, and the solution of this system provides the symmetries.

However, in the case of implicit space algebraic curves this approach is more complicated. Assume the simpler case of a space algebraic curve $\mathcal{C}$ given as the intersection of two algebraic surfaces defined by $F_1(x, y, z), F_2(x, y, z)$. If we pick a generic space symmetry $f(\boldsymbol{x}) = A\boldsymbol{x} + \boldsymbol{b}$, now the condition for f to be a symmetry of $\mathcal{C}$ is that

$$\tilde{F}_1(\boldsymbol{x}) = F_1(A\boldsymbol{x} + \boldsymbol{b}), \tilde{F}_2(\boldsymbol{x}) = F_2(A\boldsymbol{x} + \boldsymbol{b}),$$

define $\mathcal{C}$ as well, but $\tilde{F}_1, \tilde{F}_2$ need not be scalar multiples of F_1, F_2: now the condition should be $\tilde{F}_1, \tilde{F}_2 \in \mathbf{I}(\mathcal{C})$. There are certainly algorithms to check whether or not a given polynomial belongs to a certain ideal (this requires using the *Gröbner basis*, which is briefly mentioned in Section 10.3 of the last chapter, see also [45] for further information). But the algorithms are computationally expensive, and applying these algorithms with so many parameters (the entries of the generic space symmetry) would be almost impossible.

There is another, also bad, idea involving projections. If f is a symmetry of $\mathcal{C} = V(F_1, F_2)$, then $V(F_1, F_2) = V(\tilde{F}_1, \tilde{F}_2)$. Thus, projecting $V(F_1, F_2)$ and $V(\tilde{F}_1, \tilde{F}_2)$ onto any plane must yield the same planar curve. In particular, the projections Π_{xy}, Π_{xz} onto the coordinate planes xy, xz of $V(F_1, F_2)$ and $V(\tilde{F}_1, \tilde{F}_2)$ should coincide, so the resultants $\text{Res}_z(F_1, F_2), \text{Res}_z(\tilde{F}_1, \tilde{F}_2)$ and $\text{Res}_y(F_1, F_2), \text{Res}_y(\tilde{F}_1, \tilde{F}_2)$ must coincide also up to multiplication by constants. Again, imposing the equality (up to constants) of these resultants with a generic f in order to devise a brute-force algorithm for finding the symmetries of $\mathcal{C}$, is not a good idea. By proceeding this way we would have a polynomial system in a high number of unknowns, the entries of f. As for the degree of this system, if the surfaces defined by F_1, F_2 have degrees m, n, the degree of the resultant is $\leq mn$ [135]. In fact, when working with homogeneous polynomials, i.e., in projective space, one always gets the degree mn, but when passing to the affine this bound may fall in some cases; still, mn is what one gets in a generic situation. Thus, in general the projections will have a degree much bigger than the original curves, and we would be bound to a hard-to-solve polynomial system.

One might still think of this as a potential test to check whether a given f is actually a symmetry of $\mathcal{C}$, but it is not a good idea either. It might serve to rule out a certain f as a symmetry of $\mathcal{C}$, when the resultants do not coincide, but the fact that the resultants coincide is not enough to ensure that f is a symmetry of $\mathcal{C}$: what lies behind is that $V(\text{Res}_y(F_1, F_2), \text{Res}_z(F_1, F_2))$ is not necessarily equal to $V(F_1, F_2)$, because the former curve might have additional components compared to the latter curve. So definitely we need something else.

In the next chapter, after addressing algebraic surfaces, we will come back to this problem, limiting ourselves to the case when $\widehat{\mathcal{C}}$, and therefore $\mathcal{C}$, is a complete intersection (see Section 5.1 for the notion of complete intersection). For now, we leave implicit space curves and we enter the simpler case of rational space curves.

5.5 SYMMETRIES OF RATIONAL SPACE ALGEBRAIC CURVES

Here we will assume that $\mathcal{C} \subset \mathbb{R}^3$ is a real, rational space curve, rationally parametrized by $\boldsymbol{p}(t) = (x(t), y(t), z(t))$, where $\boldsymbol{p}(t)$ is a real, proper parametrization, see Section 3.1 in Chapter 3 for the notion of properness. The reason for this assumption is that, as we did for rational plane curves, in order to find $\text{Sym}(\mathcal{C})$ we want to use Theorem 45, which allows us to solve the problem in any dimension. The strategy is pretty much the same as in Chapter 3, adapted to space, so we will skip the proofs and refer to [11], where this strategy is developed, for details. In short, we compute first the Möbius function $\varphi(t)$ behind each symmetry, and then find the symmetry $f(\boldsymbol{x}) = A\boldsymbol{x} + \boldsymbol{b}$ itself from $\varphi(t)$ using the fact that

$$A\boldsymbol{p}(t) + \boldsymbol{b} = \boldsymbol{p}(\varphi(t)), \tag{5.6}$$

either solving for the entries of $A, \boldsymbol{b}$ the linear system stemming from Eq. (5.6) once φ is known, or by computing first A from its action on $\boldsymbol{p}'(0), \boldsymbol{p}''(0), \boldsymbol{p}'(0) \times \boldsymbol{p}''(0)$, then $\boldsymbol{b}$ (see Example 19 in Chapter 8 for this). We will also say here that φ is *associated* with f.

To do this, we will suppose that $\mathcal{C}$ is not contained in a plane, since in that case the algorithm in Chapter 3 suffices. Additionally, we will use the classical differential invariants well-known in the differential geometry of space curves, *curvature* and *torsion*. The curvature $\kappa_{\boldsymbol{p}}$ and the torsion $\tau_{\boldsymbol{p}}$ of a parametric curve $\boldsymbol{p}(t)$ in $\mathbb{R}^3$ are defined as

$$\kappa_{\boldsymbol{p}} = \frac{\|\boldsymbol{p}' \times \boldsymbol{p}''\|}{\|\boldsymbol{p}'\|^3}, \quad \tau_{\boldsymbol{p}} = \frac{\langle \boldsymbol{p}', \boldsymbol{p}'', \boldsymbol{p}''' \rangle}{\|\boldsymbol{p}' \times \boldsymbol{p}''\|^2}, \tag{5.7}$$

where $\times$ denotes the cross product, $\| \bullet \|$ denotes the usual norm, and $\langle \bullet \rangle$ represents the mixed product. Notice that $\kappa_{\boldsymbol{p}}$ is always non-negative and is identically zero only if $\mathcal{C}$ is a line, therefore planar, which we are excluding; however, $\tau_{\boldsymbol{p}}$ can be positive, negative or zero, although it can be identically zero only, also, in the case of planar curves. If $\boldsymbol{p}(t)$ is rational, one can see that $\kappa_{\boldsymbol{p}}$ is in general not rational, but $\kappa_{\boldsymbol{p}}^2$ is; because of this, as we did in Chapter 3, we will replace κ_p by

$$\mathrm{k}_{\boldsymbol{p}} = \kappa_{\boldsymbol{p}}^2 = \frac{\|\boldsymbol{p}' \times \boldsymbol{p}''\|^2}{\|\boldsymbol{p}'\|^6}. \tag{5.8}$$

However, $\tau_{\boldsymbol{p}}$ is always rational.

The discussion is analogous to that in Chapter 3. First, the properties of the cross product (see Eq. (3.9) in Chapter 3) and the fact that the dot product is preserved under orthogonal transformations, allow us to prove the following result, analogous to Lemma 48 in Chapter 3, and which can be proven in a similar way. The $\pm$ in the case of the torsion comes from the $\pm$ in Eq. (3.9), when analyzing the mixed product: for orientation-preserving symmetries the sign is $+$, and for orientation-reversing symmetries, the sign is $-$.

Lemma 97. *Let $\mathcal{C} \subset \mathbb{R}^3$ be a space curve properly parametrized by $\boldsymbol{p}(t)$, and let $f(\boldsymbol{x}) = A\boldsymbol{x} + \boldsymbol{b}$ be a Euclidean symmetry of $\mathcal{C}$. Then*

$$\mathrm{k}_{\boldsymbol{p}} = \mathrm{k}_{f\circ\boldsymbol{p}}, \; \tau_{\boldsymbol{p}} = \pm\tau_{f\circ\boldsymbol{p}} \tag{5.9}$$

In order to compute φ also using, as in Chapter 3, bivariate factoring, the following result is needed. This result, analogous to Lemma 48 in Chapter 3 and which can be proven in a similar way making use of the Chain rule, guarantees that $\mathrm{k}_{\boldsymbol{p}}, \tau_{\boldsymbol{p}}$ commute with Möbius transformations.

Lemma 98. *Let $\boldsymbol{p}(t)$ be a space parametrization, and let φ be a Möbius transformation. Then*

$$\mathrm{k}_{\boldsymbol{p}\circ\varphi} = \mathrm{k}_{\boldsymbol{p}} \circ \varphi, \; \tau_{\boldsymbol{p}\circ\varphi} = \tau_{\boldsymbol{p}} \circ \varphi \tag{5.10}$$

Thus, we get the following corollary.

Corollary 99. *Let $\mathcal{C} \subset \mathbb{R}^3$ be a space curve properly parametrized by $\boldsymbol{p}(t)$, and let φ be a Möbius transformation associated with a Euclidean symmetry of $\mathcal{C}$. Then*

$$\mathrm{k}_{\boldsymbol{p}} = \mathrm{k}_{\boldsymbol{p}} \circ \varphi, \; \tau_{\boldsymbol{p}} = \pm\tau_{\boldsymbol{p}} \circ \varphi \tag{5.11}$$

Corollary 99 leads to an algorithm to compute the symmetries using bivariate polynomial factoring. Calling

$$\mathrm{k}_{\boldsymbol{p}} = \mathrm{k}_{\boldsymbol{p}}(t) = \frac{P(t)}{Q(t)}, \tau_{\boldsymbol{p}} = \tau_{\boldsymbol{p}}(t) = \frac{R(t)}{S(t)},$$

we let

$$K(t,\omega) = P(t)Q(\omega) - P(\omega)Q(t), \; T^{\pm}(t,\omega) = R(t)S(\omega) \mp R(\omega)S(t).$$

It cannot happen that K and T^+ are, both, identically zero. Indeed, in that case then $\kappa_{\boldsymbol{p}}$ and $\tau_{\boldsymbol{p}}$ are themselves constant: however, since by our hypotheses neither $\kappa_{\boldsymbol{p}}$ nor $\tau_{\boldsymbol{p}}$ can be identically zero, the fact that both $\kappa_{\boldsymbol{p}}, \tau_{\boldsymbol{p}}$ are nonzero constants implies that $\mathcal{C}$ is a circular helix [56], which is not an algebraic curve. It can happen, though, that just one of either K or T^+ is a nonzero constant: there are examples (*Salkowski curves*) of rational space curves with either constant curvature, or constant torsion, see [90].

We need to find the Möbius-like factors $F(t,\omega)$ (see Eq. (3.13) in Chapter 3) which are common to $K(t,\omega)$ and $T^{\pm}(t,\omega)$, i.e., which divide

$$M^{\pm}(t,\omega) = \gcd(K(t,\omega), T^{\pm}(t,\omega)),$$

where we need to separately address M^+ and M^-. Thus, we arrive at the following theorem, analogous to Theorem 50 in Chapter 3 and which can be proven in a similar way. To prove it we need to use the Fundamental Theorem of Space Curves, which states that, when parametrized by *arc-length*, $\kappa_{\boldsymbol{p}}, |\tau_{\boldsymbol{p}}|$ completely define the curve up to rigid motions, i.e., Euclidean symmetries.

Theorem 100. *Let $\mathcal{C} \subset \mathbb{R}^3$ be a planar rational curve, properly parametrized by $\mathbf{p}(t)$. The curve $\mathcal{C} \subset \mathbb{R}^3$ is invariant under a Euclidean symmetry $f(\boldsymbol{x}) = A\boldsymbol{x} + \boldsymbol{b}$ if and only if there exists a Möbius transformation $\varphi(t)$, as in Eq. (5.8), such that the polynomial $M^{\pm}(t,\omega)$ has a Möbius-like factor, and*

$$\|\boldsymbol{p}'\| = \|(\boldsymbol{p} \circ \varphi)'\|. \tag{5.12}$$

Furthermore, φ is associated with f.

Thus, we have the following algorithm `Rational Space Symmetries`, to compute the symmetries of a space rational curve, properly parametrized.

Example 9. *Let $\mathcal{C}$ be the* crunode *space curve parametrized by*

$$\boldsymbol{x} : t \longmapsto \left(\frac{t}{t^4+1}, \frac{t^2}{t^4+1}, \frac{t^3}{t^4+1} \right).$$

Applying Algorithm `Rational Space Symmetries` *we get $M^+(t,\omega) = (t-\omega)(t+\omega)$. The first factor corresponds to the identity map*

Algorithm 3 Rational Space Symmetries

Require: A proper parametrization $\boldsymbol{p}(t)$ of a space curve $\mathcal{C} \subset \mathbb{R}^3$, not contained in any plane.

Ensure: $\mathrm{Sym}(\mathcal{C})$.

compute $K(t,\omega)$, $T^{\pm}(t,\omega)$
compute $M^{\pm}(t,\omega) = \gcd(K(t,\omega), T^{\pm}(t,\omega))$
factor $M^{\pm}(t,\omega)(t,\omega)$ over the reals
for each Möbius-like factor $F(t,\omega)$ of $M^{\pm}(t,\omega)(t,\omega)$ **do**
  solve $F(t,\omega) = 0$ for ω to find $\varphi(t)$
  check whether or not $\|\boldsymbol{p}'\| = \|(\boldsymbol{p} \circ \varphi)'\|$
  if the above condition holds **then**
    find $A, \boldsymbol{b}$ from Eq. (5.6)
    return $f(\boldsymbol{x}) = A\boldsymbol{x} + \boldsymbol{b}$
  end if
end for

$\varphi_1(t) = t$ and the trivial symmetry $f_1(\boldsymbol{x}) = \boldsymbol{x}$. The second factor corresponds to the Möbius transformation $\varphi_2(t) = -t$. A direct computation shows that $\varphi_2(t)$ satisfies Eq. (5.12)*. By using Eq.* (5.6)*, we deduce that $\mathcal{C}$ is invariant under $f_2(\boldsymbol{x}) = A_2\boldsymbol{x}$, where*

$$A_2 = \begin{bmatrix} -1 & 0 & 0 \\ 0 & 1 & 0 \\ 0 & 0 & -1 \end{bmatrix}.$$

This is an axial symmetry about the y-axis.

Additionally, $M^-(t,\omega) = (t\omega - 1)(t\omega + 1)$, whose factors correspond to the Möbius transformations $\varphi_3(t) = 1/t$ and $\varphi_4(t) = -1/t$. Again, a direct computation shows that $\varphi_3(t)$ and $\varphi_4(t)$ satisfy Eq. (5.12)*, and they correspond to reflections $f_3(\boldsymbol{x}) = A_3\boldsymbol{x}$ and $f_4(\boldsymbol{x}) = A_4\boldsymbol{x}$, with*

$$A_3 = \begin{bmatrix} 0 & 0 & 1/2 \\ 0 & 1 & 0 \\ 2 & 0 & 0 \end{bmatrix}, \qquad A_4 = \begin{bmatrix} 0 & 0 & -1/2 \\ 0 & 1 & 0 \\ -2 & 0 & 0 \end{bmatrix}$$

These symmetries are reflections on the planes $\Pi_3 : z - 2x = 0$ and $\Pi_4 : z + 2x = 0$, which intersect in the axis of the crunode; see Fig. 5.3.

FIGURE 5.3 The crunode curve from Example 9, together with its axis of symmetry and its reflection planes.

5.6 RATIONAL INVARIANTS

The square of the curvature or the torsion are examples of *rational invariants*. Given a parametrization $\boldsymbol{u}(t)$ in $\mathbb{K}^n$, with $\mathbb{K}$ a field, a rational invariant is an expression, rational in the derivatives of several orders of $\boldsymbol{u}(t)$, which is invariant under a certain group of interest. In the case of the curvature and the torsion, the group is the group of Euclidean symmetries. However, if we change to a different group, we can create invariants for that group.

To give an idea of how this works, we will consider not the usual group in this book but the group of affine mappings, $f(\boldsymbol{x}) = A\boldsymbol{x}+\boldsymbol{b}$ where A is a non-singular $n \times n$ matrix, not necessarily orthogonal (of course, Euclidean symmetries are a subcase of this more general problem). Now let $\boldsymbol{u}(t), \boldsymbol{v}(t)$ be two rational parametrizations in $\mathbb{K}^n$ such that $A\boldsymbol{u}(t) + \boldsymbol{b} = \boldsymbol{v}(t)$, which for brevity we will write just as $A\boldsymbol{u} + \boldsymbol{b} = \boldsymbol{v}$. Repeatedly differentiating this equation with respect to t yields $AD(\boldsymbol{u}) = D(\boldsymbol{v})$ where

$$D(\boldsymbol{u}) = \left[\boldsymbol{u}', \boldsymbol{u}'', \ldots, \boldsymbol{u}^{(n)}\right], \; D(\boldsymbol{v}) = \left[\boldsymbol{v}', \boldsymbol{v}'', \ldots, \boldsymbol{v}^{(n)}\right],$$

i.e., $D(\boldsymbol{u}), D(\boldsymbol{v})$ are matrices whose columns consist of the first

n derivatives of $\boldsymbol{u}, \boldsymbol{v}$ with respect to t. Whenever $\boldsymbol{u}, \boldsymbol{v}$ are not contained in hyperplanes, $D(\boldsymbol{u}), D(\boldsymbol{v})$ are invertible [122]. Thus, we can write $A = D(\boldsymbol{u})(D(\boldsymbol{v}))^{-1}$. Differentiating this equality with respect to t, and taking into account that A is a constant matrix, we get that

$$\frac{d(D(\boldsymbol{u})(D(\boldsymbol{v}))^{-1})}{dt} = 0.$$

Expanding the derivative in the left-hand side of the above equation we arrive at

$$(D(\boldsymbol{u}))^{-1}\frac{dD(\boldsymbol{v})}{dt} = (D(\boldsymbol{v}))^{-1}\frac{dD(\boldsymbol{u})}{dt}. \tag{5.13}$$

Denoting

$$U = (D(\boldsymbol{u}))^{-1}\frac{dD(\boldsymbol{u})}{dt}, \quad V = (D(\boldsymbol{u}))^{-1}\frac{dD(\boldsymbol{v})}{dt},$$

one can check that

$$U = \begin{bmatrix} 0 & 0 & \cdots & 0 & \dfrac{\|\boldsymbol{u}^{(n+1)}, \boldsymbol{u}'', \cdots, \boldsymbol{u}^{(n)}\|}{\|\boldsymbol{u}', \boldsymbol{u}'', \cdots, \boldsymbol{u}^{(n)}\|} \\ 0 & 1 & \cdots & 0 & \dfrac{\|\boldsymbol{u}', \boldsymbol{u}^{(n+1)}, \cdots, \boldsymbol{u}^{(n)}\|}{\|\boldsymbol{u}', \boldsymbol{u}'', \cdots, \boldsymbol{u}^{(n)}\|} \\ \vdots & \vdots & \ddots & \vdots & \vdots \\ 0 & 0 & \cdots & 1 & \dfrac{\|\boldsymbol{u}', \boldsymbol{u}'', \cdots, \boldsymbol{u}^{(n+1)}\|}{\|\boldsymbol{u}', \boldsymbol{u}'', \cdots, \boldsymbol{u}^{(n)}\|} \end{bmatrix},$$

and

$$V = \begin{bmatrix} 0 & 0 & \cdots & 0 & \dfrac{\|\boldsymbol{v}^{(n+1)}, \boldsymbol{v}'', \cdots, \boldsymbol{v}^{(n)}\|}{\|\boldsymbol{v}', \boldsymbol{v}'', \cdots, \boldsymbol{v}^{(n)}\|} \\ 0 & 1 & \cdots & 0 & \dfrac{\|\boldsymbol{v}', \boldsymbol{v}^{(n+1)}, \cdots, \boldsymbol{v}^{(n)}\|}{\|\boldsymbol{v}', \boldsymbol{v}'', \cdots, \boldsymbol{v}^{(n)}\|} \\ \vdots & \vdots & \ddots & \vdots & \vdots \\ 0 & 0 & \cdots & 1 & \dfrac{\|\boldsymbol{v}', \boldsymbol{v}'', \cdots, \boldsymbol{v}^{(n+1)}\|}{\|\boldsymbol{v}', \boldsymbol{v}'', \cdots, \boldsymbol{v}^{(n)}\|} \end{bmatrix}.$$

Next let us define

$$\begin{aligned} A_i(\boldsymbol{u}) &:= \|\boldsymbol{u}', \ldots, \boldsymbol{u}^{(i-1)}, \boldsymbol{u}^{(n+1)}, \boldsymbol{u}^{i+1}, \ldots, \boldsymbol{u}^{(n)}\|, \\ \Delta(\boldsymbol{u}) &:= \|\boldsymbol{u}', \boldsymbol{u}'', \ldots, \boldsymbol{u}^{n}\|. \end{aligned} \tag{5.14}$$

Thus, $A_i(\boldsymbol{u})$ is the result of replacing $\boldsymbol{u}^{(i)}$ in $\|\boldsymbol{u}'\boldsymbol{u}''\cdots\boldsymbol{u}^{(n)}\|$ by $\boldsymbol{u}^{(n+1)}$. Finally, for $i = 1, \ldots, n$, let

$$I_i(\boldsymbol{u}) := \frac{A_i(\boldsymbol{u})}{\Delta(\boldsymbol{u})}, \tag{5.15}$$

which correspond to the entries of the last column of U; since $\boldsymbol{u}$ does not belong to a hyperplane, $D(\boldsymbol{u})$ is invertible and therefore $\Delta(\boldsymbol{u})$ is not identically zero.

By Eq. (5.13), U, V are equal and therefore their last columns coincide. Thus, $I_i(\boldsymbol{u}) = I_i(\boldsymbol{v})$ for $i = 1, \ldots, n$, so the I_i are examples of rational invariants with respect to the affine group: they are rational expressions depending on the derivatives of the parametrization, that are invariant under the group of affine transformations $f(\boldsymbol{x}) = A\boldsymbol{x} + \boldsymbol{b}$.

In order to see how to use this in practice, assume that we want to decide whether two curves $\mathcal{C}_1, \mathcal{C}_2 \subset \mathbb{K}^n$, properly parametrized by $\boldsymbol{p}(t), \boldsymbol{q}(t)$, are affinely equivalent, i.e., whether or not there exists an affine mapping $f(\boldsymbol{x}) = A\boldsymbol{x} + \boldsymbol{b}$, $f : \mathbb{K}^n \to \mathbb{K}^n$, such that $f(\mathcal{C}_1) = \mathcal{C}_2$. Using the invariants in Eq. (5.15) and Theorem 45, it is clear that a necessary condition for $\mathcal{C}_1, \mathcal{C}_2$ to be affinely equivalent is the existence of a Möbius function φ such that

$$I_i(\boldsymbol{p}) = I_i(\boldsymbol{q} \circ \varphi) \tag{5.16}$$

for $i = 1, \ldots, n$. In fact, one can prove that the condition is also sufficient (see for instance Theorem 7 in [63]). Now we wonder if we can also apply here the strategy for Euclidean symmetries, i.e., whether or not we can find the affine equivalences from the I_i, by using bivariate factoring. The answer, however, is negative, and the reason is that the I_i do not commute with Möbius transformations, i.e., $I_i(\boldsymbol{q}\circ\varphi) \neq I_i(\boldsymbol{q})\circ\varphi$. For instance, in the case $n = 3$, expanding $I_i(\boldsymbol{q}\circ\varphi)$, which for simplicity we write as $I_i(\boldsymbol{q}(\varphi))$, we get (see [20])

$$\begin{aligned}
\varphi'^3 I_1(\boldsymbol{q}(\varphi)) &= 3\varphi''^3 + \frac{3}{2}\varphi'^2\varphi''^2 I_3(\boldsymbol{q})(\varphi) - \varphi'^4\varphi'' I_2(\boldsymbol{q})(\varphi) + \varphi'^6 I_1(\boldsymbol{q})(\varphi)\\
\varphi'^2 I_2(\boldsymbol{q}(\varphi)) &= -9\varphi''^2 + \varphi'^4 I_2(\boldsymbol{q})(\varphi) - 3\varphi'^2\varphi'' I_3(\boldsymbol{q})(\varphi)\\
\varphi' I_3(\boldsymbol{q}(\varphi)) &= 6\varphi'' + \varphi'^2 I_3(\boldsymbol{q})(\varphi),
\end{aligned} \tag{5.17}$$

where for brevity $I_i(\boldsymbol{q})(\varphi)$ denotes $I_i(\boldsymbol{q})\circ\varphi$, and φ', φ'' are the first and second derivatives of the Möbius function φ.

What we can do, though, is to get *new* rational invariants, J_i, from the I_i, that do commute with Möbius transformations, i.e., such that $J_i(\boldsymbol{p}\circ\varphi) = J_i(\boldsymbol{p})\circ\varphi$. In the case $n = 3$, the rough idea for doing this is to eliminate, in Eq. (5.17), the derivatives φ', φ''. By doing this (see [20] for further information), we reach, for instance,

$$\begin{aligned}&\frac{[36I_1(\boldsymbol{q}(\varphi)) + 6I_2(\boldsymbol{q}(\varphi))I_3(\boldsymbol{q}(\varphi)) + I_3(\boldsymbol{q}(\varphi))^3]^2}{[4I_2(\boldsymbol{q}(\varphi)) + I_3(\boldsymbol{q}(\varphi))^2]^3}\\ &= \frac{[36I_1(\boldsymbol{q})(\varphi) + 6I_2(\boldsymbol{q})(\varphi)I_3(\boldsymbol{q})(\varphi) + I_3(\boldsymbol{q})(\varphi)^3]^2}{[4I_2(\boldsymbol{q})(\varphi) + I_3(\boldsymbol{q})(\varphi)^2]^3},\end{aligned} \tag{5.18}$$

so that

$$J_1 = \frac{[36I_1 + 6I_2I_3 + I_3^3]^2}{[4I_2 + I_3^2]^3} \tag{5.19}$$

is a Möbius-commuting rational invariant, because $J_1(\boldsymbol{q}\circ\varphi) = J_1(\boldsymbol{q})\circ\varphi$. Another invariant with the same property is

$$J_2 = \frac{6I_0 - 5I_3^2}{4I_2 + I_3^2}, \tag{5.20}$$

where

$$I_0 = \frac{\|\boldsymbol{u}', \cdots, \boldsymbol{u}^{(n-1)}, \boldsymbol{u}^{(n+2)}\|}{\Delta(\boldsymbol{u})}.$$

Since the J_i depend rationally on the I_i, if $\mathcal{C}_1, \mathcal{C}_2$ are affinely equivalent we also have $J_i(\boldsymbol{p}) = J_i(\boldsymbol{q}\circ\varphi)$; but now, additionally, we have $J_i(\boldsymbol{q}\circ\varphi) = J_i(\boldsymbol{q})\circ\varphi$, and we can apply bivariate factoring also here. For a generic value of n, the idea is similar: expanding $I_i(\boldsymbol{q}\circ\varphi)$, we will get rational expressions involving $I_(\boldsymbol{q})\circ\varphi$ and the derivatives of φ. By eliminating these derivatives, we can get Möbius-commuting invariants. In fact, one can get closed expressions for the affine invariants in any dimension; see [20] for details. Notice that if we request $\boldsymbol{p} = \boldsymbol{q}$ and A orthogonal, we are just considering Euclidean symmetries of the rational curve defined by $\boldsymbol{p}$.

5.7 FINAL REMARKS AND OPEN PROBLEMS

There are a number of papers generalizing the problem for rational curves treated in this chapter. Similarities between rational space curves are addressed in [12]. Projective equivalences of rational curves in any dimension are studied, as we already mentioned in Chapter 3, in [29, 69]. The same problem for space rational curves is approached in [63] using rational invariants. The strategy in [63] is fully developed in [20], where the affine equivalence problem not only for rational curves, but for certain types of analytic, non-rational parametrizations, is solved in arbitrary dimension.

As for open problems, let us just mention, for rational space curves, the generalizations to space of the open problems 7, 8, 9, and 10 in Chapter 3. There are more open problems, related to implicit algebraic space curves, but we will present them at the end of the next chapter, after a second round on implicit space algebraic curves has been completed.

CHAPTER 6

Algebraic surfaces (I): implicit surfaces

Algebraic surfaces are typically more challenging than algebraic curves, and this is the case also when trying to compute their symmetries. In this chapter we consider general properties on the symmetries of these objects, and specifically address the case of implicit algebraic surfaces. The case of rational surfaces will be considered in the remaining chapters. Additionally, we will have a second round on implicit algebraic space curves, whose symmetries can be partially addressed with the help of surfaces.

6.1 GENERAL PROPERTIES

An algebraic surface $S \subset \mathbb{R}^3$ is the zeroset of a polynomial $F(x, y, z)$. The implicit equation of S is, therefore, $F(x, y, z) = 0$, which is unique up to non-zero constants. The notions of degree, components, irreducibility and square-freeness introduced in Chapter 2 can be directly translated to surfaces. However, at a certain point we will need the following notion, that can be also defined for implicit algebraic (plane and space) curves, although we did not need it before.

Definition 101. *We say that* $\boldsymbol{x}_0 \in S$ *is a* singular point *of the surface* S *implicitly defined by* $F(x, y, z) = 0$, *if*

$$F(\boldsymbol{x}_0) = F_x(\boldsymbol{x}_0) = F_y(\boldsymbol{x}_0) = F_z(\boldsymbol{x}_0) = 0. \tag{6.1}$$

DOI: 10.1201/9781003648710-6

The set of all the singular points of S is called the singular locus *of S, which we denote by* $\mathrm{Sing}(S)$.

If $\boldsymbol{x}_0$ is not singular, then we say that $\boldsymbol{x}_0$ is *regular*. The definition for planar curves is completely analogous, and we can put together the definition for both, curves and surfaces, by just saying that a point $\boldsymbol{x}_0$ belonging to the zeroset of a polynomial F is singular whenever the gradient ∇F, vanishes at $\boldsymbol{x}_0$. By the Implicit Function Theorem, if $\boldsymbol{x}_0$ is a regular point of an implicit curve/surface, then the curve is, locally, the graph of a univariate/bivariate function. If $\boldsymbol{x}_0$ is singular, the local behavior of the curve/surface can be much more intricate: for curves, cusps (see the deltoid curve at the right of Fig. 2.1) and self-intersections (see the lemniscate, the three-leaved rose and the Descartes' Folium in Fig. 2.1, left and middle) are always singular. For surfaces, again self-intersections (see for instance Fig. 6.4, left) and cusp-like phenomena, like the vertex of a cone (see Fig. 6.10, right), for instance, are also singular. If we move to the projective setting, singular points are defined in the same way, although now we must add to Eq. (6.1) the derivative with respect to the homogenization variable.

If $\boldsymbol{x}_0$ is a regular point of a curve/surface, we can define the *tangent line/tangent plane* of the curve/surface at $\boldsymbol{x}_0$ as the line/plane

$$\nabla F(\boldsymbol{x}_0) \cdot (\boldsymbol{x} - \boldsymbol{x}_0) = 0.$$

Thus, at a regular point the tangent line, for curves, and the tangent plane, for surfaces, are well defined. At a singular point of a curve we might also have a tangent line (although we could not find it with the above expression), but we might also have several tangent lines, as it happens with the lemniscate or the three-leaved rose. At a singular point of a surface we might even have infinitely many tangent planes, and an example is the vertex of a cone. An important difference between curves and surfaces regarding singularities is that while for curves the number of singular points is finite (whenever the polynomial defining the curve is square-free), in the case of surfaces the singular locus may have 1-dimensional components, i.e., we can have *singular curves*, which are common curves to the zerosets of F and of the components of ∇F.

What about the tangent line in the case of space algebraic curves? Here we limit ourselves to the case when $\mathcal{C} = V(F_1, F_2)$, so $\mathcal{C}$ is given as the intersection of two algebraic surfaces. Notice that at a regular point of a surface, the tangent plane is tangent to all the (space) curves contained in the surface, passing through the point. Thus, if we have a point $\boldsymbol{x}_0 \in \mathcal{C}$ which is a regular point of the surfaces S_1, S_2 defined by F_1, F_2, then the tangent direction to $\mathcal{C}$ at $\boldsymbol{x}_0$ (see Fig. 6.1) should belong to both the tangent plane to S_1 at $\boldsymbol{x}_0$, and the tangent plane to S_2 at $\boldsymbol{x}_0$. The first tangent plane is normal to $\nabla F_1(\boldsymbol{x}_0)$, and the second tangent plane is normal to $\nabla F_2(\boldsymbol{x}_0)$. Thus, the tangent line to $\mathcal{C}$ at $\boldsymbol{x}_0$ should be parallel to

$$\nabla F_1(\boldsymbol{x}_0) \times \nabla F_2(\boldsymbol{x}_0). \tag{6.2}$$

In particular, we can define a *singular point* of a space curve $\mathcal{C} = V(F_1, F_2)$ as a point where the cross product in Eq. (6.1) is zero. Notice that according to this definition, the singular points of either F_i belonging to $\mathcal{C}$ are singular points of $\mathcal{C}$; furthermore, if the surfaces defined by F_1, F_2 are tangent at a point $\boldsymbol{x}_0$, then $\boldsymbol{x}_0 \in \mathcal{C}$ is also a singular point of $\mathcal{C}$. Additionally, it can happen that all the points of $\mathcal{C}$ are singular: this happens when the surfaces intersect along a singular curve, or are tangent to each other along a curve. Notice that this geometric idea of a singular point can be translated into Algebra by saying that the singular points are the points where the rank of the Jacobian matrix, i.e., the 3×2 matrix whose columns are the derivatives of F_1, F_2, is less than two. We can use this idea to generalize the notion of singular point to the projective setting.

After the above digression on singularities of planar and space curves and surfaces, which we will need afterwards, we go back to surfaces. From now on we will assume that the polynomial F implicitly defining the surface is square-free, i.e., that F has no multiple factors. The reason is that, under this assumption, the following lemma, analogous to Lemma 20 in Chapter 2 and which can be proven in the same way, holds.

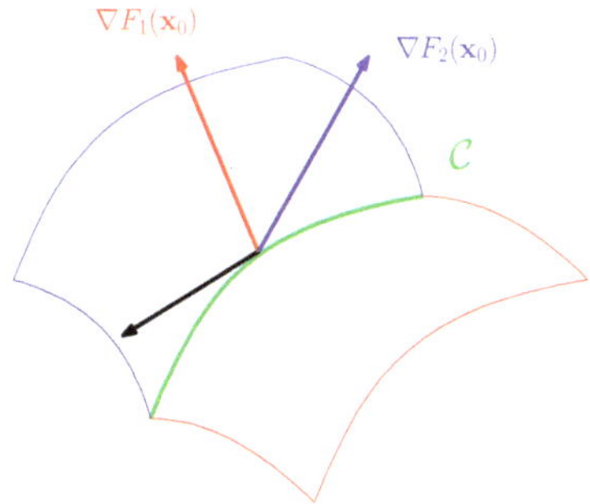

FIGURE 6.1 Tangent direction at a point $\boldsymbol{x}_0$ of an algebraic space curve $\mathcal{C} = V(F_1, F_2)$

Lemma 102. *Let $f(\boldsymbol{x}) = A\boldsymbol{x} + \boldsymbol{b}$ be a symmetry of a surface $S \subset \mathbb{R}^3$, defined by a square-free polynomial $F(x, y, z)$. Then there exists $\lambda \in \mathbb{R}$ such that $F \circ f = \lambda F$.*

Our interest of course is to study the symmetries of S; we will denote by $\mathrm{Sym}(S)$ the group of symmetries of S. As we also did with planar curves and space curves, the first task is to guarantee that $\mathrm{Sym}(S)$ is finite, or rather to isolate the cases when this is not true. Once we do this, we will discover that the constant λ appearing in Lemma 102 must be $\lambda = \pm 1$, as it also happened with curves. Furthermore, by simplicity we will suppose that F is irreducible.

To guarantee that $\mathrm{Sym}(S)$ is finite we will need to exclude two types of surfaces, namely *cylindrical surfaces* and *surfaces of revolution*. A *cylindrical surface* is a particular case of a *ruled surface*, which is a surface made of lines, i.e., with at least one line through each point; these lines are called *rulings* (see Fig. 6.2; some rulings are plotted in black). Cylindrical surfaces are ruled surfaces where all the rulings are parallel, and therefore are invariant under any translation parallel to the direction of the rulings. Thus, they have infinitely many symmetries.

Proposition 103. *If S is invariant under any of the following transformations:*

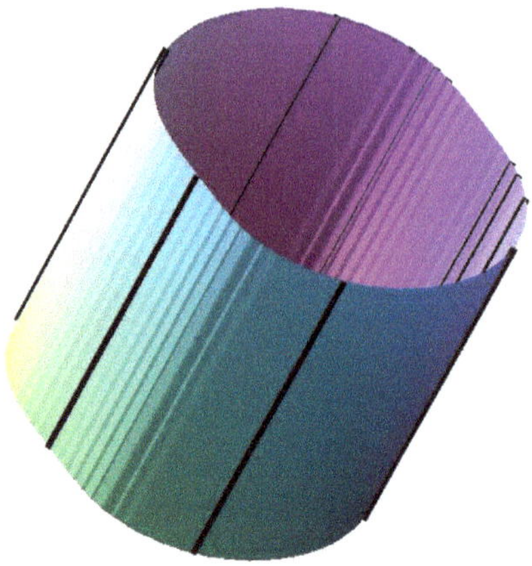

FIGURE 6.2 Cylindrical surface; some rulings are shown in black

(i) a translation by a nonzero vector $\mathbf{v} \in \mathbb{R}^3$,

(ii) a screw motion about an axis ℓ, *by a nonzero vector* $\mathbf{v}$ *(parallel to* ℓ*)*,

(iii) a glide reflection about a plane Π, *by a nonzero vector* $\mathbf{v}$ *(parallel to* Π*)*,

then S *is a cylindrical surface and its rulings are parallel to* $\mathbf{v}$.

Proof. For (i) or (iii), we observe that for every $P \in S$ the line $\mathcal{L}_{P,\mathbf{v}} = P + \lambda\mathbf{v}$ has infinitely many points in common with S. Thus, the polynomial in λ obtained by substituting the parametrization of $\mathcal{L}_{P,\mathbf{v}}$ into the polynomial $F(x, y, z)$, $F(P + \lambda\mathbf{v})$, has infinitely many roots, so it must be the zero polynomial. Thus, $\mathcal{L}_{P,\mathbf{v}}$ must be contained in S, and the result follows.

So let us address (ii). Suppose that S is invariant under a screw motion f about an axis $\mathcal{L}$ by an angle θ, with a translation vector $\mathbf{v}$ (parallel to $\mathcal{L}$). If S is a union of circular cylinders of axis $\mathcal{L}$, then S is cylindrical, with rulings parallel to $\mathbf{v}$, and we have finished. Otherwise, consider, for $R > 0$, the circular cylinder $C_{\mathcal{L},R}$ with axis $\mathcal{L}$, and let $S \cap C_{\mathcal{L},R} = \mathcal{C}_R$, where $\mathcal{C}_R$ is an algebraic space curve. Thus, we have that

$$S = \bigcup_{R>0} \mathcal{C}_R.$$

Next the curve $\mathcal{C}_R$ is also invariant under f, so by Lemma 93 the curve $\mathcal{C}_R$ should be planar. But then the only option is that $\mathcal{C}_R$ is a line, and S itself a cylinder, in which case S is also a cylindrical surface. □

The next result shows that if we work with a non-cylindrical surface, we can be sure that there is at most one central symmetry; here, and in several other coming results, we invoke Lemma 18 in Chapter 1.

Proposition 104. *If S is not cylindrical, the symmetry center of S, if it exists, is unique.*

Proof. Assume that S has two symmetry centers, i.e., that S is invariant under two different central symmetries. By part (1) of Lemma 18, S is invariant under a translation. Therefore by Proposition 103 S must be cylindrical, which cannot happen by hypothesis. □

A surface is a surface of revolution if there exists a line $\mathcal{L}$, called the *axis of revolution* of S, such that S is invariant under any rotation about $\mathcal{L}$, by an angle $\theta \in [0, 2\pi)$; thus, these surfaces also have infinitely many symmetries. Fig. 6.3, left, shows a surface of revolution, where the axis of revolution is plotted in red; notice that the sections of S by planes normal to the axis of revolution are circles.

Lemma 105. *Let S be an algebraic surface of degree d, invariant under a rotation about an axis $\mathcal{L}$, by a non-trivial angle θ (i.e., $\theta \neq 2k\pi$, with $k \in \mathbb{Z}$). If $\mathcal{L}$ is not an axis of revolution of S, then $\theta = \frac{2\pi}{N}$, where $0 < N \leq 2d$, $N \in \mathbb{N}$.*

Proof. Since by hypothesis $\mathcal{L}$ is not an axis of revolution, there exists a plane Π, normal to $\mathcal{L}$, such that the intersection curve $\mathcal{C} = \Pi \cap S$ is not a product of circles centered at $P = \Pi \cap \mathcal{L}$. Now $\mathcal{C}$ is an algebraic planar curve of degree at most d. Furthermore, since S has rotational symmetry about $\mathcal{L}$ the curve $\mathcal{C}$ is invariant under the (planar) rotation about the point $P = \Pi \cap \mathcal{L}$, by the angle θ. But then the result follows from Lemma 24 in Chapter 2. □

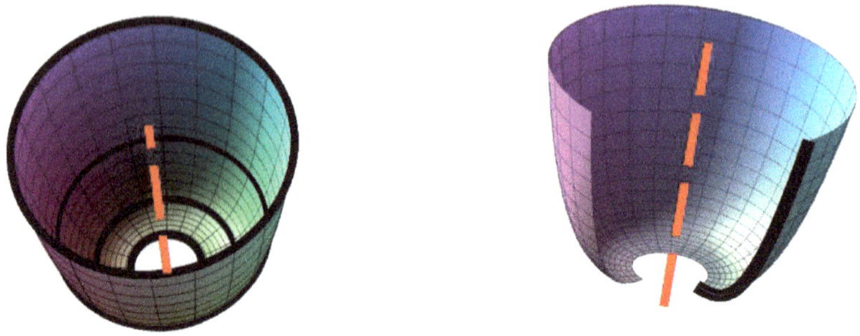

FIGURE 6.3 A surface of revolution (left), with the axis of revolution shown in red; the same surface (right) showing the directrix curve, plotted in black thick line

Lemma 105 provides the following corollary.

Corollary 106. *If S is not a surface of revolution, then the number of angles $\theta \in [0, 2\pi)$ such that S is invariant under some rotation $\mathcal{R}_{\mathcal{L},\theta}$ is finite.*

Notice that Corollary 106 does not imply that S has finitely many axes of rotation, but only that the rotation angles coming up in the rotations leaving the surface invariant exhibit finitely many values. Now we want to prove that the number of axes of rotation is also finite. In order to do this, we first need the following result.

Lemma 107. *If S is not a plane, then S cannot have two different parallel axes of rotation.*

Proof. Let $\mathcal{L}$ be an axis of rotation of S. For a given plane Π normal to $\mathcal{L}$, a rotation of S about $\mathcal{L}$ induces a rotation of the planar curve $S \cap \Pi$ around the point $\mathcal{L} \cap \Pi$. Now if S has another axis of rotation $\mathcal{L}' \neq \mathcal{L}$, parallel to $\mathcal{L}$, then $S \cap \Pi$ exhibits rotational symmetry around two different centers of rotation (the intersections of Π with $\mathcal{L}, \mathcal{L}'$). However, the center of rotation of an algebraic planar curve other than a line, if any, is unique (see Lemma 23 in Chapter 2). So $S \cap \Pi$ must be a line. Since this must happen for any plane Π normal to $\mathcal{L}$, we deduce that S is a plane, which is excluded. □

Now we can get closer to what we want.

Proposition 108. *Let S be a non-cylindrical algebraic surface.*

(1) All the axes of rotation of S intersect at a point.

(2) If S is not a surface of revolution, then S has finitely many axes of rotation.

Proof. (1) Suppose that S has several axes of rotation. By Lemma 107, no two of them can be parallel. By part (3) of Lemma 18, the composition of two rotations whose axes do not intersect is a screw motion. Hence if S has two axes of rotation that do not intersect, then S is invariant under two rotations with non-concurrent axes, and therefore it is invariant by a screw motion. Therefore, by Proposition 103 the surface S is cylindrical, which cannot happen by hypothesis.

(2) Suppose that S has infinitely many axes of rotation, all of which, by the first part of the statement, share a point P. Therefore, these axes form infinitely many different angles with each other. Since S has infinitely many axes of rotation, S is invariant under infinitely many rotations $\mathcal{R}_{\mathcal{L},\theta}$. Let $\mathcal{R}_{\mathcal{L}_1,\alpha}$, $\mathcal{R}_{\mathcal{L}_2,\beta}$, where $\mathcal{L}_1 \neq \mathcal{L}_2$, $\mathcal{L}_1 \cap \mathcal{L}_2 = P$, be any two of these rotations, and let Φ be the angle between $\mathcal{L}_1, \mathcal{L}_2$. By part (3) of Lemma 18, the composition of $\mathcal{R}_{\mathcal{L}_1,\alpha}$, $\mathcal{R}_{\mathcal{L}_2,\beta}$ is another rotation $\mathcal{R}_{\mathcal{L}_3,\gamma}$, where $\mathcal{L}_3$ is concurrent with $\mathcal{L}_1, \mathcal{L}_2$ (i.e., $\mathcal{L}_3$ also goes through P), and γ is related with α, β, Φ according to Equation (1.16). However, by Corollary 106, α, β, γ can have just finitely many values. This yields finitely many values for Φ too, and we get a contradiction. □

Notice that if S is a surface of revolution, then we can certainly have infinitely many axes of rotation. For instance, if S is an ellipsoid of revolution generated by rotating the ellipse

$$\left\{\frac{y^2}{a^2} + \frac{z^2}{b^2} = 1, x = 0\right\}$$

about the z-axis, any line contained in the plane $z = 0$, through the origin, is a symmetry axis of S, and therefore an axis of rotation (the rotation angle is π, in that case). Additionally, since every axial symmetry of a surface is a rotation of angle π, we get the following corollary.

Corollary 109. *Let S be a non-cylindrical real algebraic surface.*

(1) All the axes of symmetry of S intersect at a point.

(2) If S is not a surface of revolution, then S has finitely many axial symmetries.

If S is a surface of revolution, then it can have infinitely many axes of symmetry, as it happens in the case of the ellipsoid of revolution. However, not every surface of revolution has infinitely many axes of symmetry. For instance, the paraboloid $x^2 + y^2 = z$ has just one symmetry axis, namely the z-axis.

Finally, let us address reflections.

Proposition 110. *Let S be a non-cylindrical algebraic surface.*

(1) If S is not a surface of revolution, then S has finitely many symmetry planes, which intersect in at least in one point.

(2) S is a surface of revolution if and only if S has infinitely many symmetry planes.

Proof. (1) By part (5) of Lemma 18, the composition of two planar symmetries is either a translation, if the symmetry planes are parallel, or a rotation about their common line, if the symmetry planes are concurrent. Since by hypothesis S is not cylindrical, by Proposition 103 every two symmetry planes must intersect at a line, which is an axis of rotation of S. Now by Proposition 108, either the number of symmetry planes is finite, or there are infinitely many symmetry planes intersecting at a certain line $\mathcal{L}$. However, in this second case these planes determine infinitely many different angles with each other. Since S is invariant under the rotations about $\mathcal{L}$ by any of these angles, by Corollary 106 the surface S must be a surface of revolution, which cannot happen by hypothesis. Finally, notice that all the symmetry planes must intersect because every two symmetry planes determine an axis of rotation of S, and by Proposition 108 all these axes intersect.

(2) ($\Leftarrow$) follows from statement (1). ($\Rightarrow$) If S is a surface of revolution, then any plane containing the axis of revolution is a symmetry plane. □

Therefore, throughout this section we have proven the following result.

Theorem 111. *Let $S \subset \mathbb{R}^3$ be an algebraic surface. Then* $\mathrm{Sym}(S)$ *is finite if and only if S is neither a cylindrical surface nor a surface of revolution.*

Finally, Theorem 111 provides the following corollary on the constant λ of Lemma 102. The proof is analogous to Corollary 27 in Chapter 2.

Corollary 112. *Let $f(\boldsymbol{x}) = A\boldsymbol{x} + \boldsymbol{b}$ be a symmetry of an irreducible algebraic surface $S \subset \mathbb{R}^3$, not cylindrical, not a surface of revolution, defined by a square-free polynomial $F(x, y, z)$. Then*

$$F \circ f = \pm F. \tag{6.3}$$

6.2 CYLINDRICAL SURFACES AND SURFACES OF REVOLUTION

We have seen that algebraic cylindrical surfaces and algebraic surfaces of revolution are the only algebraic surfaces with infinitely many symmetries. For completeness, in this section we address the symmetries of these surfaces. Notice first that one can algorithmically decide whether or not a given surface belongs to one of these classes: for cylindrical surfaces, one can use the algorithm in [4], which also allows us to find the direction $\mathbf{v}$ of the rulings; for surfaces of revolution, one can use the algorithms in [4, 129].

A cylindrical surface S is a union of infinitely many copies of $\mathcal{C} = S \cap \Pi$, where Π is a plane normal to the rulings of S. Thus, S is obviously invariant under any translation by a vector parallel to its rulings, and symmetric with respect to any plane normal to its rulings. The remaining symmetries of S follow from the symmetries of $\mathcal{C}$, plus their composition with the translations leaving the surface invariant. So we focus on the, more interesting, symmetries of surfaces of revolution. First we recall the following result, called *Study's Lemma*, that will be used in the proof of the next result and also in other parts of the book. We will state this lemma for multivariate polynomials, although it can formulated in a more general way; see Section 6.13 of [58] for a proof (and a more general statement). Observe that for $n = 2$, the result is a direct consequence of Bezout's Theorem.

Lemma 113. [Study's Lemma] *Let $F, G \in \mathbb{C}[x_1, \dots, x_n]$. If F is irreducible and the varieties $V(F), V(G)$ satisfy that $V(F) \subset V(G)$, then F divides G, i.e., F is a factor of G.*

Now we can prove the following proposition.

Proposition 114. *Let S be an irreducible algebraic surface of revolution, not a plane or a sphere. Then the axis of revolution of S is unique.*

Proof. Assume that S has two different axes of revolution, $\mathcal{L}_1$ and $\mathcal{L}_2$, which cannot be parallel because of Lemma 107. If $\mathcal{L}_1$ and $\mathcal{L}_2$ are skew, then by statement (4) of Lemma 18 the surface S is invariant under a screw motion, so S is cylindrical; in that case, since S is a surface of revolution it must be a circular cylinder, and the statement follows. If S is not cylindrical, then $\mathcal{L}_1$ and $\mathcal{L}_2$ intersect at a point P. Since $\mathcal{L}_2$ is an axis of revolution of S, whenever S is not a plane we can find a plane Π normal to $\mathcal{L}_2$ such that $S \cap \Pi$ contains a circle C. For every point of C, the distance to P is the same; let this distance be r. Now since $\mathcal{L}_1$ is also an axis of revolution of S, by rotating C around $\mathcal{L}_1$ we generate a 2-dimensional piece of a sphere $\mathrm{Sph}_{P,r}$ centered at P, with radius equal to r. Since the intersection of S and $\mathrm{Sph}_{P,r}$ is a two-dimensional subset and since $S, \mathrm{Sph}_{P,r}$ are irreducible and algebraic, then by Study's Lemma $\mathrm{Sph}_{P,r} = S$, i.e., S is a sphere. □

If S is a surface of revolution which is not a union of spheres, the axis of revolution $\mathcal{L}$ can be computed with the algorithms in [4, 129]. It is clear, by construction, that the axis of revolution is an axis of symmetry of S, and that any plane containing $\mathcal{L}$ is a symmetry plane of S. Now the remaining symmetries of S also follow, by construction, from the symmetries of a section of S with any plane Π containing $\mathcal{L}$; such a section, that we denote by $\mathcal{D}$, is called a *directrix* curve of S. In fact, the surface S is generated by rotating $\mathcal{D}$ around $\mathcal{L}$. Fig. 6.3 shows a surface of revolution and its directrix curve (right), plotted in black thick line.

More precisely, we have the following results. In all the propositions that follow, which relate the symmetries of S with those of $\mathcal{D}$, the implication ($\Leftarrow$) is a consequence of the fact that S is constructed by rotating $\mathcal{D}$ about $\mathcal{L}$. So we focus on ($\Rightarrow$) in all the propositions.

Proposition 115. *Let S be a surface of revolution, not a plane, and let $\mathcal{L}$ be the axis of revolution. S has central symmetry iff $\mathcal{D}$ has central symmetry with respect to a point of $\mathcal{L}$.*

Proof. We prove ($\Rightarrow$). Suppose that S is symmetric with respect to a point P. If $P \notin \mathcal{L}$, then let Π be the plane containing P and $\mathcal{L}$. Then the planar curve $S \cap \Pi$ inherits central symmetry with respect to P; additionally, the planar curve $S \cap \Pi$ also has axial symmetry with respect to $\mathcal{L}$. Now let $P' \in S \cap \Pi$ be the symmetric point of P with respect to $\mathcal{L}$. Since $\mathcal{L}$ is an axis of symmetry of $S \cap \Pi$, we deduce that $S \cap \Pi$ is also symmetric with respect to P'. Therefore $S \cap \Pi$ has two different centers of symmetry, and hence $S \cap \Pi$ must be a line containing P, P'. Since S is a surface of revolution generated by this line, S must be a plane, which is forbidden by hypothesis. So $P \in \mathcal{L}$. Since the intersection curve of S with any plane containing $\mathcal{L}$ inherits central symmetry, the result follows. □

Proposition 116. *Let S be a surface of revolution, not a sphere or a plane, and let $\mathcal{L}$ be the axis of revolution. A line $\tilde{\mathcal{L}} \neq \mathcal{L}$ is an axis of symmetry of S iff $\mathcal{L}, \tilde{\mathcal{L}}$ intersect, and $\tilde{\mathcal{L}}$ lies in a plane Π normal to $\mathcal{L}$ which also contains an axis of symmetry of $\mathcal{D}$.*

Proof. We prove ($\Rightarrow$). If S is cylindrical, then S is a union of circular cylinders with the same axis, and the result follows. So assume that S is not cylindrical. From Corollary 109, $\tilde{\mathcal{L}}$ and $\mathcal{L}$ must intersect. Now suppose that $\tilde{\mathcal{L}}$ is not perpendicular to $\mathcal{L}$, and let

$\Phi \neq 0, \frac{\pi}{2}$ be the angle between $\tilde{\mathcal{L}}, \mathcal{L}$. Since $\mathcal{L}$ is an axis of revolution of S, we deduce that S is invariant under every rotation $\mathcal{R}_{\mathcal{L},\theta}$, with $\theta \in [0, 2\pi)$. Furthermore, since $\tilde{\mathcal{L}}$ is an axis of symmetry, S is invariant under the rotation $\mathcal{R}_{\tilde{\mathcal{L}},\pi}$. So S is invariant under the composition of $\mathcal{R}_{\mathcal{L},\theta}$, for any θ, and $\mathcal{R}_{\tilde{\mathcal{L}},\pi}$. By statement (3) of Lemma 18, the composition of these two rotations is another rotation about an axis L concurrent with $\mathcal{L}, \tilde{\mathcal{L}}$, by an angle γ corresponding to Eq. (1.16). Now if $\Phi \neq \frac{\pi}{2}$, then since θ can take any value we get infinitely many values for γ too. But then from Lemma 105 we deduce that L is another axis of revolution, implying that S has two different axes of revolution. However, by Proposition 114 this means that S is either a sphere or a plane, which is forbidden by hypothesis. Therefore, $\tilde{\mathcal{L}}$ is perpendicular to $\mathcal{L}$. Finally, since $\tilde{\mathcal{L}}$ is an axis of symmetry of S, the section $\mathcal{G}$ of S with the plane containing both $\tilde{\mathcal{L}}, \mathcal{L}$ inherits the axial symmetry with respect to $\tilde{\mathcal{L}}$, i.e., $\mathcal{G}$ is symmetric with respect to $\tilde{\mathcal{L}}$. But $\mathcal{G}$ is the result of rotating $\mathcal{D}$ around $\mathcal{L}$ by a rotation angle α. So $\mathcal{D}$ must also be symmetric with respect to an axis ℓ such that $\mathcal{R}_{\mathcal{L},\alpha}(\ell) = \tilde{\mathcal{L}}$.
□

Proposition 117. *Let S be a surface of revolution, not a sphere or a plane, let $\mathcal{L}$ be the axis of revolution of S, and let Π be a plane not containing $\mathcal{L}$. The plane Π is a symmetry plane of S iff Π is normal to $\mathcal{L}$, and $\mathcal{D}$ is symmetric with respect to the intersection line of Π and the plane containing $\mathcal{D}$.*

Proof. We prove ($\Rightarrow$). If S is cylindrical, then S is a union of circular cylinders with the same axis, and the result follows. So assume that S is not cylindrical. If Π is parallel to $\mathcal{L}$, then we can find another symmetry plane $\Pi^\star$, containing $\mathcal{L}$, which is parallel to Π. Therefore S is symmetric with respect to two parallel planes, namely Π and $\Pi^\star$, which by statement (5) of Lemma 18 and Proposition 103 implies that S is cylindrical. Since we are assuming that we are not in this case, Π and $\mathcal{L}$ intersect. Now the intersection of Π with any plane $\tilde{\Pi}$ containing $\mathcal{L}$ yields a line $\tilde{\mathcal{L}}$. Furthermore, also by statement (5) of Lemma 18, S is invariant under a rotation about $\tilde{\mathcal{L}}$ by an angle equal to twice the angle between Π and $\tilde{\Pi}$. If we pick $\tilde{\Pi}$ such that the angle θ between Π and $\tilde{\Pi}$ is not $\theta = \frac{\pi}{N}$, with $N \in \mathbb{N}$, by Lemma 105 we deduce that the line $\tilde{\mathcal{L}} = \Pi \cap \tilde{\Pi}$ is an axis of revolution of S. Since $\tilde{\mathcal{L}} \neq \mathcal{L}$, this implies that S has two different axes of revolution, and therefore S is either a plane or a sphere, which is forbidden by hypothesis. Therefore, Π is perpendicular to $\mathcal{L}$. Now the section of S with any plane $\widehat{\Pi}$ containing $\mathcal{L}$ yields a curve which is symmetric with respect to the intersection of Π and $\widehat{\Pi}$. Thus, the rest of the implication follows. □

Corollary 118. *Let S be a surface of revolution, not a sphere or a plane.*

(1) If $\mathcal{D}$ has no axis of symmetry perpendicular to $\mathcal{L}$, then S has just one axis of symmetry (the axis of revolution), and the symmetry planes of S are the planes containing $\mathcal{L}$.

(2) S has either one axis of symmetry (the axis of revolution) or infinitely many axes of symmetry.

(3) S has infinitely many axes of symmetry iff S has some symmetry plane not containing the axis of revolution. Furthermore, the axes of symmetry of S are the axis of revolution, and the intersections of the planes containing the axis of revolution with other symmetry planes not containing the axis of revolution.

(4) If S has infinitely many axes of symmetry and S is not cylindrical, all the symmetry axes of S but one (the axis of revolution) lie on one plane, which is a symmetry plane of S.

So we deduce that the symmetries of a surface of revolution can be computed from the symmetries of the directrix curve $\mathcal{D}$, which is a planar algebraic curve. Thus, the symmetries of $\mathcal{D}$ can be found by using the techniques in Chapter 2.

6.3 COMPUTATION OF SYMMETRIES

Let us consider the computation of the symmetries of an implicit algebraic surface S, implicitly defined by $F(x, y, z) = 0$, where F is irreducible, not a cylindrical surface or a surface of revolution; we will be following the ideas of [15], from where we will, additionally, borrow examples and pictures. In order to do this it is just natural to try to copy the strategy in Chapter 2: indeed, we can take advantage of the fact that Laplacians commute with orthogonal transformations (see Theorem 28 in Chapter 2) also here, take Laplacians in Eq. (6.3) and get a Laplacian chain

$$F \to \Delta F \to \Delta^2 F \to \cdots \Delta^k F \to \cdots \tag{6.4}$$

and in turn a chain of symmetry groups

$$\mathrm{Sym}(F) \subset \mathrm{Sym}(\Delta F) \subset \mathrm{Sym}(\Delta^2 F) \subset \cdots \mathrm{Sym}(\Delta^k F) \subset \cdots \tag{6.5}$$

Whenever we end up with a polynomial defining a surface with finitely many symmetries, we can extract the symmetries of the original surface by testing whether each symmetry satisfies Eq. (6.3).

There are several things that can go wrong here, though. In the case of curves, the Laplacian chain ended when we reached either a harmonic polynomial, or a polynomial of degree two, or a polynomial of degree one. In the harmonic case we could compute the symmetries of the curve making use of complex numbers and Cauchy-Riemann relations, and in the other cases we could, essentially, reduce the computation to either the harmonic case, or to Linear Algebra. In the case of surfaces, however, harmonic polynomials do not offer any special advantage because we cannot switch to complex numbers anymore. If we reach a polynomial of degree two, i.e., a quadric, which in turn requires $F(x, y, z)$ to have even degree, then whenever the quadric is not degenerate and is not a surface of revolution, the symmetry group of the quadric is finite, and can be computed making use of the matrix defining the quadric, i.e., using Linear Algebra methods. But in any other case, we do not have enough information to find $\mathrm{Sym}(S)$.

Thus, instead of a general method like the one we presented in Chapter 2, here we will present a different method where rotational symmetries, reflectional symmetries and central symmetries are addressed separately, making extensive use of Eq. (6.3). We can still take advantage of the Laplacian chain in Eq. (6.4) to replace $F(x, y, z)$ by the last polynomial in Eq. (6.4), not constant and with finitely many symmetries. By simplicity, we will also call $F(x, y, z)$ to the implicit equation of such surface, that we will also denote by S, whose degree, denoted by n, is necessarily smaller than the degree of the original surface. Also, sometimes we will write $\mathrm{Sym}(F)$ to denote the set of symmetries of the surface defined by S, i.e., $\mathrm{Sym}(S)$.

To do this we need a closer look to Eq. (6.3). The problem with Eq. (6.3) is that the sign $\pm$ is difficult to predict. However, let us see that both the parity of $n = \deg(F)$ and the type of symmetry that S exhibits, have some consequences on this sign. We will focus on reflections, axial symmetries and central symmetries. Similar results for rotations which are not axial symmetries are not so easy to produce, so for now we will leave general rotations aside. It will be useful to write

$$F(x, y, z) = F_n(x, y, z) + F_{n-1}(x, y, z) + \cdots + F_0(x, y, z), \quad (6.6)$$

where for $i = 0, 1, \ldots, n$, $F_i(x, y, z)$ is a homogeneous polynomial of degree i. In particular, we will call $F_n(x, y, z)$ the *form of highest degree* of $F(x, y, z)$.

Now for reflections and axial symmetries we observe that any symmetry of this type can be decomposed (see Chapter 1) as

$$f = \psi \circ \xi \circ \psi^{-1},$$

where ψ, ξ are rigid motions and $\xi(\boldsymbol{x}) = A\boldsymbol{x}$, with A a matrix of the type

$$\begin{bmatrix} 1 & 0 & 0 \\ 0 & 1 & 0 \\ 0 & 0 & -1 \end{bmatrix}, \quad \begin{bmatrix} -1 & 0 & 0 \\ 0 & -1 & 0 \\ 0 & 0 & 1 \end{bmatrix}, \quad (6.7)$$

the left-hand matrix corresponding to reflections, the right-hand matrix corresponding to axial symmetries. Furthermore, substituting $f = \psi \circ \xi \circ \psi^{-1}$ into Eq. (6.3) and calling $\tilde{F} = F \circ \psi$, which defines the surface corresponding to $\psi(S)$, we get that

$$\tilde{F} \circ \xi = \pm \tilde{F}, \quad (6.8)$$

so the sign $\pm$ is the same in Eqs. (6.3) and (6.8). However, in Eq.

(6.8) we can argue more easily the sign at the right-hand side. Indeed, since ξ fixes the origin, we also have that $\tilde{F}_n \circ \xi = \tilde{F}_n$, where $\tilde{F}_n$ is the form of highest degree of $\tilde{F}$. Let us write

$$\tilde{F}_n(x,y,z) = a_0(x,y)z^n + a_1(x,y)z^{n-1} + \cdots + a_{n-1}(x,y)z + a_n(x,y), \tag{6.9}$$

where for $i = 0, \ldots, n$, $a_i(x,y)$ is a homogeneous polynomial of total degree i.

Let us consider reflections. Since $\xi(x,y,z) = (x,y,-z)$, we deduce that $(\tilde{F}_n \circ \xi)(x,y,z) = \tilde{F}_n(x,y,-z)$. We need to distinguish the cases when n is even, and when n is odd. If n is even, then

$$\begin{aligned}(\tilde{F}_n \circ \xi)(x,y,z) &= \tilde{F}_n(x,y,-z)\\ &= a_0(x,y)z^n - a_1(x,y)z^{n-1} + \cdots - a_{n-1}(x,y)z + a_n(x,y).\end{aligned}$$

Thus, Eq. (6.8) is satisfied iff either the sign in Eq. (6.8) is +, and $a_1(x,y) = a_3(x,y) = \cdots = a_{n-1}(x,y) = 0$, or the sign in Eq. (6.8) is −, and $a_0(x,y) = a_2(x,y) = \cdots = a_n(x,y) = 0$. However, in the last case $\tilde{F}_n$ is not irreducible and has z as a factor, which corresponds to the reflection plane $z = 0$. From here we deduce that in this last case the highest-order form of the original polynomial F also contains an irreducible linear factor defining a plane parallel to the reflection plane. If n is odd we can argue in a similar way. So we have the following lemma.

Lemma 119. *Let $n = \deg(F)$, and let $f \in \mathrm{Sym}(F)$ be a reflection on a plane Π. If n is even and the sign in Eq. (6.8) is −, then the form F_n of highest degree of F contains a linear factor defining a plane parallel to Π. If n is odd and the sign in Eq. (6.8) is +, then F_n contains a linear factor defining a plane parallel to Π.*

For axial symmetries the discussion is analogous, although now we must take into account that $\xi(x,y,z) = (-x,-y,z)$, and the fact that the $a_i(x,y)$ are homogeneous. Then we arrive at the following result.

Lemma 120. *Let $n = \deg(F)$, and let $f \in \mathrm{Sym}(F)$ be an axial symmetry about a line $\mathcal{L}$. If n is even and the sign in Eq. (6.8) is −, then the form F_n of highest degree of F contains a linear factor* *defining a plane perpendicular to* $\mathcal{L}$. *If* n *is odd and the sign in Eq.* (6.8) *is* $+$, *then* F_n *contains a linear factor defining a plane parallel to* Π.

For central symmetries we have the following result, which again follows by analyzing the form of highest degree, taking into account that the form of highest degree is homogeneous.

Lemma 121. *Let* $n = \deg(F)$, *and let* $f \in \mathrm{Sym}(F)$ *be a central symmetry. Then the sign in Eq.* (6.8) *is* $+$ *when* n *is even and is* $-$ *when* n *is odd.*

Remark 9. *Notice that if we have several symmetries of the same type (say axial, reflectional, etc.), not all of them need to be associated with the same sign* $\pm$ *in Eq.* (6.3). *Thus, in general we will need to look for symmetries of each type associated to the* $+$ *sign, and to the* $-$ *sign. We will refer to these two cases as "Case* $+$*" and "Case* $-$*". This does not apply to central symmetries, though, because* S *can have at most one central symmetry and from Lemma 121 we just need to check either the Case* $+$ *or the Case* $-$, *depending on whether the degree of* F *is even or odd. Nevertheless, Lemma 119 and Lemma 120 allow us to rule out the existence of symmetries in the cases covered in these results when the form of highest degree does not have linear factors.*

The following observation will also be useful. If $f(\boldsymbol{x}) = A\boldsymbol{x} + \mathbf{b}$ is a symmetry of S, from Eq. (6.8) we deduce that $\widehat{f}(\boldsymbol{x}) = A\boldsymbol{x}$ is a symmetry of the surface defined by F_n, the form of highest degree of F. Furthermore, the sign, plus or minus, in Eq. (6.3) is the same for F, f and $F_n, \widehat{f}$. Moreover, if f is a rotation about a line $\mathcal{L}$, then $\widehat{f}$ is a rotation about a line parallel to $\mathcal{L}$ through the origin. If f is a reflection on a plane Π, then $\widehat{f}$ is a reflection on a plane parallel to Π through the origin. In particular, if $\widehat{f}$ is known, the direction of $\mathcal{L}$ and the rotation angle, when f is a rotation, or the normal vector to Π, when f is a reflection, are known, and therefore the matrix A (not the vector $\boldsymbol{b}$) is also known. Notice that in the cases covered in Lemma 119 and Lemma 120, we can find the directions of the tentative reflection planes or symmetry axes, and therefore the tentative matrices A, by just factoring out the form of highest degree of F.

Additionally, taking gradients in Eq. (6.8) we get that

$$\nabla F(A\boldsymbol{x}+\boldsymbol{b}) = \pm A\nabla F(\boldsymbol{x}), \tag{6.10}$$

where the sign $\pm$ in Eq. (6.10) coincides with the sign $\pm$ in Eq. (6.8). Thus, using the terminology of Chapter 4, if the sign in Eq. (6.8) is $+$, $f(\boldsymbol{x}) = A\boldsymbol{x}+\boldsymbol{b}$ is also a symmetry of the gradient field ∇F, which is a space vector field; if the sign in Eq. (6.8) is $-$, f is an antisymmetry of ∇F. Although most of the ideas and results of Chapter 4 are not applicable to space vector fields, Proposition 80 is certainly applicable. This leads to the following result concerning rotational symmetries, and therefore axial symmetries, in particular; the proof is analogous to that of Proposition 80.

Proposition 122. *Let $\mathbf{V}(\boldsymbol{x})$ be a polynomial space vector field, and let $f(\boldsymbol{x}) = A\boldsymbol{x}$ be a space rotation about a line $\mathcal{L}$ through the origin. If f is a symmetry of $\mathbf{V}$, then $\mathcal{L}$ is a common line to the surfaces defined by the components of $\mathbf{V}(\boldsymbol{x}) \times \boldsymbol{x}$.*

As a final observation, taking norms in Eq. (6.10) and since A is orthogonal we get that

$$\|\nabla F(A\boldsymbol{x}+\boldsymbol{b})\|^2 = \|\nabla F(\boldsymbol{x})\|^2. \tag{6.11}$$

Thus, any symmetry of S is also a symmetry of the algebraic surface defined by $G(\boldsymbol{x}) = \|\nabla F(\boldsymbol{x})\|^2$; furthermore, since $G(\boldsymbol{x})$ is non-negative the sign in Eq. (6.8) is necessarily $+$.

So let us address the computation of the different symmetry types. For all of them, with the exception of central symmetries, we will distinguish between "Case $-$" and "Case $+$", see Remark 9. Certainly, an option is to compute, for each type of symmetry, the symmetries associated with each case. As we have just observed, another option, to the price of working with a polynomial of higher degree, is to address the symmetries of $G(\boldsymbol{x}) = \|\nabla F(\boldsymbol{x})\|^2$, where we can only have Case $+$.

6.3.1 Rotational symmetries

Although rotational symmetries include axial symmetries, some specific aspects of axial symmetries will we addressed in the next subsection.

Case −

Let $f(\boldsymbol{x}) = A\boldsymbol{x} + \boldsymbol{b}$ be a rotational symmetry of S satisfying that $F \circ f = -F$; from Eq. (6.10), we have that

$$\nabla F(A\boldsymbol{x} + \boldsymbol{b}) = -A\nabla F(\boldsymbol{x}). \tag{6.12}$$

Let us consider first the axis of rotation $\mathcal{L}$. Since $\mathcal{L}$ is the set of fixed points of f, $\forall \boldsymbol{x} \in \mathcal{L}$ we have $f(\boldsymbol{x}) = A\boldsymbol{x} + \boldsymbol{b} = \boldsymbol{x}$. Substituting this into $F \circ f = -F$ we get that $F(\boldsymbol{x}) = 0$, so we obtain the following result.

Lemma 123. *Let $f(\boldsymbol{x}) = A\boldsymbol{x} + \boldsymbol{b}$ be a rotational symmetry of S satisfying that $F \circ \phi = -F$, and let $\mathcal{L}$ be the axis of rotation of f. Then $\mathcal{L}$ is a line contained in S.*

Furthermore, from Eq. (6.12) the points $\boldsymbol{x} \in \mathcal{L}$ also satisfy that $(A + I)\nabla F(\boldsymbol{x}) = 0$. If f is not an axial symmetry (which will be discussed in the next section), $\lambda = -1$ is not an eigenvalue of $A + I$, and therefore we deduce that $\nabla F(\boldsymbol{x}) = 0$ for all the points $\boldsymbol{x} \in \mathcal{L}$. Taking Lemma 123 into account, we deduce that $\mathcal{L}$ is part of the singular locus of S (see Definition 101), i.e., it is a *singular line* of S.

Theorem 124. *Let $f(\boldsymbol{x}) = A\boldsymbol{x} + \boldsymbol{b}$ be a rotational symmetry of S, not an axial symmetry, satisfying that $F \circ \phi = -F$, and let $\mathcal{L}$ be the axis of rotation of f. Then $\mathcal{L}$ is a singular line of S.*

Theorem 124 provides an algorithm to find the rotational symmetries of S which are not axial symmetries in the case $F \circ \phi = -F$: indeed, we compute the singular lines of S from the system in Eq. (6.1), and for each line $\mathcal{L}$, we consider the intersection of S with a plane Π normal to $\mathcal{L}$. This yields a planar algebraic curve $\mathcal{C}$. If S is rotationally symmetric about the line $\mathcal{L}$, $\mathcal{C}$ must exhibit rotational symmetry about the point $\Pi \cap \mathcal{L}$. This can be checked with the results in Chapter 2, which also allow us to find the rotation angle.

Case +

Now let $f(\boldsymbol{x}) = A\boldsymbol{x}+\boldsymbol{b}$ be a rotational symmetry of S satisfying that $F \circ f = F$. Then $\widehat{f}(\boldsymbol{x}) = A\boldsymbol{x}$ is a rotational symmetry of the surface defined by F_n, the form of highest degree of F, and the axis of rotation $\widehat{\mathcal{L}}$ of $F_n(x, y, z) = 0$ is a line parallel to the axis of rotation $\mathcal{L}$ of $F(x, y, z) = 0$ through the origin. Furthermore, $\widehat{f}$ satisfies that $F_n \circ \widehat{f} = F_n$. Taking Eq. (6.10) into account, we deduce that

$$\nabla F_n(A\boldsymbol{x}) = A\nabla F_n(\boldsymbol{x}), \tag{6.13}$$

so $\widehat{f}$ is a symmetry of the space vector field ∇F_n. Then we have the following result.

Theorem 125. *Let $f(\boldsymbol{x}) = A\boldsymbol{x} + \boldsymbol{b}$ be a rotational symmetry of S with axis of rotation $\mathcal{L}$, satisfying that $F \circ f = F$, and let $\widehat{\mathcal{L}}$ be the axis of rotation of $\widehat{f}(\boldsymbol{x}) = A\boldsymbol{x}$, parallel to $\mathcal{L}$ through the origin.*

(1) $\widehat{\mathcal{L}}$ is a common line to the surfaces defined by the components of the space vector field $\nabla F_n(\boldsymbol{x}) \times \boldsymbol{x}$.

(2) $\mathcal{L}$ is the solution to $(A - I)\nabla F(\boldsymbol{x}) = \boldsymbol{0}$.

Proof. The statement (1) follows from Eq. (6.12) and Proposition 122 applied to the field $\mathbf{V} = \nabla F_n$. For statement (2), observe that from Eq. (6.10) we get that $\nabla F(A\boldsymbol{x} + \boldsymbol{b}) = A\nabla F(\boldsymbol{x})$. Since the points $\boldsymbol{x} \in \mathcal{L}$ are fixed points of f, they satisfy that $A\boldsymbol{x} + \boldsymbol{b} = \boldsymbol{x}$. Substituting this into $\nabla F(A\boldsymbol{x} + \boldsymbol{b}) = A\nabla F(\boldsymbol{x})$, we deduce that the points of $\mathcal{L}$ fulfill $(A - I)\nabla F(\boldsymbol{x}) = 0$. □

Remark 10. *Let us note that Theorem 125 covers also the situation when S possesses a singular line being simultaneously an axis of symmetry for Case +. In other words, the singular lines as axes of symmetry are not only reserved for Case −.*

Theorem 125 provides an algorithm to find the rotational symmetries of S in the case $F \circ f = F$, *whenever* the components of the space vector field $\nabla F_n(\boldsymbol{x}) \times \boldsymbol{x}$ have finitely many lines in common. Indeed, if this condition holds we can compute the tentative $\widehat{\mathcal{L}}$'s as the (finitely many) common lines of the components of $\nabla F_n(\boldsymbol{x}) \times \boldsymbol{x}$. For each tentative $\widehat{\mathcal{L}}$, the rotation angle can be computed by intersecting the surface S (or alternatively the surface defined by F_n) with a plane normal to $\widehat{\mathcal{L}}$, and studying the rotational symmetries of the resulting curve with the algorithm in Chapter 2. The direction of $\widehat{\mathcal{L}}$ and the rotation angle allow us to compute the matrix A, and finally the axis of rotation $\mathcal{L}$ of S is computed by solving $(A - I)\nabla F(\boldsymbol{x}) = 0$. Of course, if the components of the space vector field $\nabla F_n(\boldsymbol{x}) \times \boldsymbol{x}$ have infinitely many common lines this is not applicable, because we have infinitely many tentative $\widehat{\mathcal{L}}$'s. This is a special case that we will treat in the next section.

Example 10. Let S be the algebraic surface defined by the cubic polynomial

$$F(x, y, z) = x^3 - 3xy^2 + 3x^2 - 3y^2 + 3x - z + 1,$$

see Fig. 6.4, left. A direct computation confirms that S has no singular lines and thus we can leave Case $-$ aside, and move to Case $+$. For $F_3(x, y, z) = x^3 - 3xy^2$ we get the vector field $\nabla F_3(\boldsymbol{x}) \times \boldsymbol{x} = (-6xyz, 3y^2z - 3x^2z, 9x^2y - 3y^3)$. Using Theorem 125, we obtain the tentative $\widehat{\mathcal{L}}$'s through the origin, parallel to the vectors

$$(0, 0, 1),\ (1, 0, 0), (1, \pm\sqrt{3}, 0),$$

as common lines to the surfaces defined by the components of $\nabla F_3(\boldsymbol{x}) \times \boldsymbol{x}$. Studying the intersections of S and the planes normal to (the direction vector of) $\widehat{\mathcal{L}}$ we find, using the method from Chapter 2, that the only axis that comes into consideration is given by $(0, 0, 1)$. At the same time we get that the rotation angles are $2\pi/3$ and $4\pi/3$, see Fig. 6.4, left. Next we determine the corresponding matrix A in the form

$$\begin{pmatrix} -\frac{1}{2} & -\frac{\sqrt{3}}{2} & 0 \\ \frac{\sqrt{3}}{2} & -\frac{1}{2} & 0 \\ 0 & 0 & 1 \end{pmatrix}, \text{ or } \begin{pmatrix} -\frac{1}{2} & \frac{\sqrt{3}}{2} & 0 \\ -\frac{\sqrt{3}}{2} & -\frac{1}{2} & 0 \\ 0 & 0 & 1 \end{pmatrix}$$

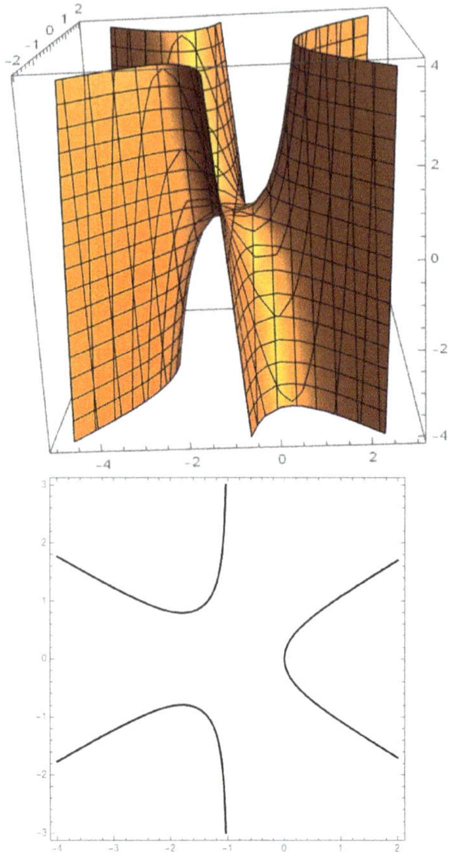

FIGURE 6.4 Left: The cubic surface S from Example 10. Right: The cubic curve $x^3-3xy^2+3x^2-3y^2+3x$ as the intersection of S with the plane $z=1$; this planar curve exhibits a rotational symmetry around the point $(1,0,1)$ by angles $2\pi/3$ and $4\pi/3$.

The final computation, cf. Theorem 125, gives that the sought axis $\mathcal{L}$ parallel to $\widehat{\mathcal{L}}$ goes through the point $(-1,0,0)$.

Example 11. Consider the algebraic surface S defined by the polynomial

$$F(x,y,z)=xyz-x^2+2xz+y^2-yz+2x+4y-2z+3,$$

see Fig. 6.5, left. A direct computation reveals that S has just one

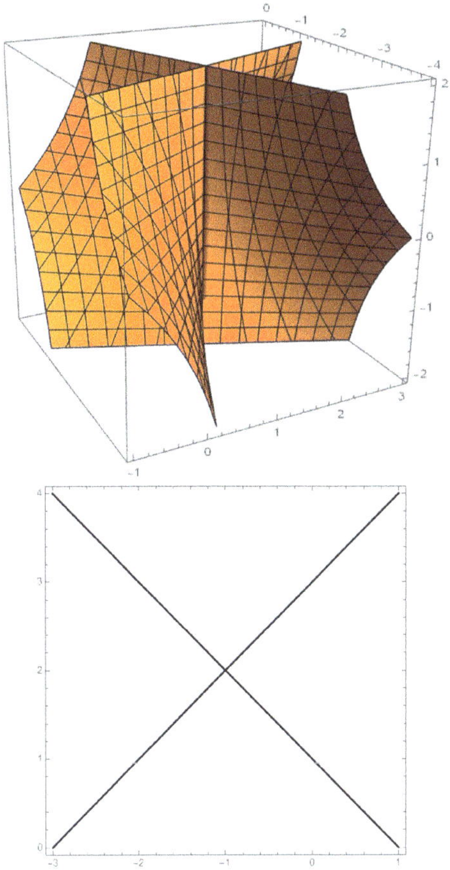

FIGURE 6.5 Left: The cubic surface S from Example 11. Right: The curve $\mathcal{C}$, obtained as the intersection of S with the plane $z = 0$; this planar curve exhibits rotational symmetry around the point $(-1, 2, 0)$ by the angles $\pi/2$, π and $3\pi/2$.

singular line, namely the line $\mathcal{L}$ through the point $(1, -2, 0)$, parallel to the vector $(0, 0, 1)$. Hence, here we need to explore Case $-$. The intersection of S with a plane normal to $\mathcal{L}$, e.g. $z = 0$, yields the planar algebraic curve $\mathcal{C}$ defined by $-(x-y+3)(x+y-1) = 0$. This curve is rotationally symmetric with angles of rotation $\pi/2$, π, $3\pi/2$. It can be verified that the angles $\pi/2$ and $3\pi/2$ correspond to Case $-$. The rotational symmetry with rotation angle π belongs to Case $+$ (in fact, it is an axial symmetry).

6.3.2 Axial symmetries

Axial symmetries are a particular instance of rotational symmetries. As such, Case + has already been covered in the previous subsection. Moreover, recall from Lemma 120 that in Case +, n even, the form of highest degree must contain a linear factor defining a plane perpendicular to the direction of the symmetry axis.

However, axial symmetries are excluded from Theorem 124, which allows us to compute rotational symmetries in Case −, so we need to investigate this case. If n is even, we can use Lemma 120. Indeed, using Lemma 120 we check whether or not F_n has some linear factor; each linear factor of F_n provides the direction of a tentative symmetry axis $\mathcal{L}$, and therefore the matrix A. Finally, the symmetry axis $\mathcal{L}$ itself, which is the set of fixed points of the axial symmetry, can be computed, using Eq. (6.10) and arguing as in the preceding paragraphs, as the solution set of $(A+I)\nabla F(\boldsymbol{x}) = 0$.

If n is odd, then we can still use Lemma 123, which is useful unless S contains infinitely many real lines, i.e., unless S is a real ruled surface. If that happens, we address the axial symmetries of the surface defined by $G(\boldsymbol{x}) = \|\nabla F(\boldsymbol{x})\|^2$, which contain as a subset the axial symmetries of the original surface; observe that here we necessarily fall into Case +.

Example 11 (continued). The term of highest degree $F_3(x, y, z) = xyz$ possesses three linear factors. Hence, the tentative axes of symmetry for Case + (as $n = 3$ is odd), cf. Lemma 120, are perpendicular to $x = 0$, $y = 0$, $z = 0$. Thus, we confirm what we knew from previous computations, i.e., that the line $\mathcal{L}$ through the point $(1, -2, 0)$ and parallel to the vector $(0, 0, 1)$ is the axis of symmetry of S.

Example 12. Consider the algebraic surface S determined by the polynomial

$$F(x, y, z) = 2x^3 - 2x^2y + x^2z + 2xy^2 + 2xz^2 - 2y^3 + y^2z$$
$$-2yz^2 + z^3 + 2x^2 - 8xz - 2y^2 - 8yz - 8x + 8y - 4z.$$

see Fig. 6.6. Since $n = 3$ is odd and the term of highest degree

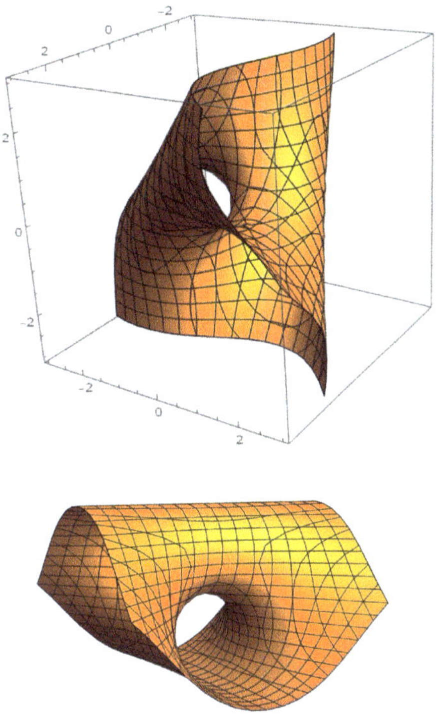

FIGURE 6.6 Left: The axially symmetric surface S from Example 12. Right: The same surface from a different viewpoint – the vector $(2,-2,1)$ is pointing upwards.

$F_3(x, y, z)$ possesses a linear factor,

$$F_3 = (2x - 2y + z)(x^2 + y^2 + z^2),$$

we arrive at the direction of a tentative axis of symmetry $(2,-2,1)$ assuming Case $+$. The final computation shows that the sought axis $\mathcal{L}$ with this direction vector and going through the origin.

Example 13. Let S be the algebraic surface (see Fig. 6.7) defined by the polynomial

$$F(x, y, z) = x^5 + 3xy^4 + z.$$

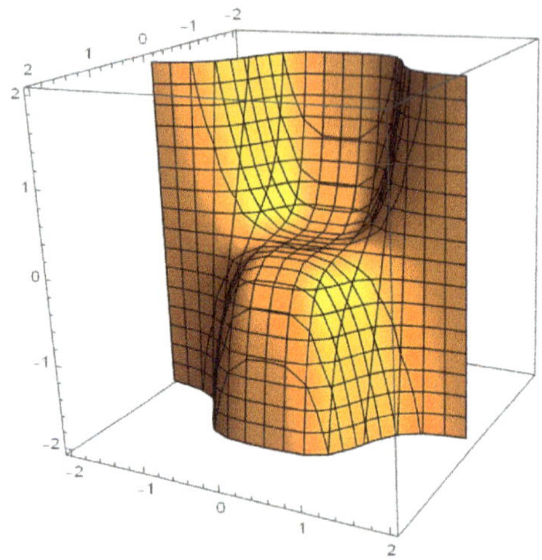

FIGURE 6.7 The axially symmetric surface S from Example 13.

Since $n = 5$ is odd and the term of highest degree $F_5(x, y, z)$ possesses a linear factor,

$$F_5 = x(x^4 + 3y^4),$$

we arrive at the direction of a tentative axis of symmetry $(1, 0, 0)$ assuming Case +. However, the next computations show that this direction does not yield the axis. So, if some axis exists, then it must belong to Case −. This means that the only tool we have at our disposal is Lemma 123. Thus, we can check that S contains only one real line, i.e., the line $\mathcal{L}$ going through the origin and with the direction vector $(0, 1, 0)$. This non-singular line is the sought axis of symmetry.

6.3.3 Reflectional symmetries

Here we distinguish again between Case − and Case +.

Case −

If n is even, from Lemma 119 we know that F_n must have a linear factor, defining a plane parallel to the reflection plane. However, we can say something more general. In order to do this, let $f(\boldsymbol{x}) = A\boldsymbol{x} + \boldsymbol{b}$ be a rotational symmetry of S satisfying that $F \circ f = -F$. If Π is the reflection plane of f, the points $\boldsymbol{x} \in \Pi$ are fixed points of f, so they satisfy that $A\boldsymbol{x} + \boldsymbol{b} = \boldsymbol{x}$. Plugging this into $F \circ \phi = -F$, we deduce that $F(\boldsymbol{x}) = 0$ for $\boldsymbol{x} \in \Pi$. Thus, we get the following result.

Lemma 126. *Let $f(\boldsymbol{x}) = A\boldsymbol{x} + \boldsymbol{b}$ be a reflectional symmetry of S satisfying that $F \circ f = -F$, and let Π be the reflection plane of f. Then Π corresponds to a linear factor of F.*

Since by hypothesis S is not a plane, we deduce the following corollary.

Corollary 127. *If S is an irreducible surface, not a plane, admiting reflectional symmetry, then $F \circ f = F$.*

Case +

If n is odd, from Lemma 119 we know that F_n must have a linear factor defining a plane $\widehat{\Pi}$ parallel to the reflection plane Π. By intersecting S with a plane normal to $\widehat{\Pi}$, the intersection curve $S \cap \widehat{\Pi}$ must exhibit reflectional symmetry with respect to the intersection line $\Pi \cap \widehat{\Pi}$, and from here we can derive Π. However, Lemma 119 does not say anything about the case when n is even, so we need a different strategy for that case. The strategy follows from the next lemma, which connects reflectional symmetries $f(\boldsymbol{x}) = A\boldsymbol{x} + \boldsymbol{b}$ of S, satisfying that $F \circ \phi = F$, with axial symmetries of the surface defined by F_n, which we have addressed in the previous subsection. Before giving the result, we recall *conical surfaces*: a conical surface is a ruled surface, i.e., a surface made of lines, where the lines contained in the surface all intersect at one point of the surface, called the *vertex*; in particular, any homogeneous polynomial in x, y, z defines a conical surface with vertex at the origin.

Lemma 128. *Let $\widehat{f}(\boldsymbol{x}) = A\boldsymbol{x}$ be a reflectional symmetry of the surface defined by F_n, associated to Case +. Then $\widehat{g}(\boldsymbol{x}) = -A\boldsymbol{x}$ is an axial symmetry of the surface defined by F_n. Furthermore, if n is odd, then $\widehat{g}$ is associated with Case −, i.e., $F_n \circ \widehat{g} = -F_n$, and if n is even, then $\widehat{g}$ is associated with Case +, i.e., $F_n \circ \widehat{g} = F_n$.*

Proof. We begin with the first part. Since F_n is homogeneous, F_n defines a conical surface with vertex at the origin. Thus, $\widehat{\varphi}(\boldsymbol{x}) = -\boldsymbol{x}$, which corresponds to a central symmetry with respect to the origin, is a symmetry of F_n. Since by hypothesis $\widehat{f}(\boldsymbol{x}) = A\boldsymbol{x}$ is also a symmetry of F_n, and since symmetries form a group, $\widehat{g} = \widehat{\varphi} \circ \widehat{f}$ is also a symmetry of F_n. Since the composition of a central symmetry and a reflection is an axial symmetry, the result follows.

In order to check whether $\widehat{g}$ is associated with Case + or Case −, and since $F_n(A\boldsymbol{x}) = F_n(\boldsymbol{x})$ because F_n is by hypothesis associated with Case +, we observe that

$$(F_n \circ \widehat{g})(\boldsymbol{x}) = F_n(-A\boldsymbol{x}) = F_n(A(-\boldsymbol{x})) = F_n(-\boldsymbol{x}).$$

Since F_n is a homogeneous polynomial of total degree n, $F_n(-\boldsymbol{x}) = (-1)^n F_n(\boldsymbol{x})$. Thus, n even implies Case +, and n odd implies Case −. □

Lemma 128, jointly with previous results applied to the surface defined by F_n, leads to an algorithm to compute the reflectional symmetries in Case +. If n is odd the algorithm will surely succeed; if n is even, the only danger is that the components of the space vector field $\nabla F_n(\boldsymbol{x}) \times \boldsymbol{x}$ have infinitely many lines in common, a case that will be treated in the next section.

Example 14. Let S be the algebraic surface defined by the polynomial

$$F(x, y, z) = x^2y^2 + x^2z^2 + y^2z^2 - 4xyz,$$

see Fig. 6.8 (left). Our goal is to find the reflectional symmetries of S. To start, the methods from Sections 6.3.1 and 6.3.2 allow us to efficiently determine rotational and axial symmetries of S. We arrive at 8 rotational symmetries and 3 axial symmetries (all belonging to Case +); all the axes pass through the origin and the angles of rotation are determined using the algorithm from Chapter 2 applied on the intersection curves of S and the planes normal to axis of rotation, see Fig. 6.8 (right). In particular, we obtain

- rotational symmetries about the axes determined by $(1, 1, 1)$, $(1, 1, -1)$, $(1, -1, 1)$, or $(-1, 1, 1)$, by angles $\pm\frac{2\pi}{3}$,
- axial symmetries about the axes determined by $(1, 0, 0)$, $(0, 1, 0)$, or $(0, 0, 1)$.

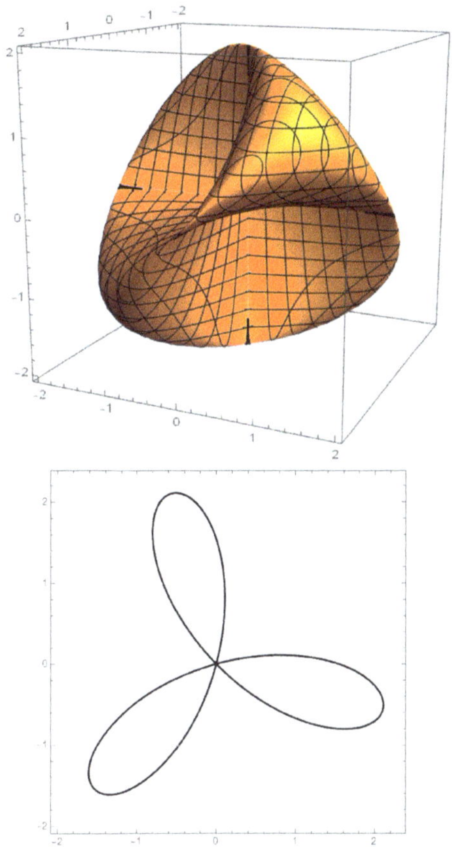

FIGURE 6.8 Left: The quartic surface S from Example 14. Right: The intersection curve of S with the plane $x+y-z=0$ (which is the plane of view); this planar curve exhibits rotational symmetry around the point $(0,0,0)$ with the angles $2\pi/3$ and $4\pi/3$.

Now we move to reflections. Since F is irreducible, by Corollary 127 only Case + is possible. Hence, we must look for axial symmetries of the surface defined by the polynomial $F_4(x,y,z) = x^2y^2 + x^2z^2 + y^2z^2$. All the tentative directions of the axes of symmetry of $F_4(x,y,z)$, gained via the approach from Theorem 125, are

$$(1,1,1), \quad (1,1,-1), \quad (1,-1,1), \quad (-1,1,1),$$
$$(1,\pm 1,0), \quad (1,0,\pm 1), \quad (0,\pm 1,1),$$
$$(1,0,0), \quad (0,1,0), \quad (1,0,0).$$

In fact, these directions are already determined during the calculation of the rotations and axial symmetries. The subsequent computations yield 6 reflectional symmetries of S given by the planes

$$\Pi_{1,2} : x \pm y = 0, \quad \Pi_{3,4} : x \pm z = 0, \quad \Pi_{5,6} : y \pm z = 0.$$

6.3.4 Central symmetries

Using Lemma 121 we can guess the sign in Eq. (6.3), which in this case corresponds to

$$F(-\boldsymbol{x} + \boldsymbol{b}) = \pm F(\boldsymbol{x}). \tag{6.14}$$

The above equality provides a polynomial system $\mathcal{S}$ in the components of $\boldsymbol{b}$. In general, we can extract from $\mathcal{S}$ a linear system of the components of $\boldsymbol{b}$; in some cases we may need equations of higher order.

Example 15. Consider the algebraic surface S defined by the polynomial

$$\begin{aligned}F(x,y,z) = {} & 3x^4 - 4x^3z + 6x^2y^2 - 12x^2yz + 12x^2z^2 + 12xy^2z + 3y^4 \\ & + 4y^3z + 12y^2z^2 + 2z^4 - 4x^3 + 36x^2y - 48x^2z - 36xy^2 + 48xyz \\ & - 24xz^2 + 4y^3 - 48y^2z + 24yz^2 - 16z^3 + 66x^2 - 120xy + 120xz \\ & + 66y^2 - 120yz + 66z^2 - 156x + 156y - 152z + 136,\end{aligned}$$

see Fig. 6.9. Since $n = 4$ is even by Lemma 121 we just focus on Case +. From $F(-\boldsymbol{x} + \boldsymbol{b}) = F(\boldsymbol{x})$, cf. (6.14), we extract the form of highest degree (cubic in this case) to get

$$\begin{aligned}& x^2z(12b_1 + 12b_2 - 24b_3 + 96) + y^2z(-12b_1 - 12b_2 - 24b_3 + 96) \\ & + xyz(24b_1 - 24b_2 - 96) + x^3(-12b_1 + 4b_3 + 8) \\ & + xy^2(-12b_1 - 12b_3 + 72) + (48 - 24b_1)xz^2 \\ & + x^2y(-12b_2 + 12b_3 - 72) + y^3(-12b_2 - 4b_3 - 8) \\ & + (-24b_2 - 48)yz^2 + (32 - 8b_3),\end{aligned}$$

and set it equal to the zero polynomial. This leads to a system of linear equations for the coefficients. The solution is $b_1 = 2, b_2 = -2, b_3 = 4$, and we verify that S is centrally symmetric around the point $(1, -1, 2)$.

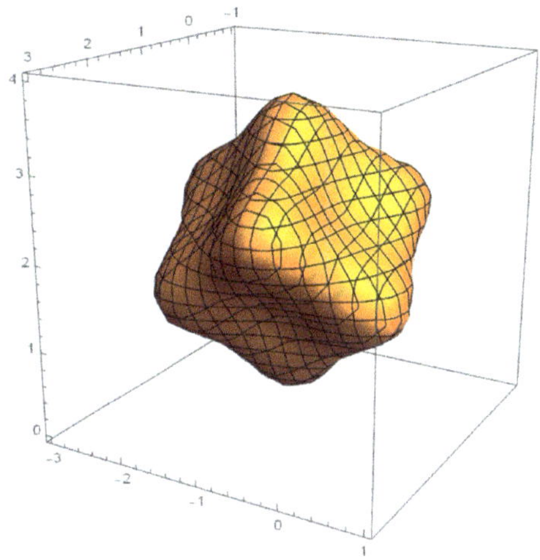

FIGURE 6.9 The quartic surface S from Example 15.

6.3.5 Special cases

From the previous subsections it remains to cover the case when $F \circ f = F$ and the components of the space vector field $\nabla F_n(\boldsymbol{x}) \times \boldsymbol{x}$ have infinitely many lines in common. We will split this into two subcases, the subcase $\nabla F_n(\boldsymbol{x}) \times \boldsymbol{x} = 0$ and the subcase $\nabla F_n(\boldsymbol{x}) \times \boldsymbol{x} \neq 0$. Furthermore, we will denote $\mathbf{W}(\boldsymbol{x}) = \nabla F_n(\boldsymbol{x}) \times \boldsymbol{x}$. Here we will just go in detail through the first subcase; for the second subcase, whose analysis is longer and rather technical, we refer to Section 4.2 of [15]. Let us also observe that the first subcase includes an important class of surfaces, *D'Arboux cyclides*: these are quartic algebraic surfaces, whose implicit equation has the form

$$F(x, y, z) = \lambda(x^2+y^2+z^2)^2+(x^2+y^2+z^2)L(x, y, z)+Q(x, y, z) = 0, \tag{6.15}$$

where λ is a constant, L is a linear polynomial, and Q is a polynomial of degree at most two. These surfaces were studied by Kummer [80] and D'Arboux [50], and have the remarkable property of containing from two to six real circles through almost every point [87, 106, 123]; conversely, every algebraic surface with at least 3 circles through each point must be Möbius equivalent (see the end of Chapter 1 for the notion of Möbius equivalence) to either a sphere or a D'Arboux cyclide [86]. Furthermore, the fact that these surfaces are made of circles makes them attractive in Architecture. For different properties related to D'Arboux cyclides, their history and applications we refer to [137], where the morphologies arising in this family of surfaces are studied.

Now we begin by identifying what the condition $\mathbf{W}(\boldsymbol{x}) = 0$ means.

Lemma 129. *Let $F(x, y, z)$ define an algebraic surface S, and let F_n be the form of highest degree of F. Then $\mathbf{W}(\boldsymbol{x}) = 0$ if and only if $F_n = k(x^2 + y^2 + z^2)^m$, $k \in \mathbb{R} - \{0\}$. In particular, if $\deg(F)$ is odd, then $\mathbf{W}(\boldsymbol{x}) \neq 0$.*

Proof. The condition $\mathbf{W}(\boldsymbol{x}) = 0$ holds iff the normal line at every regular point of the surface $F_n(\boldsymbol{x}) = 0$ goes through the origin. By Theorem 5 in [5], a surface is a surface of revolution about a line ℓ if and only if all the normals at the regular points of the surface intersect ℓ. Using this, $\mathbf{W}(\boldsymbol{x}) = 0$ occurs iff F_n defines a surface of revolution about *any* line through the origin. Since by Proposition 114 in Section 6.2 the axis of a surface of revolution is unique except for spheres, we deduce that $\mathbf{W}(\boldsymbol{x}) = 0$ happens iff F_n defines a sphere with center at the origin. Since F_n is homogeneous, the result follows. □

From Lemma 129 we observe that the situation we are analyzing can only happen for even degrees. Now let us go back to the Laplacian chain in Eq. (6.4), and to the chain of symmetry groups in Eq. (6.5), derived from the Laplacian chain. Let us see that in our case, the sequence in Eq. (6.4) must contain a sphere. In order to show this, we need the following instrumental lemma, which can be easily checked.

Lemma 130. *Let $F(x, y, z) = (x^2 + y^2 + z^2)^m$, $m \in \mathbb{N}$. Then $\Delta F = (4m^2 + 2m)(x^2 + y^2 + z^2)^{m-1}$; in particular, $\Delta F \neq 0$.*

Writing F as a sum of homogeneous forms, we get that

$$\Delta F = \Delta F_n + \Delta F_{n-1} + \cdots + \Delta F_0.$$

Then we have the following result.

Proposition 131. *Let F define a surface S, let F_n be the form of highest degree of F, and assume that $F_n(x, y, z) = k(x^2 + y^2 + z^2)^m$. Then the chain in Eq. (6.4) contains a nonzero quadratic polynomial defining a sphere* $\mathbf{S}$.

Proof. By Lemma 130, the Laplacian ΔF_n is not identically zero, and has degree $2m-2$. Since the degree of ΔF_i for $i \neq n$ is strictly less than $2m-2$, ΔF cannot be zero either, and the form of highest degree of ΔF is of the type $k(x^2+y^2+z^2)^{m-1}$. The same situation arises every time we take the Laplacian in Eq. (6.4), so we must end with a quadratic polynomial whose form of highest degree is of the type $k(x^2 + y^2 + z^2)$. □

The sphere $\mathbf{S}$ in the statement of Proposition 131 can be computed by just repeatedly taking the Laplacian of F. Let $\mathbf{p}$ be the center of this sphere; furthermore, notice that $\mathbf{p}$ is always real, although the sphere might not be real.

Theorem 132. *Let F define a surface S, let F_n be the form of highest degree of F, and assume that $F_n(x, y, z) = k(x^2+y^2+z^2)^m$. If $f \in \mathrm{Sym}(F)$, then $\mathbf{p}$ is a fixed point of f.*

Proof. Because of the chain in Eq. (6.5), $f \in \mathrm{Sym}(\mathbf{S})$, where $\mathbf{S}$ is the sphere in the statement of Proposition 131. Since all the symmetries of $\mathbf{S}$ fix $\mathbf{p}$, then $\mathbf{p}$ must also be a fixed point of f. □

Therefore, by applying a translation τ moving $\mathbf{p}$ to the origin, in order to compute the symmetries of S we can restrict to the case of symmetries fixing the origin. By an abuse of notation, let F denote also the polynomial defining $\tau(S)$, and let f be a symmetry fixing the origin. Then,

$$F \circ f = (F_n \circ f) + (F_{n-1} \circ f) + \cdots + (F_0 \circ f).$$

Thus, $F \circ f = F$ if and only if $F_i \circ f = F_i$ for $i = 0, 1, \ldots, n$, so we can compute the symmetries of F from the symmetries of homogeneous forms F_i of small degree, using the ideas in the previous subsections.

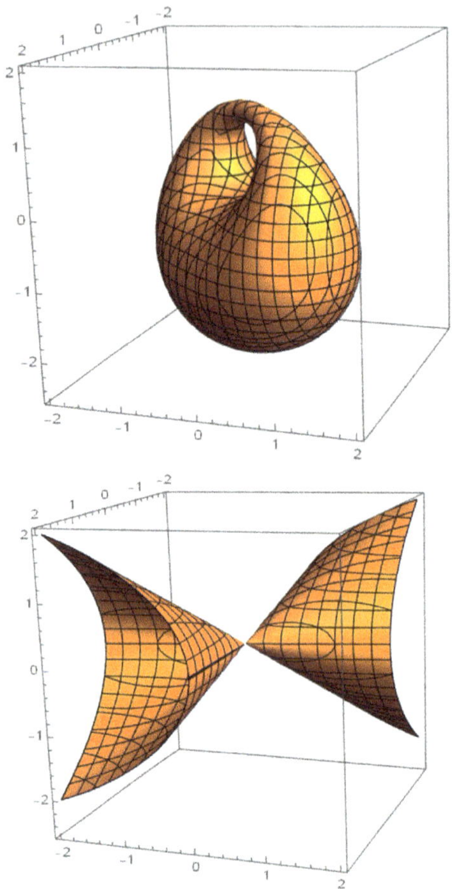

FIGURE 6.10 Left: The surface S from Example 16. Right: The surface (quadratic cone) described by the homogeneous form F_2.

Example 16. Let S be the algebraic surface defined by the polynomial

$$F(x, y, z) = (x^2 + y^2 + z^2)^2 - y^2 - 4z^2 - 5xy + 3z,$$

see Fig. 6.10 (left), which corresponds to a D'Arboux cyclide. Certainly this surface belongs among the class studied in this subsection, and $F_4(x, y, z) = (x^2 + y^2 + z^2)^2$. Following Proposition 131, the sphere contained in the chain (6.4) is described by the polynomial $2x^2 + 2y^2 + 2z^2 - 1$, i.e., the center $\mathbf{p}$ is at the origin. Hence, no translation τ is needed in this case. Then, we try to compute the symmetries of F from the symmetries of $F_2(x,y,z) = -y^2 - 4z^2 - 5xy$. This is a quadric, in particular a cone with the vertex at the origin, see Fig. 6.10 (right), which has three axes and three axial planes determined by the vectors

$$(0,0,1),\ (\sqrt{26} \pm 1, \mp 5, 0).$$

However, not all of them are symmetries of F. One can check that F has just one axial symmetry about the line $\mathcal{L}$ determined by the vector $(0,0,1)$, and two reflectional symmetries on the orthogonal planes $\Pi_{1,2} : \left(\sqrt{26}x \pm 1\right) x \mp 5y = 0$.

6.4 IMPLICIT ALGEBRAIC SPACE CURVES (II)

At this point we know how to compute the symmetries of an algebraic surface, so let us see how this can help us to solve, under certain hypotheses, the problem of computing the symmetries of an algebraic space curve, something that we promised in the previous chapter. Here we will be switching between the affine and the projective settings, partially because we will need some results holding in the projective setting, and partially because at some points we will be discussing ad-hoc arguments for homogeneous polynomials.

We limit ourselves to non-planar space curves $\mathcal{C} \subset \mathbb{R}^3$ given as the intersection of two algebraic surfaces implicitly defined by F_1, F_2, so $\mathcal{C} = V(F_1, F_2)$, with F_1, F_2 square-free. We will also assume that the (projective) space curve $V(\widehat{F}_1, \widehat{F}_2)$, where $\widehat{F}_1, \widehat{F}_2$ are the result of homogenizing F_1, F_2 with respect to a homogenization variable ω, is a *transversal* intersection. The notion of transversality is projective, and appears in the context of Algebraic Geometry (see for instance Chapter 18 of [68]); in our case, it means that no component of $V(\widehat{F}_1, \widehat{F}_2)$ is a singular curve of the surfaces defined by either $\widehat{F}_1$ or $\widehat{F}_2$, and that the surfaces defined by $\widehat{F}_1, \widehat{F}_2$ are not tangent along any component of $V(\widehat{F}_1, \widehat{F}_2)$. By Proposition 17.18 of [68], under the transversality assumption $V(\widehat{F}_1, \widehat{F}_2)$ is a complete intersection (see Section 5.1), meaning that the ideal of this curve is $\langle \widehat{F}_1, \widehat{F}_2 \rangle$. Also, in that case $\mathcal{C} = V(F_1, F_2)$ is a complete intersection too, and $\widehat{\mathcal{C}} = V(\widehat{F}_1, \widehat{F}_2)$. Moreover, we will also request that $\mathcal{C}$, the affine curve, is irreducible; $\widehat{\mathcal{C}}$ might have other components at infinity, though.

First we consider the case when F_1, F_2 are polynomials of different degrees, m, n, where $m < n$. Under the hypothesis of transversality, we can guarantee that the *degree* of $\widehat{\mathcal{C}}$, which we reviewed in the previous chapter and that corresponds to the number of intersection points between the curve and a generic plane (the notion for projective curves is analogous), is equal to mn (see Theorem 18.3 in [68]); furthermore, $\deg(\widehat{\mathcal{C}}) = \deg(\mathcal{C})$. Let us also mention that without the hypothesis of transversality, we just might ensure that the degree is less or equal than mn (see Corollary 18.5 in [68]). Then we get the following result.

Theorem 133. *Let $\mathcal{C} = V(F_1, F_2)$ be an irreducible algebraic space curve, where $m = \deg(F_1) < n = \deg(F_2)$. Under the previous assumptions, every symmetry f of $\mathcal{C}$ is a symmetry of the surface defined by F_1 (the surface of least degree).*

Proof. Let f be a symmetry of $\mathcal{C}$. Since $\mathcal{C} = V(F_1, F_2)$, every point $\boldsymbol{x} \in \mathcal{C}$ satisfies that $F_1(\boldsymbol{x}) = 0$. Additionally, for every point $\boldsymbol{x} \in \mathcal{C}$, $f(\boldsymbol{x}) \in \mathcal{C}$ too; thus, $F_1(f(\boldsymbol{x})) = 0$, i.e., $(F_1 \circ f)(\boldsymbol{x}) = 0$, so $\mathcal{C} \subset V(F_1, F_1 \circ f)$. Assume that $F_1, F_1 \circ f$ define different surfaces. In that case, $V(F_1, F_1 \circ f)$ is an algebraic space curve, and since $\mathcal{C}$ is irreducible, $\mathcal{C}$ must be an irreducible component of $V(F_1, F_1 \circ f)$. Because f is linear, the degree of $F_1 \circ f$ coincides with the degree, m, of F_1. Since $\mathcal{C} \subset V(F_1, F_1 \circ f)$, we deduce that the degree of $\mathcal{C}$ is at most m^2. But $m < n$, so $m^2 < mn = \deg(\mathcal{C})$ and we get a contradiction. Therefore, $F_1, F_1 \circ f$ must define the same surface, implying that f is a symmetry of the surface defined by F_1. □

However, we still need a test to check which symmetries of the surface defined by F_1 are, actually, symmetries of $\mathcal{C}$. Again the property of being a complete intersection comes to the rescue. Because $\mathcal{C}$ is a complete intersection, $\mathbf{I}(\mathcal{C}) = \langle F_1, F_2 \rangle$. Next if f is a symmetry of $\mathcal{C}$, for any point $\boldsymbol{x} \in \mathcal{C}$, $f(\boldsymbol{x}) \in \mathcal{C}$, and since F_2 vanishes at the points of $\mathcal{C}$, F_2 vanishes at $f(\boldsymbol{x})$. Thus, $\boldsymbol{x}$ is a point of the surface defined by $F_2 \circ f$, so $F_2 \circ f \in \langle F_1, F_2 \rangle$. In a projective setup, and denoting the result of homogenizing $F_2 \circ f$ by $\widehat{F_2 \circ f}$, we have that $\widehat{F_2 \circ f} \in \langle \widehat{F}_1, \widehat{F}_2 \rangle$. So there exist homogeneous polynomials α_1, α_2 such that

$$\widehat{F_2 \circ f} = \alpha_1 \widehat{F}_1 + \alpha_2 \widehat{F}_2. \tag{6.16}$$

Lemma 134. *The homogeneous polynomial α_2 in Eq. (6.16) is a constant.*

Proof. Since f is linear, $\deg(\widehat{F_2 \circ f}) = \deg(\widehat{F}_2) = n$. Now the right-hand side of Eq. (6.16) is the (nonzero) sum of two homogeneous polynomials, so its degree is the maximum of the degrees of these polynomials. Taking into account that the degree of a product of homogeneous polynomials is the sum of their degrees, the degree of the right-hand side of Eq. (6.16) is $\max\{\deg(\alpha_1)+m, \deg(\alpha_2)+n\}$, that must be equal to n. If $\deg(\alpha_1) + m > \deg(\alpha_2) + n$, we get that $n = \deg(\alpha_1) + m > \deg(\alpha_2) + n$: since $\deg(\alpha_2) \geq 0$, this is a contradiction. Therefore $\deg(\alpha_1)+m \leq \deg(\alpha_2)+n$, so we conclude that $n = \deg(\alpha_2) + n$, and therefore $\deg(\alpha_2) = 0$, implying that α_2 is a constant. □

Dehomogenizing, we get that $F_2 \circ f = \tilde{\alpha}_1 F_1 + \alpha_2 F_2$, where $\tilde{\alpha}_1$ is a polynomial in x, y, z, and α_2 is a constant. Furthermore, recall that because of Theorem 133, $F_1 \circ f = \pm F_1$. Putting these two conditions together, we get that

$$\begin{cases} F_1 \circ f &= \pm F_1 \\ F_2 \circ f &= \tilde{\alpha}_1(x, y, z) F_1 + \alpha_2 F_2 \end{cases}$$

where we have made explicit the fact that $\tilde{\alpha}_1 = \tilde{\alpha}_1(x, y, z)$ is, in general, a polynomial (while $\alpha_2 \in \mathbb{R}$). These equations admit a more elegant formulation: calling

$$\mathbf{V} = \begin{bmatrix} F_1 \\ F_2 \end{bmatrix}, \; Q = \begin{bmatrix} \pm 1 & 0 \\ \tilde{\alpha}_1(x, y, z) & \alpha_2 \end{bmatrix},$$

we have

$$\mathbf{V} \circ f = Q \cdot \mathbf{V} \tag{6.17}$$

with Q a numeric matrix, which is reminiscent of other conditions found before, like Eq. (2.2), Eq. (4.4) or Eq. (6.3). Notice that different f's have different associated matrices Q's, which form a matrix group; furthermore, since $\mathrm{Sym}(\mathcal{C})$ is finite, we can deduce that for each Q there exists $N \in \mathbb{N}$ such that $Q^N = I$. In particular, Q is regular, $|Q| = \pm 1$, and the eigenvalues of Q must be N-th roots of the unity.

Lemma 135. *The constant α_2 is either $\alpha_2 = 1$ or $\alpha_2 = -1$.*

Proof. Since $|Q| = \pm\alpha_2$ and $|Q| = 1$ or $|Q| = -1$, the lemma follows. □

Thus, the possible values for the trace $\mathrm{Tr}(Q)$ of the matrices Q are $2, -2$, when the elements in the main diagonal of Q have the same sign, or 0, when the elements in the main diagonal of Q have opposite sign. We have yet one more consequence of Lemma 135 and Eq. (6.17).

Proposition 136. *Let f be a symmetry of $\mathcal{C}$, and let Q be its associated matrix according to Eq. (6.17). If $\mathrm{Tr}(Q) = 0$, then ϕ is an involution. Furthermore, if f is not an involution, then f must be a common symmetry of the surfaces defined by F_1, F_2.*

Proof. Repeatedly applying Eq. (6.17) we get that $\mathbf{V} \circ f^k = Q^k \cdot \mathbf{V}$. If $\mathrm{Tr}(Q) = 0$, then one can check that $Q^2 = I$, so f^2 is the identity and f is an involution. Now if $\mathrm{Tr}(Q) = \pm 2$ we get that

$$Q^k = \begin{bmatrix} 1 & 0 \\ \pm k\tilde{\alpha}_1(x, y, z) & 1 \end{bmatrix}.$$

Since there exists $N \in \mathbb{N}$ so that $Q^N = I$, we have no other possibility but $\tilde{\alpha}_1 = 0$, so $F_2 \circ f = \pm F_2$ and f is a symmetry of the surface defined by F_2. We already know that f must be a symmetry of the surface defined by F_1, so the result follows. □

Now we can provide a final theorem for the case of different degrees,

Theorem 137. *Under the previous assumptions and with* $\deg(F_1) < \deg(F_2)$*,* f *is a symmetry of* $\mathcal{C}$ *if and only if* f *is a symmetry of the surface defined by* F_1 *and* $(F_2 \circ f) \pm F_2 \equiv \text{mod } F_1$*. Furthermore, if* f *is not an involution, then* f *is a symmetry of* $\mathcal{C}$ *if and only if* f *is a common symmetry of the surfaces defined by* F_1, F_2*.*

Proof. For the first part, $(\Rightarrow)$ is a consequence of Theorem 133 plus Lemma 135; for $(\Leftarrow)$, the condition $(F_2 \circ f) \pm F_2 \equiv \text{mod } F_1$ implies that $F_2 \circ f = \tilde{\alpha}_1 F_1 \pm F_2$, and since f is a symmetry of F_1, $F_1 \circ f = \pm F_1$. Thus $V(F_1 \circ f, F_2 \circ f) \subset \mathcal{C}$, and because $\mathcal{C}$ is irreducible, $V(F_1 \circ f, F_2 \circ f) = \mathcal{C}$, so $f(\mathcal{C}) = \mathcal{C}$, i.e., f is a symmetry of $\mathcal{C}$. For the second part, $(\Rightarrow)$ follows from Proposition 136, and $(\Leftarrow)$ can be proven as before. □

Thus, whenever the surface defined by F_1 has finitely many symmetries, i.e., whenever $F_1 = 0$ is not either cylindrical or a surface of revolution, we can find them using Theorem 137. If $F_1 = 0$ has infinitely many symmetries, we can still find the non-involutional symmetries with the help of the same theorem whenever $F_2 = 0$ has finitely many symmetries.

Let us move now to the case when F_1, F_2 are polynomials of the same degree, n. A first result here is that, again taking advantage of the property of being a complete intersection, we can also get a simple condition to test a potential symmetry. First we work in a projective setting: if f is a symmetry of the curve, then $\widehat{F_1 \circ f}, \widehat{F_2 \circ f} \in \mathbf{I}(\widehat{\mathcal{C}}) = \langle \widehat{F}_1, \widehat{F}_2 \rangle$. Thus, there exist homogeneous polynomials a, b, c, d, where, $a = a(x, y, z, \omega)$, etc., such that

$$\widehat{F_1 \circ f} = a\widehat{F}_1 + b\widehat{F}_2, \quad \widehat{F_2 \circ f} = c\widehat{F}_1 + d\widehat{F}_2. \tag{6.18}$$

Lemma 138. *The polynomials* a, b, c, d *in Eq.* (6.18) *are constant.*

Proof. Let us see that a, b are constant; for c, d, the reasoning is analogous. Because f is a linear mapping, $\deg(\widehat{F_1 \circ f}) = \deg(\widehat{F}_1)$. Furthermore, the degree of $a\widehat{F}_1 + b\widehat{F}_2$, taking into account that all the polynomials intervening are homogeneous and that the degrees of $\widehat{F}_1, \widehat{F}_2$ coincide, is equal to $\max\{\deg(a), \deg(b)\} + \deg(\widehat{F}_1)$. Since this must be equal to $\deg(\widehat{F}_1)$, we get that $\max\{\deg(a), \deg(b)\} = 0$, so a, b are constant. □

Therefore, dehomogenizing Eq. (6.18) we conclude that

$$F_1 \circ f = aF_1 + bF_2, \;\; F_2 \circ f = cF_1 + dF_2 \tag{6.19}$$

where the coefficients a, b, c, d are constants. By writing

$$\mathbf{V} = \begin{bmatrix} F_1 \\ F_2 \end{bmatrix}, \; Q = \begin{bmatrix} a & b \\ c & d \end{bmatrix},$$

we recover the expression in Eq. (6.17),

$$\mathbf{V} \circ f = Q \cdot \mathbf{V}. \tag{6.20}$$

Again, $|Q| = ad - bc = \pm 1$.

Proposition 139. *Let $\mathcal{C} = V(F_1, F_2)$ be a space algebraic curve, not planar, where $\deg(F_1) = \deg(F_2) = n$. Under the previous assumptions, f is a symmetry of $\mathcal{C}$ if and only if Eq. (6.20) holds.*

Proof. The implication ($\Rightarrow$) is a consequence of previous considerations. For ($\Leftarrow$), notice that from Eq. (6.19) we can write F_1, F_2 as linear combinations of $F_1 \circ f, F_2 \circ f$. Thus, every common zero of $F_1 \circ f, F_2 \circ f$ is also a common zero of F_1, G_1, so, invoking irreducibility, $V(F_1, F_2) = V(F_1 \circ f, F_2 \circ f)$, i.e., $f(\mathcal{C}) = \mathcal{C}$. □

But there is a further consequence of Proposition 139, that allows us to compute the symmetries of $\mathcal{C}$.

Theorem 140. *Let $\mathcal{C} = V(F_1, F_2)$ a space algebraic curve. Under the previous assumptions, every symmetry f of $\mathcal{C}$ is a symmetry of the surface defined by $\|\nabla F_1 \times \nabla F_2\|^2$.*

Proof. Recall that $f(\boldsymbol{x}) = A\boldsymbol{x} + \boldsymbol{b}$, with A orthogonal. Applying the Chain Rule,

$$\nabla(F_1 \circ f) = \nabla F_1(A\boldsymbol{x} + \boldsymbol{b}) = A^T(\nabla F_1) \circ f, \;\; \nabla(F_2 \circ f) = A^T(\nabla F_2) \circ f.$$

Next using the identity for the cross product in Eq. (3.8), and since A is orthogonal, we deduce that

$$\begin{aligned}\nabla(F_1 \circ f) \times \nabla(F_2 \circ f) &= A^T(\nabla F_1) \circ f \times A^T(\nabla F_2) \circ f \\ &= |A| A^T(\nabla F_1 \times \nabla F_2) \circ f.\end{aligned} \tag{6.21}$$

Additionally, from Eq. (6.19), we get that $\nabla(F_1 \circ f) = a\nabla F_1 + b\nabla F_2$, $\nabla(F_2 \circ f) = c\nabla F_1 + d\nabla F_2$. Thus,

$$\nabla(F_1 \circ f) \times \nabla(F_2 \circ f) = (ad - bc)(\nabla F_1 \times \nabla F_2). \tag{6.22}$$

Taking norms in Eq. (6.21), Eq. (6.22), since $|A| = \pm 1$, $ad - bc = \pm 1$, and since orthogonal matrices preserve the norm, we conclude that

$$\|(\nabla F_1 \times \nabla F_2) \circ f\|^2 = \|\nabla F_1 \times \nabla F_2\|^2,$$

which implies that f is a symmetry of the surface defined by $\|\nabla F_1 \times \nabla F_2\|^2$. □

Theorem 140 implies that we can reduce the computation to the symmetries of a surface, namely the surface defined by $\|\nabla F_1 \times \nabla F_2\|^2$. Thus, if this surface has finitely many symmetries, we can compute them with the algorithms in this chapter, and then check which ones are symmetries of $\mathcal{C}$ by using Eq. (6.20), which amounts to linear system solving.

Remark 11. *From Eqs.* (6.21) *and* (6.22) *we also get that*

$$(\nabla F_1 \times \nabla F_2) \circ f = \pm A(\nabla F_1 \times \nabla F_2). \tag{6.23}$$

This means that f is either a symmetry or an anti-symmetry (recall this notion from Chapter 4) of the tangent vector field of $\mathcal{C}$, i.e., the space vector field defined by $\nabla F_1 \times \nabla F_2$, which is an algebraic space vector field. In particular, if the singular locus of $\mathcal{C}$ is nonempty, f is a symmetry of the singular locus of $\mathcal{C}$, which was expectable.

Example 17. *Let $\mathcal{C} = V(F_1, F_2)$, where*

$$F_1(x, y, z) = 2x^2 + 4y^2 + z^2 - 1, \;\; F_2(x, y, z) = 2x^2 + y^2 + z^2 - z - 1.$$

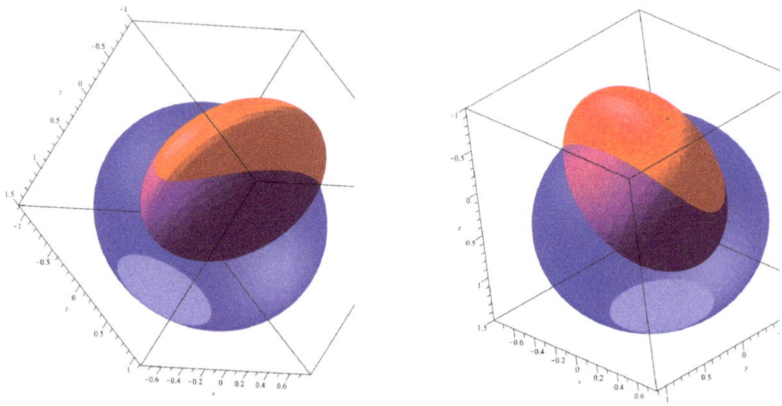

FIGURE 6.11 Two views of the space curve in Example 17, given as the intersection of two ellipsoids (in blue and red)

The space curve $\mathcal{C}$ is the intersection of two ellipsoids: this curve is not rational, and in fact it is an example of a space elliptic curve; two different views of $\mathcal{C}$ are shown in Fig. 6.11.

The computation of $\|\nabla F_1 \times \nabla F_2\|^2$ yields

$$\|\nabla F_1 \times \nabla F_2\|^2 = 576x^2y^2 + 144y^2z^2 - 192y^2z + 16x^2 + 64y^2.$$

The surface defined by $\|\nabla F_1 \times \nabla F_2\|^2 = 0$ has 7 non-trivial symmetries, 3 of which are symmetries of $\mathcal{C}$: the symmetries with respect to the coordinate planes $y = 0$, $x = 0$, plus the axial symmetry about the z-axis. The symmetries that we get are common symmetries of the surfaces defined by F_1, F_2, so in all the cases Eq. (6.20) is satisfied, with Q the identity matrix.

6.5 FINAL REMARKS AND OPEN PROBLEMS

In this chapter we have presented results to compute the symmetries of an implicit algebraic surface. A full algorithm requires to consider different types of symmetries, and distinctions according to the even or odd degree of the surface; in [15], one can find illustrative flow charts with the full procedures. A more global algorithm to solve the problem, that does not require to distinguish cases, is provided in [64], where the more general question of finding the isometries, i.e., congruences, between two implicit algebraic surfaces is addressed. In order to do this, the authors in [64] identify two points on the surface that must be mapped to each other, related to the gradients and Hessians of the surfaces. Thus, the algorithm in [64] proceeds by reducing the problem to finding isometries fixing the origin, which in turn is solved making use of elimination tools and polynomial factoring, in a manner similar to [6]. Also, symmetries of algebraic surfaces are studied in [29]; here the main question is how to visualize the group of symmetries of the surface, which, excluding cylindrical surfaces and surfaces of revolution, must be either trivial, or cyclic, or dihedral, or a group of symmetries of a platonic solid, see Section 4.4 of [29].

Additionally, we have also presented results for computing the symmetries of an implicit algebraic space curve given as the intersection of two algebraic surfaces, under certain conditions. We begin by suggesting some open problems related to space curves.

Open Problem 15. *Compute the symmetries of an implicit space algebraic curve, with fewer restrictions.* The methods we have provided distinguish between intersecting surfaces with different or equal degrees, are only applicable for space curves given as the intersection of two surfaces, and require irreducibility, transversality and complete intersection. Furthermore, these methods ultimately rely on determining the symmetries of some surface, which must be tested to extract, from them, the symmetries of the space curve. Thus, if this surface has infinitely many symmetries (i.e., if it is a cylindrical surface, or a surface of revolution), the method does not work well. It would be good to have a more global algorithm with relaxed hypotheses, that could work in a more general setting. Also, the simplest examples of symmetric space algebraic curves that one can produce are space curves whose symmetries are common symmetries of the two intersecting surfaces, since any such common symmetry is, surely, a symmetry of the intersection curve. Thus, one may wonder under which conditions the only symmetries of a space algebraic curve are like that, i.e., common symmetries of the intersecting surfaces.

Open Problem 16. *Compute symmetries, affine and projective equivalences of elliptic and hyperelliptic space curves.* This is the generalization to space of the problem treated in [8]. This class of curves includes notable families, like the non-rational curves resulting from intersecting two quadric curves, or a quadric and a ruled surface; a more complete account of examples of situations when these curves appear can be checked in [28]. A good starting point would be to devise an algorithm for computing the symmetries of the space curves which result from intersecting two quadrics: certainly the algorithm presented in this chapter is available for some cases, but there may be extra advantages that allow us to find a simpler solution. The interested reader can check the different possibilities occurring when intersecting two quadric surfaces in, for instance, [133].

Now we consider open problems regarding surfaces.

Open Problem 17. *Compute the symmetries of cubic implicit surfaces.* Since surfaces are harder than curves, in general, typically one selects certain types of surfaces where the problem at hand may be solved more efficiently. In this sense, a natural choosing would be cubic surfaces. A remarkable property of these surfaces is that they contain lines: G. Salmon [110], after correspondence with A. Cayley, proved that projective smooth cubic surfaces contain exactly 27 (projective, complex and real) straight lines, some of them at infinity. Schläffi [113] proved later that the number of real straight lines must be 3, 7, 15 or 27. If the cubic is singular [25], the number of straight lines goes down to 21. These properties could be of use to find the symmetries of a cubic surface, since obviously any symmetry of the surface must also be a symmetry of the lines contained in it. As an additional comment, the problem of computing the straight lines contained in an algebraic surface is interesting in itself: the straight lines can certainly be computed by means of a brute-force algorithm, but one might come up with more efficient methods to do this.

Open Problem 18. *Classify the symmetries of D'Arboux cyclides.* D'Arboux cyclides came up in Subsection 6.3.5, and the results given there provided an algorithm to compute the symmetries of a surface of this kind. However, the question here is more theoretical and ambitious: what kind of symmetries can these surfaces have? What are the conditions on λ, L, Q in Eq. (6.15) for these symmetries to arise? A special case of D'Arboux cyclides, *Dupin cyclides*, will be studied in Chapter 8, but the general case seems to be still open.

Open Problem 19. *Compute affine, projective and Möbius equivalences between two implicit algebraic surfaces.* The generalization of the problem considered here to affine and projective equivalences is natural, and completely open. In particular, the also open problem of finding the projective equivalences between two implicit planar algebraic curves is equivalent, working in a projective setup, to the computation of the affine equivalences between two conical surfaces. Special cases of these transformations, like similarities, for instance, are also interesting.

Open Problem 20. *Generating surfaces with symmetries; approximate symmetries.* In Chapter 2, we addressed the general form of planar algebraic curves with rotational symmetry, and observed that these results can be used to find *approximate* rotational symmetries of implicitly given curves in the case when the coefficients of the implicit equations are floating point numbers, i.e., are known up to finite precision. Analogous problems for surfaces are completely open. Of course, complex numbers are now out of the picture. One might wonder whether *quaternions* [56, 62] could be used instead; quaternion multiplication, though, is not commutative, which is a complication.

CHAPTER 7

Rational surfaces (I): generalities; ruled surfaces

For plane or space rational curves (in fact, for curves in any dimension) one could reduce the problem to the parameter space, whenever the parametrization of the curve was proper, and benefit from the fact that Möbius transformations are the birational mappings of the (real or complex, affine or projective) line. For rational surfaces there is a big change: one can also reduce the problem to the parameter space, but the mappings that we have there, birational planar mappings, also called *Cremona transformations*, do not have, unlike the birational mappings of the line, a closed form. So unless we impose additional requirements on the parametrization to work with, we do not know, a priori, what kind of transformation we must look for in the parameter space. We will be able, though, to give a general algorithm to find the symmetries, but the fact that we do not know what kind of Cremona transformation we have behind also opens up the possibility of addressing certain types of surfaces where the form of the Cremona transformation can be guessed. For special surfaces, like ruled surfaces, treated in this chapter, and other surfaces that will be addressed in the coming chapters, there is a specially advantageous option as well, namely reducing the problem to that of space rational curves.

DOI: 10.1201/9781003648710-7

7.1 GENERAL NOTIONS

We start with the following definition, analogous to Definition 41 in Chapter 3.

Definition 141. *Let $S \subset \mathbb{R}^3$ be a surface. We say that S is* rational *if S admits a parametrization $\boldsymbol{x}(t,s) = (x(t,s), y(t,s), z(t,s))$ where $x(t,s)$, $y(t,s)$, $z(t,s)$ are rational functions.*

If $F(x,y,z) = 0$ is the implicit equation of S, then $F(\boldsymbol{x}(t,s)) = 0$ for $(t,s) \in \mathbb{C}^2$. In that case we say that $\boldsymbol{x}(t,s)$ is a *rational parametrization* of S, and that S is a rational surface; furtheremore, rational surfaces are irreducible [45]. As in the case of curves, we say that $\boldsymbol{x}(t,s)$ is *proper* if it is generically injective, i.e., if for almost all $(t_0,s_0) \in \mathbb{C}^2 \backslash \mathcal{C}$, where $\dim(\mathcal{C}) \leq 1$, setting $\boldsymbol{x}(t_0,s_0) = \boldsymbol{x}(t,s)$ implies that $(t,s) = (t_0,s_0)$. Thus, a proper parametrization covers the surface just once, with perhaps the exception of a finite set of points or, at most, one space curve, containing the *self-intersections* of the surface. To check exactly, i.e., symbolically, whether or not a rational surface parametrization is proper, and to reparametrize it in the case it is not, is a much more challenging problem compared to curves; the interested reader can check [97, 98, 100] for more information on this question. However, one can always check properness using a probabilistic test, namely picking a random pair $(t_0,s_0) \in \mathbb{C}^2$, and solving $\boldsymbol{x}(t_0,s_0) = \boldsymbol{x}(t,s)$; if the solution is $(t,s) = (t_0,s_0)$, then with probability almost one the parametrization is proper.

If $\boldsymbol{x}(t,s)$ is proper, then $\boldsymbol{x}^{-1}$ exists and is a rational mapping. This allows us to replicate Theorem 45 in Chapter 3 for the case of surfaces. Thus, we have the following result, that can be proven as we did with Theorem 45.

Theorem 142. *Let $S_i \subset \mathbb{C}^3$, with $i = 1,2$, be two rational surfaces defined by rational, proper parametrizations $\boldsymbol{x}_i(t,s) = (x_i(t,s), y_i(t,s), z_i(t,s))$. If the surfaces S_i are related by a real birational mapping f, i.e., if there exists a rational mapping $f : \mathbb{R}^3 \to \mathbb{R}^3$ with rational inverse f^{-1} such that $f(S_1) = S_2$, then there also exists a birational mapping $\psi : \mathbb{C}^2 \to \mathbb{C}^2$ such that*

$$f \circ \boldsymbol{x}_1 = \boldsymbol{x}_2 \circ \psi, \tag{7.1}$$

i.e., making commutative the following diagram:

$$\begin{array}{ccc} S_1 & \xrightarrow{f} & S_2 \\ \uparrow{\scriptstyle x_1} & & \uparrow{\scriptstyle x_2} \\ \mathbb{C}^2 & \underset{\psi}{\dashrightarrow} & \mathbb{C}^2 \end{array} \tag{7.2}$$

Furthermore, if f_1, f_2*,* $f_1 \neq f_2$*, satisfy Eq.* (7.1) *for birational mappings* ψ_1, ψ_2*, then* $\psi_1 \neq \psi_2$*.*

Birational mappings of the plane are called *Cremona transformations*, and the group of Cremona transformations is called the *Cremona group*. Compared to the group of birational mappings of the line, consisting of the Möbius transformations, the structure of the Cremona group is much more complicated and is not fully understood yet. Since Noether and Castelnuovo [38] the Cremona group is known to be generated by the *standard quadratic transformation*, which can be written as

$$(x, y) \longmapsto \left(\frac{1}{x}, \frac{1}{y}\right),$$

and the *projective linear transformations*,

$$(x, y) \longmapsto \left(\frac{a_{11}x + a_{12}y + a_{13}}{a_{31}x + a_{32}y + a_{33}}, \frac{a_{21}x + a_{22}y + a_{23}}{a_{31}x + a_{32}y + a_{33}}\right). \tag{7.3}$$

Furthermore, the finite subgroups of the Cremona group have also been studied [54]; notice that by Theorem 111 in the previous chapter, whenever S is not either cylindrical or a surface of revolution its symmetry group is finite, and therefore the group consisting of the ψ satisfying Theorem 142 must also be finite, and in particular a finite subgroup of the Cremona group. However, the classification of finite subgroups of the Cremona group is not considered to be entirely completed; also, some of these subgroups are complicated to deal with from a computational point of view.

Although in this case we do not necessarily know a priori what kind of transformation we have at the bottom of Eq. (7.2), despite of this in the next section we will able to provide a pretty general strategy to solve the problem. There are cases, though, where we can guess the shape of ψ, and which lead to special algorithms. Here we will present two of these cases; other cases are associated with special types of surfaces, and will be addressed later, either in this chapter or in coming chapters.

For the first case, addressed in [70], we need to move to a projective setup. In order to do this, given a rational parametrization $\boldsymbol{x}(t,s) = (x(t,s), y(t,s), z(t,s))$ we write

$$x(t,s) = \frac{A(t,s)}{D(t,s)},\ y(t,s) = \frac{B(t,s)}{D(t,s)},\ z(t,s) = \frac{C(t,s)}{D(t,s)}, \tag{7.4}$$

where $\gcd(A(t,s), B(t,s), C(t,s), D(t,s)) = 1$. We choose a homogenization variable ω, and consider the mapping

$$t \to \frac{t}{\omega},\ s \to \frac{s}{\omega}.$$

Plugging this mapping into $[A(t,s) : B(t,s) : C(t,s) : D(t,s)]$ and clearing denominators, we reach a projective parametrization of the projective closure of S,

$$\widehat{\boldsymbol{x}}(t,s,\omega) = [\widehat{A}(t,s,\omega) : \widehat{B}(t,s,\omega) : \widehat{C}(t,s,\omega) : \widehat{D}(t,s,\omega)],$$

where all the components are homogeneous polynomials in t, s, ω of the same degree. The *base points* of $\widehat{\boldsymbol{x}}(t,s,\omega)$ are the points $[t : s : \omega] \in \mathbb{P}^2(\mathbb{C})$ where

$$\widehat{A}(t,s,\omega) = 0,\ \widehat{B}(t,s,\omega) = 0,\ \widehat{C}(t,s,\omega) = 0,\ \widehat{D}(t,s,\omega) = 0.$$

Notice that $\widehat{\boldsymbol{x}}(t,s,\omega)$ has no (projective, real or complex) base points iff the solution to the above polynomial system is just $t = 0, s = 0, \omega = 0$, since $[0 : 0 : 0]$ does not represent any point in $\mathbb{P}^2(\mathbb{C})$. The case that we consider now, precisely, is that of rational surfaces without projective base points.

For these surfaces we have the following result, whose proof can be found in [70] (see Theorem 4 in [70]).

Theorem 143. *Let $S_i \subset \mathbb{R}^3$, with $i = 1, 2$, be two rational surfaces defined by rational, proper parametrizations $\boldsymbol{x}_i(t,s) = (x_i(t,s), y_i(t,s), z_i(t,s))$, and assume that the corresponding projective parametrizations do not have projective base points. Then the mapping ψ in Theorem 142 is a projective linear transformation.*

Theorem 143 is used in [70] to solve the problem of finding the projective transformations between two rational surfaces given by parametrizations matching the hypotheses in Theorem 143, which in particular solves the problem of computing the symmetries of a given rational surface without base points. The strategy used, analogous to the strategy in [69] for rational curves, is, essentially, to derive a polynomial system from Eq. (7.2) to solve for the coefficients of the transformation ψ. Although there are notable examples of parametrizations without projective base points, for instance several classes of quadratically parametrizable surfaces [44, 70], base points, however, are frequent.

The second case regards *polynomially parametrized* surfaces, i.e., the case when S is defined by a parametrization as in Eq. (7.4) with $D(t,s)$ a constant. If we have a polynomial parametrization which is surjective (also called *normal* in several references, see for instance [102, 115, 116]) we have the following result; see Theorem 16 in [7] for a proof.

Theorem 144. *Let $S \subset \mathbb{R}^3$ be a rational surface defined by means of a proper, polynomial, surjective parametrization $\boldsymbol{x}(t,s) = (x(t,s), y(t,s), z(t,s))$, and let $f : \mathbb{R}^3 \to \mathbb{R}^3$ be a symmetry of S. Then the mapping ψ in Theorem 142 is linear affine, i.e., it has the form*

$$\psi(t,s) = (a_{11}t + a_{12}s + a_{13}, a_{21}t + a_{22}s + a_{23}). \tag{7.5}$$

If we are just interested in involutions, which is the problem addressed in [7] (for the case of rational surfaces fulfilling the hypotheses in Theorem 144), we can express the coefficients in Eq. (7.5) in terms of at most two coefficients, exploiting two facts: (i) if $f : \mathbb{R}^3 \to \mathbb{R}^3$ is an involution of S, then ψ is an involution of $\mathbb{R}^2$;

(ii) if f is a symmetry of S, the *first fundamental form* (see the next section) of the surfaces defined by $\boldsymbol{x}$ and $f \circ \boldsymbol{x}$ must coincide at corresponding points. Therefore, in this particular case we can find the involutions of the surface by just using bivariate system solving.

The shape of ψ can be also guessed in the case of *ruled surfaces*, that will be addressed later.

Remark 12. *We should also mention that the problem of computing the symmetries of a rational surface, and in fact the computation of the projective isomorphisms between two rational surfaces, has also been addressed without using Theorem 142, i.e., without making use of the mapping ψ. Such an approach is considered in [75], where the problem is reduced to five cases, two of which are solved in [75], involving surfaces covered by either lines or conics. However, the approach in [75] requires deep elements of Algebraic Geometry and falls beyond the reach of this book.*

Remark 13. *Of course, an alternative for finding the symmetries of a rational surface is to compute the implicit equation of the surface, and apply the results in the previous chapter. Nevertheless, computing the implicit equation of the surface from the parametrization is not an easy question either, and efficient methods to do this are in fact the object of still ongoing research.*

7.2 A PRETTY GENERAL ALGORITHM

In order to present this algorithm, developed in [21], we need to recall some notions on surfaces from Differential Geometry; see [53, 65, 120] for more information on this topic. First, we say that a parametrization $\boldsymbol{x}(t,s)$ (not necessarily rational) of a surface S is *regular* at (t_0, s_0) if the partial derivatives $\boldsymbol{x}_t, \boldsymbol{x}_s$ are linearly independent at the point; otherwise, we say that $\boldsymbol{x}(t,s)$ is *singular* at (t_0, s_0). At regular points $P = \boldsymbol{x}(t,s)$ the *tangent plane* to S is well defined and corresponds to

$$T_P(S) = P + \mathcal{L}(\boldsymbol{x}_t, \boldsymbol{x}_s),$$

where $\mathcal{L}(\boldsymbol{x}_t, \boldsymbol{x}_s)$ denotes the linear variety spanned by the vectors $\boldsymbol{x}_t, \boldsymbol{x}_s$. The normal direction to the tangent plane is therefore that of $\boldsymbol{x}_t \times \boldsymbol{x}_s$; the unitary *normal vector* is defined as

$$\boldsymbol{N}_{\boldsymbol{x}}(t,s) = \frac{\boldsymbol{x}_t \times \boldsymbol{x}_s}{||\boldsymbol{x}_t \times \boldsymbol{x}_s||}. \tag{7.6}$$

Since $\boldsymbol{N}_{\boldsymbol{x}}(t,s)$ is unitary one can also see $\boldsymbol{N}_{\boldsymbol{x}}$ as a mapping from the surface S onto the unit sphere $\mathbb{S}^2$, i.e., $\boldsymbol{N}_{\boldsymbol{x}} : S \to \mathbb{S}^2$; this mapping is called the *Gauss map*.

The *first fundamental form* and the *second fundamental form* are two symmetric quadratic forms defined on each tangent plane $T_P(S)$ of S, with P regular. The first fundamental form captures the metric of the surface and is defined by the matrix

$$\mathbf{I}_{\boldsymbol{x}} = \begin{bmatrix} E & F \\ F & G \end{bmatrix} = \begin{bmatrix} \boldsymbol{x}_t \cdot \boldsymbol{x}_t & \boldsymbol{x}_t \cdot \boldsymbol{x}_s \\ \boldsymbol{x}_t \cdot \boldsymbol{x}_s & \boldsymbol{x}_s \cdot \boldsymbol{x}_s \end{bmatrix}. \tag{7.7}$$

The second fundamental form captures the local shape of the surface around the point and is defined by the matrix

$$\mathbf{II}_{\boldsymbol{x}} = \begin{bmatrix} L & M \\ M & N \end{bmatrix} = \begin{bmatrix} \boldsymbol{x}_{tt} \cdot \boldsymbol{N} & \boldsymbol{x}_{ts} \cdot \boldsymbol{N} \\ \boldsymbol{x}_{ts} \cdot \boldsymbol{N} & \boldsymbol{x}_{ss} \cdot \boldsymbol{N} \end{bmatrix}. \tag{7.8}$$

Finally, the *Weingarten map* $\mathcal{W}_{\boldsymbol{x}}$ is the (linear) mapping $\mathcal{W}_{\boldsymbol{x}} : T_P(S) \to T_P(S)$ defined as $\mathcal{W}_{\boldsymbol{x}} = -d\boldsymbol{N}_{\boldsymbol{x}}$, i.e., the differential of the Gauss map, with a changed sign. The matrix defining $\mathcal{W}_{\boldsymbol{x}}$ is

$$\mathbf{W}_{\boldsymbol{x}} = \mathbf{I}_{\boldsymbol{x}}^{-1}\mathbf{II}_{\boldsymbol{x}}. \tag{7.9}$$

The determinant of $\mathbf{W}_{\boldsymbol{x}}$ is the *Gauss curvature* of the surface, $\mathbf{K}_{\boldsymbol{x}}$, and the trace of $\mathbf{W}_{\boldsymbol{x}}$ is the *mean curvature*, $\mathbf{H}_{\boldsymbol{x}}$. In terms of the entries of the matrices $\mathbf{I}_{\boldsymbol{x}}, \mathbf{II}_{\boldsymbol{x}}$ (see for instance Section 3.4 of [95]),

$$\mathbf{K}_{\boldsymbol{x}} = \frac{LN - M^2}{EG - F^2}, \quad \mathbf{H}_{\boldsymbol{x}} = \frac{EN + GL - 2FM}{2(EG - F^2)}. \tag{7.10}$$

The above two formulas follow from a more meaningful interpretation of the Gauss and mean curvatures: by *Meusnier's theorem*, given a regular point $P \in S$ and a direction $\mathbf{u}$ on the tangent plane

$T_P(S)$, the curvature κ of every curve through P, contained in S, and tangent to $\mathbf{u}$, is the same at the point P. The *principal directions* at P are the directions $\mathbf{u}_{min}$ and $\mathbf{u}_{max}$ where the curvature κ at P reaches its minimum κ_{min} and maximum κ_{max}, respectively; such values are called *principal curvatures* (the curves contained in S tangent to the principal directions are called *lines of curvature*, and form a family covering the surface). But the principal curvatures are nothing else than the eigenvalues of the matrix defining the Weingarten map, so $\mathbf{K}_{\boldsymbol{x}} = \kappa_{min} \cdot \kappa_{max}$, and $\mathbf{H}_{\boldsymbol{x}} = \kappa_{min} + \kappa_{max}$. Taking this into account, Eq. (7.10) follows.

Now if $f(\boldsymbol{x}) = A\boldsymbol{x} + \boldsymbol{b}$ is a symmetry of a rational surface S properly parametrized by $\boldsymbol{x}(t,s)$, then from Eq. (7.1) we have that $f \circ \boldsymbol{x} = \boldsymbol{x} \circ \psi$ for some Cremona transformation ψ. In order to find the symmetries of S, we need to study first the relationship between $\mathbf{K}_{\boldsymbol{x}}, \mathbf{H}_{\boldsymbol{x}}, \mathbf{K}_{f\circ\boldsymbol{x}}, \mathbf{H}_{f\circ\boldsymbol{x}}$ and $\mathbf{K}_{\boldsymbol{x}\circ\psi}, \mathbf{H}_{\boldsymbol{x}\circ\psi}$. To do this, we must see how $\boldsymbol{N}_{\boldsymbol{x}}, \mathbf{I}_{\boldsymbol{x}}, \mathbf{II}_{\boldsymbol{x}}$ are affected when moving to $f \circ \boldsymbol{x}$ or $\boldsymbol{x} \circ \psi$. We start with $f \circ \boldsymbol{x}$. The following result follows easily from Eqs. (7.6), (7.7), and (7.8) by replacing $\boldsymbol{x} \to A\boldsymbol{x} + \boldsymbol{b}$, taking into account that A is an orthogonal matrix, so inner products are preserved; Eq. (3.9) is also needed for the statement regarding the normal vector, which in turn must be used to prove the statement regarding the second fundamental form.

Lemma 145. *Let $\boldsymbol{x}(t,s)$ be a rational parametrization of a surface S, and let $f(\boldsymbol{x}) = A\boldsymbol{x} + \boldsymbol{b}$ be a Euclidean symmetry.*

(1) $\mathbf{I}_{f\circ\boldsymbol{x}} = \mathbf{I}_{\boldsymbol{x}}$.

(2) $\boldsymbol{N}_{f\circ\boldsymbol{x}} = \det(A) A \boldsymbol{N}_{\boldsymbol{x}}$.

(3) $\mathbf{II}_{f\circ\boldsymbol{x}} = \det(A) \mathbf{II}_{\boldsymbol{x}}$.

The following corollary follows from Lemma 145 and Eq. (7.9), taking also into account that the Gauss curvature and the mean curvature are the determinant and trace of the matrix defining the Weingarten map.

Corollary 146. *Let $\boldsymbol{x}(t,s)$ be a rational parametrization of a surface S, and let $f(\boldsymbol{x}) = A\boldsymbol{x} + \boldsymbol{b}$ be a Euclidean symmetry. Then*

$$\mathbf{W}_{f\circ\boldsymbol{x}} = \det(A)\mathbf{W}_{\boldsymbol{x}}, \ \mathbf{K}_{f\circ\boldsymbol{x}} = \mathbf{K}_{\boldsymbol{x}}, \ \mathbf{H}_{f\circ\boldsymbol{x}} = \det(A)\mathbf{H}_{\boldsymbol{x}}. \quad (7.11)$$

In the language of Section 5.6, Corollary 146 shows that $\mathbf{K}_{\boldsymbol{x}}$ is an invariant for the group of space Euclidean symmetries; furthermore, from Eq. (7.10) we observe that $\mathbf{K}_{\boldsymbol{x}}$ is a rational expression of t, s. The mean curvature $\mathbf{H}_{\boldsymbol{x}}$ is "almost" an invariant, since according to Corollary 146 there might be a change of sign when moving from $\boldsymbol{x}$ to $f \circ \boldsymbol{x}$. However, Eq. (7.10) shows that for $\boldsymbol{x}(t, s)$ rational, in general $\mathbf{H}_{\boldsymbol{x}}$ is not rational, since there might be a square-root coming from the denominator in Eq. (7.6). But $\widetilde{\mathbf{H}}_{\boldsymbol{x}} = \mathbf{H}_{\boldsymbol{x}}^2$ is certainly rational and is an invariant.

Now let us analyze $\boldsymbol{x} \circ \psi$, which is just a reparametrization of the surface; we call $\boldsymbol{y} := \boldsymbol{x} \circ \psi$, $\psi(t, s) = (\psi_1(t, s), \psi_2(t, s))$. Using the Chain rule,

$$\begin{aligned} \boldsymbol{y}_t &= \boldsymbol{x}_t(\psi)\frac{\partial \psi_1}{\partial t} + \boldsymbol{x}_s(\psi)\frac{\partial \psi_2}{\partial t}, \\ \boldsymbol{y}_s &= \boldsymbol{x}_t(\psi)\frac{\partial \psi_1}{\partial s} + \boldsymbol{x}_s(\psi)\frac{\partial \psi_2}{\partial s}, \end{aligned} \tag{7.12}$$

where $\boldsymbol{y}_t, \boldsymbol{y}_s$ are evaluated at (t, s), and $\boldsymbol{x}_t(\psi), \boldsymbol{x}_s(\psi)$ denote that $\boldsymbol{x}_t, \boldsymbol{x}_s$ must be evaluated at $\psi(t, s)$. In matrix notation, we have that

$$\begin{bmatrix} \boldsymbol{y}_t \\ \boldsymbol{y}_s \end{bmatrix} = J^T \cdot \begin{bmatrix} \boldsymbol{x}_t(\psi) \\ \boldsymbol{x}_s(\psi) \end{bmatrix},$$

where J represents the Jacobian of ψ. Taking Eq. (7.7) into account, we get that

$$\mathbf{I}_{\boldsymbol{y}} = J^T \cdot \mathbf{I}_{\boldsymbol{x}}(\psi) \cdot J,$$

where, again, the entries of the first fundamental form at the left-hand side must be evaluated at (t, s), while at the right-hand side the entries of the form are evaluated at $\psi(t, s)$. To relate $\boldsymbol{N}_{\boldsymbol{y}}$ and $\boldsymbol{N}_{\boldsymbol{x}}$ we also use Eq. (7.12). Finally, to relate $\mathbf{II}_{\boldsymbol{y}}$ and $\mathbf{II}_{\boldsymbol{x}}$ we need to differentiate Eq. (7.11) again, applying the Chain rule; although the calculations are lengthy, we get the same result that we obtained for the first fundamental form. So we have the following lemma.

Lemma 147. *Let $\boldsymbol{x}(t, s)$ be a rational parametrization of a surface S, let ψ be a planar rational mapping, and let J be the Jacobian of ψ.*

(1) $\mathbf{I}_{\boldsymbol{x} \circ \psi} = J^T \cdot \mathbf{I}_{\boldsymbol{x}}(\psi) \cdot J$.

(2) $\boldsymbol{N}_{\boldsymbol{x}\circ\psi} = \det(J) A \boldsymbol{N}_{\boldsymbol{x}}(\psi)$.

(3) $\mathbf{II}_{\boldsymbol{x}\circ\psi} = J^T \cdot \mathbf{II}_{\boldsymbol{x}}(\psi) \cdot J$.

Then we derive the following corollary.

Corollary 148. *Let $\boldsymbol{x}(t, s)$ be a rational parametrization of a surface S, let ψ be a planar birational mapping, and let J be the Jacobian of ψ. Then*

$$\mathbf{W}_{\boldsymbol{x}\circ\psi} = J^{-1} \cdot \mathbf{W}_{\boldsymbol{x}}(\psi) \cdot J, \tag{7.13}$$

and

$$\mathbf{K}_{\boldsymbol{x}\circ\psi} = \mathbf{K}_{\boldsymbol{x}}(\psi), \ \ \mathbf{H}_{\boldsymbol{x}\circ\psi} = \mathbf{H}_{\boldsymbol{x}}(\psi). \tag{7.14}$$

Proof. From Eq. (7.9) and statements (1), (3) of Lemma 147,

$$\mathbf{W}_{\boldsymbol{x}\circ\psi} = \left(J^T \cdot \mathbf{I}_{\boldsymbol{x}}(\psi) \cdot J\right)^{-1} \left(J^T \cdot \mathbf{II}_{\boldsymbol{x}}(\psi) \cdot J\right),$$

and Eq. (7.13) is proved. Since from Eq. (7.13) the matrices $\mathbf{W}_{\boldsymbol{x}\circ\psi}$, $\mathbf{W}_{\boldsymbol{x}}(\psi)$ are similar, they have the same determinant and trace, and Eq. (7.14) follows. □

Finally, we reach the following theorem.

Theorem 149. *Let $\boldsymbol{x}(t, s)$ be a proper parametrization of a surface S, and let $f(\boldsymbol{x}) = A\boldsymbol{x} + \boldsymbol{b}$ be a Euclidean symmetry of S. Then there exists a birational planar mapping ψ satisfying that $f \circ \boldsymbol{x} = \boldsymbol{x} \circ \psi$, and*

$$\mathbf{K}_{\boldsymbol{x}} = \mathbf{K}_{\boldsymbol{x}}(\psi), \ \ \mathbf{H}_{\boldsymbol{x}} = \det(A)\mathbf{H}_{\boldsymbol{x}}(\psi). \tag{7.15}$$

Proof. From Theorem 142, there exists a birational mapping ψ such that $f \circ \boldsymbol{x} = \boldsymbol{x} \circ \psi$. By Corollary 146, $\mathbf{K}_{f\circ\boldsymbol{x}} = \mathbf{K}_{\boldsymbol{x}}$, and by Corollary 7.14, $\mathbf{K}_{\boldsymbol{x}\circ\psi} = \mathbf{K}_{\boldsymbol{x}}(\psi)$. Since $f \circ \boldsymbol{x} = \boldsymbol{x} \circ \psi$, we get the result for $\mathbf{K}_{\boldsymbol{x}}$. For $\mathbf{H}_{\boldsymbol{x}}$ the idea is the same. □

We can extract an algorithm from Theorem 149 by proceeding as we did with rational curves. Using $\widetilde{\mathbf{H}}_{\boldsymbol{x}} = \mathbf{H}_{\boldsymbol{x}}^2$, let us write Eq. (7.15) as

$$\mathbf{K}_{\boldsymbol{x}}(t, s) = \mathbf{K}_{\boldsymbol{x}}(u, v), \ \ \widetilde{\mathbf{H}}_{\boldsymbol{x}}(t, s) = \widetilde{\mathbf{H}}_{\boldsymbol{x}}(u, v),$$

which is clearly fulfilled, under the hypotheses of Theorem 149, for $u = \psi_1(t,s)$, $v = \psi_2(t,s)$. Now $\mathbf{K}_{\boldsymbol{x}}$ and $\widetilde{\mathbf{H}}_{\boldsymbol{x}}$ are rational functions, so after clearing denominators, the above equations lead to

$$\xi_1(t,s,u,v) = 0,\ \ \xi_2(t,s,u,v) = 0, \tag{7.16}$$

where ξ_1, ξ_2 are two polynomials in the variables t, s, u, v, the first one coming from $\mathbf{K}_{\boldsymbol{x}}$, the second one from $\widetilde{\mathbf{H}}_{\boldsymbol{x}}$. Using resultants (see Section 4.2), if t, s, u, v satisfy Eq. (7.16), then t, s, u satisfy

$$\eta_1(t,s,u) = 0,\ \text{ with } \eta_1 := \text{Res}_v(\xi_1, \xi_2), \tag{7.17}$$

and t, s, v satisfy

$$\eta_2(t,s,v) = 0,\ \text{ with } \eta_2 := \text{Res}_u(\xi_1, \xi_2). \tag{7.18}$$

Next let

$$\psi_1(t,s) = \frac{\psi_{1,n}(t,s)}{\psi_{1,d}(t,s)},\ \psi_2(t,s) = \frac{\psi_{2,n}(t,s)}{\psi_{2,d}(t,s)}, \tag{7.19}$$

where $\psi_{i,n}, \psi_{i,d}$, for $i = 1, 2$, are polynomials, and $\gcd(\psi_{i,n}, \psi_{i,d}) = 1$ for $i = 1, 2$. Then we have the following result.

Theorem 150. *Let $\boldsymbol{x}(t,s)$ be a proper parametrization of a surface S, and let $f(\boldsymbol{x}) = A\boldsymbol{x} + \boldsymbol{b}$ be a Euclidean symmetry of S. Then there exists a birational planar mapping $\psi = (\psi_1, \psi_2)$, with components as in Eq. (7.19), such that $\psi_{1,d}(t,s)u - \psi_{1,n}(t,s)$ is a factor of $\eta_1(t,s,u)$, and $\psi_{2,d}(t,s)v - \psi_{2,n}(t,s)$ is a factor of $\eta_2(t,s,v)$.*

Proof. From Theorem 142, there exists a birational mapping ψ such that $f \circ \boldsymbol{x} = \boldsymbol{x} \circ \psi$. Then for any (t,s) we have that $(t, s, \psi_1(t,s), \psi_2(t,s))$ satisfies Eq. (7.16), and therefore $(t, s, \psi_1(t,s))$ fulfills Eq. (7.17), and $(t, s, \psi_2(t,s))$ fulfills Eq. (7.18). Hence, all the points of the surface (in the (t,s,u) space) defined by $\psi_{1,d}(t,s)u - \psi_{1,n}(t,s) = 0$, are also points of the surface $\eta_1(t,s,u) = 0$. Furthermore, since $\gcd(\psi_{1,n}, \psi_{1,d}) = 1$ the polynomial $\psi_{1,d}(t,s)u - \psi_{1,n}(t,s)$ is irreducible, so by Study's Lemma (see Lemma 113), $\psi_{1,d}(t,s)u - \psi_{1,n}(t,s)$ divides $\eta_1(t,s,u)$. For $\psi_{2,d}(t,s)v - \psi_{2,n}(t,s)$ we argue in the same way. □

Thus, if, say, $\eta_1(t, s, u)$ is not identically zero, we can look for the factors of η_1 which are linear in u, which provides the functions $\psi_1(t, s)$, substitute $u = \psi_1(t, s)$ into $\xi_1(t, s, u, v)$, $\xi_2(t, s, u, v)$, and recover v from the linear factor in v of the gcd of the polynomials in t, s, v resulting from that substitution. We might also work with $\eta_2(t, s, v)$, but we do not need to compute both resultants. Notice that in this process it is not necessary to know a priori how $\psi(t, s)$ looks like. Furthermore, once ψ is known, we can compute f from the equality $f \circ \boldsymbol{x} = \boldsymbol{x} \circ \psi$. This leads to an algorithm, Algorithm 4, for computing the symmetries of S. The algorithm works whenever not both resultants η_1, η_2 are identically zero; if η_1, η_2 are both identically zero, then either ξ_1, ξ_2 share a factor, or some of them is identically zero. We can identify some cases where this can happen:

- *Gauss curvature being constant:* It is well-known that the only surfaces with nonzero constant Gauss curvature are isometric to the sphere, if $\mathbf{K} > 0$, or to the *pseudosphere*, if $\mathbf{K} < 0$. However, the pseudosphere is a surface of revolution whose directrix is a trascendental curve, the *tractix*, so the pseudosphere is not algebraic. This implies that the only irreducible algebraic surface with constant nonzero Gaussian curvature, is the sphere. However, there are algebraic, and in fact rational, surfaces with vanishing Gauss curvature; these surfaces are called *developable* surfaces, and will come up in the next section.

- *Mean curvature being constant:* These surfaces certainly exist, and among them we have the remarkable class of surfaces where $\mathbf{H} = 0$, called *minimal surfaces*, which will be addressed in Chapter 9.

- *Certain functional relationships between the Gauss curvature and the mean curvature:* an example is the case of *linear Weingarten surfaces* (see for instance [96]), which are surfaces where $a\mathbf{H} + b\mathbf{K} = c$, for constants a, b, c; this family includes as particular cases surfaces with constant Gauss or mean curvature.

Algorithm 4 Rational Surface Symmetries

Require: A proper parametrization $\boldsymbol{x}(t,s)$ of a surface $S \subset \mathbb{R}^3$.
Ensure: $\mathrm{Sym}(S)$.

compute $\mathbf{K}(t,s)$, $\widetilde{\mathbf{H}}(t,s)$, and the polynomials ξ_1, ξ_2
compute the resultant $\eta_1(t,s,u) = \mathrm{Res}_v(\xi_1, \xi_2)$
if η_1 is not identically zero **then**
    compute the univariate factors in u of $\eta_1(t,s,u)$, and the corresponding functions $u = \psi_1^{(j)}(t,s)$
    for each function $\psi_1^{(j)}(t,s)$ **do**
        substitute $u = \psi^{(j)}(t,s)$ into ξ_1, ξ_2
        compute the gcd of the polynomials, after the above substitution
        compute the univariate factors in v of the gcd
        compute $v = \psi_1^{(j)}(t,s)$ by solving for v in each univariate factor
    end for
else
    compute the resultant $\eta_2(t,s,v) = \mathrm{Res}_u(\xi_1, \xi_2)$
    if η_2 is not identically zero **then**
        proceed as before, with η_2
    end if
end if
if η_1, η_2 are both identically zero **then**
    return `method fails: zero resultants`
end if
for each couple $\psi^{(j)}(t,s) = (\psi_1^{(j)}(t,s), \psi_2^{(j)}(t,s))$ computed in the previous process **do**
    compute the symmetry f from $f \circ \boldsymbol{x} = \boldsymbol{x} \circ \psi^{(j)}$. and **return** f
end for

In these cases the algorithm fails, and an alternative must be used. Some of these cases, namely developable and minimal surfaces, will be studied later, developable surfaces in the next section, minimal surfaces in Chapter 9.

Example 18. *Consider the toy example of the ellipsoid parametrized by*

$$\boldsymbol{x}(t,s)=\left(\frac{2(-s^2-t^2+1)}{s^2+t^2+1},\frac{-2t}{s^2+t^2+1},\frac{8s}{s^2+t^2+1}\right).$$

In this case, Algorithm 4 provides 8 symmetries, including the trivial symmetry (the identity) corresponding to the Cremona transformations:

$$\left(\frac{t}{t^2+s^2},\frac{-s}{t^2+s^2}\right),\quad \left(\frac{t}{t^2+s^2},\frac{s}{t^2+s^2}\right),$$
$$\left(\frac{-t}{t^2+s^2},\frac{-s}{t^2+s^2}\right),\quad \left(\frac{-t}{t^2+s^2},\frac{s}{t^2+s^2}\right),$$

and

$$(-t,-s),\ (-t,s),\ (t,-s),\ (t,s).$$

For instance, the symmetry corresponding to the first Cremona transformation, $\left(\frac{t}{t^2+s^2},\frac{-s}{t^2+s^2}\right)$*, is* $f(\boldsymbol{x})=A\boldsymbol{x}$ *with*

$$A=\begin{bmatrix}-1&0&0\\0&1&0\\0&0&-1\end{bmatrix}.$$

7.3 BACKGROUND ON RULED SURFACES

We say that $S\subset\mathbb{R}^3$ is *ruled* if S is covered by lines, called the *rulings* of the surface. These surfaces are well-known in Differential Geometry; one can check classical texts like [53, 65, 105, 120] for further reading on the topic and also for several facts that we will be using in this section.

If S is rational, then we can always find [103] a rational parametrization of the form

$$\boldsymbol{x}(t,s)=\boldsymbol{u}(t)+s\boldsymbol{v}(t),\tag{7.20}$$

where $\boldsymbol{u}(t),\boldsymbol{v}(t)$ parametrize space rational curves. The curve defined by $\boldsymbol{u}(t)$ is called the *directrix*. If S is a real surface, i.e., if S

contains a 2-dimensional set of real points, we cannot always guarantee that $\boldsymbol{u}(t), \boldsymbol{v}(t)$ are real, though, and a good example is that of quadrics, all of which are ruled surfaces (some with real rulings, some with complex rulings). We refer to Eq. (7.20) as a *standard parametrization* of S, and we will assume the ruled surface S we work with to be parametrized in this way; notice that, regardless of potential reparametrizations of $\boldsymbol{u}(t), \boldsymbol{v}(t)$, a standard parametrization of S is not necessarily unique, since different curves may serve as directrices of the surface. Eq. (7.20) shows that a ruled surface is indeed covered by a family of lines, where $\boldsymbol{v}(t)$ provides the direction of the ruling through each point $\boldsymbol{u}(t)$ of the directrix. There are ruled surfaces covered by two different families of curves, called *doubly ruled* surfaces; however, the only doubly ruled surfaces are the plane, the hyperbolic paraboloid, and the single-sheeted hyperboloid, see [72]. The only triply ruled surface is the plane.

For instance, the single-sheeted hyperboloid admits the standard parametrizations

$$\boldsymbol{x}_{\pm}(t,s) = \left(\frac{1-t^2}{1+t^2} \mp s\frac{2t}{1+t^2}, \frac{2t}{1+t^2} \pm s\frac{1-t^2}{1+t^2}, s\right). \tag{7.21}$$

The directrix curve is the unit circle, in this case. In Fig. 7.1, we show two portions of the single-sheeted hyperboloid, with the directrix curve in red, and some rulings, in yellow, of each family; at the left we have some rulings of the family corresponding to $\boldsymbol{x}_+(t,s)$, and at the right, some rulings of the family corresponding to $\boldsymbol{x}_-(t,s)$.

A special type of ruled surfaces, called *developable surfaces*, is classical in Differential Geometry, and widely used in applications. Developable surfaces are the ruled surfaces S that are *isometric* to the plane, meaning that there exists a dipheomorphism $\phi : S \to \mathbb{R}^2$ where the differential $d\phi$ preserves the inner product between the vectors of the tangent plane of S at each point. Intuitively speaking, developable surfaces are the surfaces that can be "unfolded" onto the plane, so that we can find a dipheomorphic mapping $\phi : S \to \mathbb{R}^2$ preserving distances.

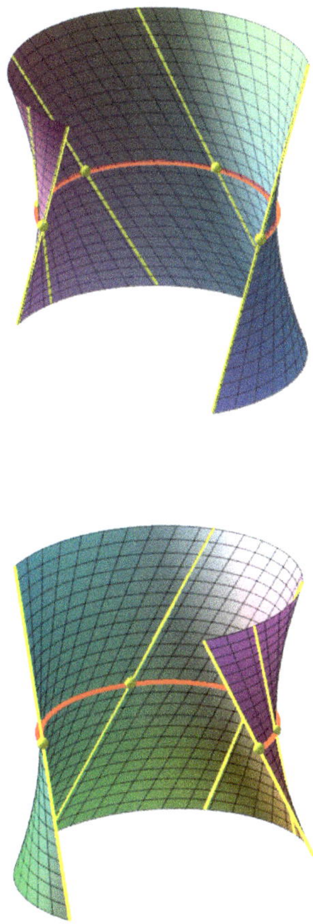

FIGURE 7.1 The single-sheeted hyperboloid; the directrix curve in red, and some rulings of the two families of rulings in yellow

Developable surfaces are the ruled surfaces with vanishing Gauss curvature. Also, if S is parametrized by Eq. (7.20), then S is developable if and only if the mixed product $[\boldsymbol{u}'(t), \boldsymbol{v}(t), \boldsymbol{v}'(t)]$ is identically zero. This allows us to classify developable surfaces in three types (the first two already appeared in Chapter 6, though):

(i) *Cylindrical surfaces*: ruled surfaces where all the rulings are parallel to a same vector. Thus, they can be parametrized as

$$\boldsymbol{x}(t, s) = \boldsymbol{u}(t) + s\boldsymbol{v}_0, \tag{7.22}$$

where $\boldsymbol{v}_0$ is a constant vector. Recall that these surfaces already appeared in Chapter 6.

(ii) *Conical surfaces*: these are ruled surfaces where all the rulings intersect at a same point, called the *vertex* of the surface. By applying if necessary a translation so that the vertex coincides with the origin, one can always parametrize these surfaces as

$$\boldsymbol{x}(t,s) = s\boldsymbol{v}(t). \tag{7.23}$$

(iii) *Tangential surfaces*: ruled surfaces consisting of the union of tangent lines to the directrix, which can be parametrized as

$$\boldsymbol{x}(t,s) = \boldsymbol{u}(t) + s\boldsymbol{u}'(t). \tag{7.24}$$

Cylindrical surfaces and conical surfaces can be recognized using the algorithm in [4]; furthermore, in those cases we can find the constant direction of the rulings, for cylindrical surfaces, and the vertex, for conical surfaces.

It is also useful to recall the notion of *line of striction* (see for instance Section 3-5 of [53] for further information on this notion, and its properties). Given a ruled surface S, parametrized as in Eq. (7.20), the line of striction is a parametric space curve $\boldsymbol{p}(t)$, contained in S, such that $\boldsymbol{p}'(t)$ is orthogonal to $\boldsymbol{v}'(t)$. The expression for the line of striction is

$$\boldsymbol{p}(t) = \boldsymbol{u}(t) - \frac{(\boldsymbol{v}(t)\times\boldsymbol{v}'(t))\cdot(\boldsymbol{v}(t)\times\boldsymbol{u}'(t))}{\|\boldsymbol{v}(t)\times\boldsymbol{v}'(t)\|^2}\boldsymbol{v}(t), \tag{7.25}$$

and can be proven to be independent of the directrix considered in Eq. (7.20). In particular, if Eq. (7.20) is rational, then $\boldsymbol{p}(t)$ is also rational. This expression does not make sense for cylindrical surfaces; also, for conical surfaces the line of striction degenerates into a point. Excluding cylindrical surfaces, we can always assume that S is parametrized as in Eq. (7.20), where the directrix is the line of striction. One can check that for tangential developable surfaces the directrix appearing in Eq. (7.24) is already the line of striction.

Finally, we recall *Plücker coordinates*, which will appear in the next section. Essentially, Plücker coordinates provide a means to see a space line as a point on a geometric object living in higher dimension; the dimension being higher seems a drawback, but the fact that the line, somehow, can be seen as a "point" seems advantageous.

Definition 151. *Let $\ell \in \mathbb{C}^3$ be a line, let $P \in \ell$ be a point on the line, and let $\boldsymbol{v} \in \mathbb{C}^3$ be a vector parallel to ℓ. The* Plücker coordinates *of ℓ are $(\boldsymbol{v}, \boldsymbol{w})$, where $\boldsymbol{w} = P \times \boldsymbol{v}$.*

Plücker coordinates do not depend on the choice of the point $P \in \ell$: indeed, if $P' \in \ell$, $P' \neq P$, then

$$P' \times \boldsymbol{v} = (P + \vec{PP'}) \times \boldsymbol{v} = (P \times \boldsymbol{v}) + (\vec{PP'} \times \boldsymbol{v}),$$

which equals $P \times \boldsymbol{v}$ because $\vec{PP'}, \boldsymbol{v}$ are parallel and therefore $\vec{PP'} \times \boldsymbol{v} = 0$. Furthermore, if we pick $\lambda \boldsymbol{v}$, $\lambda \neq 0$, instead of $\boldsymbol{v}$, we reach $\lambda(\boldsymbol{v}, \boldsymbol{w})$, which shows that Plücker coordinates have a projective structure. Finally, notice that $\boldsymbol{v}, \boldsymbol{w}$ are not independent, since $\boldsymbol{v} \cdot \boldsymbol{w} = 0$. The equation $\boldsymbol{v} \cdot \boldsymbol{w} = 0$ corresponds to the equation, in homogeneous coordinates, of a quadric in $\mathbb{P}^5(\mathbb{C})$, called the *Plücker quadric*, or the *Klein quadric*. Thus, a line in $\mathbb{C}^3$ can be identified with a point in $\mathbb{P}^5(\mathbb{C})$ belonging to the Plücker quadric.

If we consider now the parametrization in Eq. (7.20), we observe that a generic ruling of S has Plücker coordinates

$$\boldsymbol{z}(t) = (\boldsymbol{v}(t), \boldsymbol{w}(t)), \;\; \boldsymbol{w}(t) = \boldsymbol{u}(t) \times \boldsymbol{v}(t). \tag{7.26}$$

The expression in Eq. (7.26) defines a rational curve in the Plücker quadric, which we refer to as the *associated curve* of the ruled surface. Thus, every rational ruled surface parametrized in standard form is represented by a rational curve (although in higher dimension); since curves, even in higher dimension, are much more tractable than surfaces, this opens up the possibility of computing with curves instead of surfaces, when it comes to ruled surfaces.

7.4 COMPUTATION OF SYMMETRIES OF RULED SURFACES

Let us see that a simple strategy to compute the symmetries of a ruled surface S is to use the line of striction. Since for cylindrical and conical surfaces the line of striction is not defined and degenerate, respectively, we will exclude these surfaces here; we will consider these two special cases later. We suppose that S is parametrized as in Eq. (7.20), where, from the previous section, we can assume the directrix $\boldsymbol{u}(t)$ to be the line of striction. Now if $f(\boldsymbol{x}) = A\boldsymbol{x} + \boldsymbol{b}$ is a symmetry of S, then

$$\tilde{\boldsymbol{x}}(t,s) = (f \circ \boldsymbol{x})(t,s) = A\boldsymbol{x}(t,s) + \boldsymbol{b} = \underbrace{(A\boldsymbol{u}(t) + \boldsymbol{b})}_{\tilde{\boldsymbol{u}}(t)} + s\,\underbrace{A\boldsymbol{v}(t)}_{\tilde{\boldsymbol{v}}(t)} \tag{7.27}$$

also parametrizes S. Then we have the following theorem.

Theorem 152. *Let $f(\boldsymbol{x}) = A\boldsymbol{x} + \boldsymbol{b}$ be a symmetry of a ruled surface S, and let $\mathcal{D}$ be the line of striction of S. Then f is also a symmetry of $\mathcal{D}$.*

Proof. Let us see that the parametrization $\tilde{\boldsymbol{u}}(t)$ in Eq. (7.27), which parametrizes the image of $\mathcal{D}$ under f, is also a parametrization of $\mathcal{D}$. Indeed,

$$\tilde{\boldsymbol{u}}'(t) \cdot \tilde{\boldsymbol{v}}'(t) = (A\boldsymbol{u}'(t)) \cdot (A\boldsymbol{v}'(t)).$$

Since A is an orthogonal matrix, $(A\boldsymbol{u}'(t)) \cdot (A\boldsymbol{v}'(t)) = \boldsymbol{u}'(t) \cdot \boldsymbol{v}'(t) = 0$. Thus, $\tilde{\boldsymbol{u}}(t)$ satisfies the condition defining the line of striction, so $\tilde{\boldsymbol{u}}(t)$ also parametrizes $\mathcal{D}$. Thus $f(\mathcal{D}) = \mathcal{D}$, and the result follows. □

The symmetries of the line of striction can be computed using the results in Chapter 5 or Chapter 3, if the line of striction is planar. Thus, whenever the line of striction has finitely many symmetries, we just need to test which ones are also symmetries of S; we can do this with a probabilistic algorithm by checking whether or not a random point $\boldsymbol{x}(t_0, s_0) \in S$ satisfies that $f(\boldsymbol{x}(t_0, s_0)) \in S$. This is not enough, though, in the cases when the line of striction is either a line or a circle, since in those cases we have infinitely many symmetries to check. For those cases we refer to the general method in the previous section.

However, there are other approaches for ruled surfaces, developed in [17] and [29], that we will briefly review. The topic treated in [17] is the computation of the affine equivalences between two ruled surfaces, so the results in [17] are applicable to our problem. In [29] projective equivalences between ruled surfaces, among other types of surfaces, are considered. Let us summarize the ideas in each case, starting with [17]. In [17] it is required that Eq. (7.20) is a proper parametrization of a ruled surface not doubly ruled, and that $\boldsymbol{v}(t)$ is polynomial, so the components of $\boldsymbol{v}(t)$ have constant denominators; this condition can be always satisfied, without affecting the properness of the parametrization, by replacing the original $\boldsymbol{v}(t)$ by $\mu(t)\boldsymbol{v}(t)$ if necessary, where $\mu(t)$ is the common denominator of the components of $\boldsymbol{v}(t)$. Notice that there is no danger in doing so, since the $\mu(t)\boldsymbol{v}(t)$ gives the same direction for the rulings, so we are just reparametrizing the surface. Under this assumption, the form of the mapping ψ in Theorem 142 can be guessed, see the next result; we refer to Section 3 of [17] for a proof.

Theorem 153. *Let S_1, S_2 be two ruled rational surfaces properly parametrized by $\boldsymbol{x}_i(t,s) = \boldsymbol{u}_i(t) + s\boldsymbol{v}_i(t)$, $i = 1,2$ as in Eq. (142), not doubly ruled, where $\boldsymbol{v}_i(t)$ is polynomial for $i = 1,2$. Let n be the maximum of the degrees of the (polynomial) components of $\boldsymbol{v}_i(t)$. Let $f : \mathbb{R}^3 \to \mathbb{R}^3$ be an affine mapping taking S_1 onto S_2. Then the mapping ψ in Theorem 142 has the form*

$$\psi(t,s) = (\psi_1(t,s), \psi_2(t,s)) = \left(\frac{\alpha t + \beta}{\gamma t + \delta}, k(\gamma t + \delta)^n s + c(t)\right), \quad (7.28)$$

where $\alpha\delta - \beta\gamma \neq 0$, $k \neq 0$, and $c(t)$ is a rational function.

The shape of the first component of the mapping $\psi(t,s)$ in Eq. (7.28), again a Möbius transformation, has an intuitive explanation: since an affine mapping maps lines to lines, the curves $t = t_a$ in the parameter space, with t_a a constant, must be mapped onto curves $t = t_b$, with t_b a constant. Thus, the first component ψ_1 of ψ satisfies that $\psi_1(t_a, s) = t_b$ for t_a, t_b constant, and for all s, so ψ_1

must depend only on t; since the $\boldsymbol{x}_i$ are proper parametrizations, and therefore birational, ψ_1 must also be birational, and because ψ_1 depends only on t, ψ_1 must be a birational mapping of the line, therefore a Möbius transformation. Deducing the shape of ψ_2 is more elaborate and uses the (technical) condition that the $\boldsymbol{v}_i(t)$ are polynomial. To find the affine equivalences between two ruled surfaces, one substitutes Eq. (7.28) into Eq. (7.1); this yields a polynomial system, whose structure can be exploited, see Section 4 of [17] for a detailed explanation.

Remark 14. *Whenever the parametrization $\boldsymbol{x}(t,s)$ of the ruled surface is in standard form and $\boldsymbol{x}(t,s)$ is proper, i.e., leaving other hypotheses in Theorem 153 aside, by the same argument behind Theorem 153 the first component $\psi_1(t,s)$ of the Cremona transformation $\psi(t,s)$ associated with a symmetry of the surface must be a univariate Möbius transformation in the variable t.*

The approach in [29] uses Plücker coordinates, instead. Applied to our problem, the main idea is that Euclidean symmetries of $\mathbb{R}^3$ (in fact, projective isomorphisms of $\mathbb{P}^3(\mathbb{R})$) give rise to projective isomorphisms of the Plücker quadric, see Theorem 2.1.10 in [105]. Thus, a symmetry $f : \mathbb{R}^3 \to \mathbb{R}^3$ of a ruled surface S, parametrized as in Eq. (7.20), gives rise to a projective isomorphism preserving the Plücker quadric, leaving the associated curve in the Plücker quadric, parametrized by $\boldsymbol{z}(t)$ as in Eq. (7.26), invariant. Projective isomorphisms between two rational curves in n-space can be computed with the techniques of [29] or [69], where rational curves in whatever dimension are also discussed. Thus, after finding the projective isomorphisms leaving $\boldsymbol{z}(t)$ invariant, we just need to check which ones leave the Plücker quadric invariant, and correspond to Euclidean symmetries. The Plücker quadric is defined by a 6×6 symmetric real matric Q, and satisfies the equation $\boldsymbol{x}^T Q \boldsymbol{x} = 0$. Thus, a projective isomorphism $f(\mathbf{x}) = M\mathbf{x}$ of $\mathbb{P}^5(\mathbb{R})$, with M a 6×6 non-singular matrix and $\mathbf{x} \in \mathbb{P}^5(\mathbb{R})$, leaves this quadric invariant whenever $M^T Q M = \lambda Q$, with $\lambda \neq 0$.

7.4.1 Developable surfaces

Since developable surfaces have zero Gauss curvature, the general algorithm presented before is not applicable to them. Also, for two of the subfamilies of developable surfaces, cylindrical and conical surfaces, using the line of striction is not an option, either.

In Section 6.2, we already observed that the symmetries of cylindrical essentially follow from the symmetries of a normal section of the surface, i.e., the intersection of the surface with a plane Π normal to the direction $\mathbf{v}$ of the rulings. Because Eq. (7.20) is linear in s, the intersection of S with a plane Π normal to the rulings is a rational planar curve, so its symmetries can be computed using the results in Chapter 3.

For conical surfaces, we observe that any symmetry $f(\boldsymbol{x}) = A\boldsymbol{x} + \boldsymbol{b}$ must leave the vertex of the surface invariant, so assuming a parametrization as in Eq. (7.23), with the vertex at the origin, $\boldsymbol{b} = 0$, and therefore $f(\boldsymbol{x}) = A\boldsymbol{x}$. Then we can use Theorem 153: because the parametrization is that in Eq. (7.23), and taking $\boldsymbol{v}(t)$ to be polynomial, the function $c(t)$ in Eq. (7.28) must be identically zero, and thus Theorem 142 leads to the condition

$$A\boldsymbol{v}(t) = k(\gamma t + \delta)^n \boldsymbol{v}\left(\frac{\alpha t + \beta}{\gamma t + \delta}\right).$$

This provides a polynomial system in the entries of A and the parameters $k, \alpha, \beta, \gamma, \delta$ that can be solved taking advantage of its special structure; see again [17] for details. Notice that every symmetry of the curve defined by $\boldsymbol{v}(t)$ is a symmetry of the conical surface, but the converse is not true: think for instance of a cone of revolution, with $\boldsymbol{v}(t)$ parametrizing an ellipse which is the intersection of the cone with a plane not normal to the axis of revolution.

Finally, in the case of tangential developable surfaces Theorem 152 works without any problem. However, we can go a bit further.

Theorem 154. *Let S be a rational tangential developable surface. Then $f : \mathbb{R}^3 \to \mathbb{R}^3$ is a symmetry of S if and only if f is a symmetry of its line of striction.*

Proof. The implication ($\Rightarrow$) is Theorem 152, so let us address ($\Leftarrow$). Thus, let $f : \mathbb{R}^3 \to \mathbb{R}^3$, $f(\boldsymbol{x}) = A\boldsymbol{x}+\boldsymbol{b}$, be a symmetry of the line of striction of S; let $\boldsymbol{u}(t)$ be a rational parametrization of the line of striction, and let us assume without loss of generality that $\boldsymbol{u}(t)$ is proper. By Theorem 45, $A\boldsymbol{u}(t)+\boldsymbol{b} = (\boldsymbol{u}\circ\varphi)(t)$, with $\varphi(t)$ a Möbius transformation. Considering the parametrization of S provided by Eq. (7.24),

$$(f\circ\boldsymbol{x})(t,s) = (A\boldsymbol{u}(t)+\boldsymbol{b})+sA\boldsymbol{u}'(t) = (\boldsymbol{u}\circ\varphi)(t)+sA\boldsymbol{u}'(t). \quad (7.29)$$

Differentiating $A\boldsymbol{u}(t) + \boldsymbol{b} = (\boldsymbol{u} \circ \varphi)(t)$ and using the Chain Rule, we deduce that $A\boldsymbol{u}'(t) = \boldsymbol{u}'(\varphi(t))\varphi'(t)$. Substituting this into Eq. (7.29), we obtain

$$(f \circ \boldsymbol{x})(t,s) = \boldsymbol{u}(\varphi(t)) + s\boldsymbol{u}'(\varphi(t))\varphi'(t). \quad (7.30)$$

Since $\boldsymbol{u}'(\varphi(t))\varphi'(t)$ and $\boldsymbol{u}'(\varphi(t))$ are proportional, we deduce that $(f\circ\boldsymbol{x})(t,s)$ parametrizes a ruled surface which is also parametrized by

$$\tilde{\boldsymbol{x}}(t,s) = \boldsymbol{u}(\varphi(t)) + s\boldsymbol{u}'(\varphi(t)).$$

However, $\tilde{\boldsymbol{x}} = \boldsymbol{x} \circ \psi$, where $\psi = \psi(t,s) = (\varphi(t),s)$, so $\tilde{\boldsymbol{x}}$ is just a reparametrization of S. Thus, $\tilde{x}$ and therefore $f \circ \boldsymbol{x}$ parametrize S, which implies that $f(S) = S$ and therefore f is a symmetry of S. □

Observe that for tangential surfaces the line of striction $\boldsymbol{u}(t)$ cannot be planar, since in that case the surface S is a plane, which is a trivial case. Thus, the number of symmetries of $\boldsymbol{u}(t)$ must be finite.

7.5 FINAL REMARKS AND OPEN PROBLEMS

Although we have provided a general algorithm for computing the symmetries of a properly parametrized rational surface, this algorithm is not always efficient. The reason is that the algorithm in Section 7.2 requires to compute a resultant between two polynomials in four variables, which can be costly as the degree of the parametrization grows. Also, the polynomial obtained after that, which we need to factor, might be big. So there seems to be, still, space for improvement. The approach presented in [75] is general, but this approach is far from simple. Thus, it would be desirable to find alternative, simple, more efficient algorithms to solve the problem that could be accessed by a wider community.

All in all, we suggest the following open problems for this chapter.

Open Problem 20. *Find alternative, simple, efficient algorithms to compute the symmetries of rational surfaces, as general as possible.* A possibility might be to separately address different types of symmetry; a first distinction between rotational symmetries and involutions might help.

Open Problem 21. *Compute approximate symmetries of rational surfaces.* Approximate problems also appear here. In particular, the fact that the rational ruled surfaces correspond to rational curves in the Plücker quadric shows that solving approximate problems for rational curves in n-space might lead to solving approximate problems for ruled surfaces as well.

Open Problem 22. *Characterize rational surfaces associated with families of subgroups of the Cremona group.* The group of Cremona transformations associated with the symmetries of a rational surface must be a subgroup of the Cremona group. And whenever the symmetry group of the surface is finite, such subgroup must also be finite. Thus, one can wonder about the properties, general form, etc. of the rational surfaces whose symmetry groups correspond to certain subgroups of the Cremona group, e.g. the projective linear transformations, the affine linear transformations, or *de Jonquieres* transformations [54],

$$\left(\frac{at+b}{ct+d}, \frac{sr_1(t)+r_2(t)}{sr_3(t)+r_4(t)}\right), \quad ad-bc\neq 0, \;\; r_1(t)r_4(t)-r_2(t)r_3(t)\neq 0,$$

to quote a few.

Open Problem 23. *Rational invariants; affine and projective equivalence.* The notion of rational invariant introduced in Section 5.6 can be extended to rational surfaces as well. The question of finding rational invariants for surfaces that can solve the more general problems of checking equivalence with respect to other groups is challenging. Notice that one needs not only a good behavior of the invariant with respect to the group of interest (that is what makes it an invariant), but also with respect to Cremona transformations. A possibility here is to restrict oneself to cases where the form of the Cremona transformation can be guessed in advance.

CHAPTER 8

Rational surfaces (II): rational canal surfaces

Canal surfaces are well-known in Computer Aided Geometric Design. These surfaces are the envelopes of families of spheres whose centers lie on a space curve, called the *spine* curve. If the radii of these spheres is constant, then we have the subfamily of *pipe surfaces*, also called *tubular surfaces*, which are used, for instance, to give thickness to a 3D object following the shape of a space curve. In this chapter we address the symmetries of canal surfaces and also the symmetries of a very special subfamily, *Dupin cyclides*, well-known in Geometry since the XIXth century and that in recent decades have acquired some relevance in Computer Aided Geometric Design and Architecture. For these surfaces, we can reduce the study and computation of its symmetries to space curves, which allows us to find these symmetries efficiently. Here we will be following the ideas and presentation in [3].

8.1 BACKGROUND ON CANAL SURFACES

Given a 1-parameter family of surfaces, implicitly defined by $F(\boldsymbol{x}, t) = 0$, where t is the parameter, the *envelope* $\mathcal{E}$ of the family (see for instance Chapter V of [120]) is the surface tangent to all the surfaces of the family. The implicit equation of $\mathcal{E}$ can be found by eliminating the parameter t from the system

$$F(\boldsymbol{x}, t) = 0, \;\; \dot{F}(\boldsymbol{x}, t) = 0,$$

DOI: 10.1201/9781003648710-8

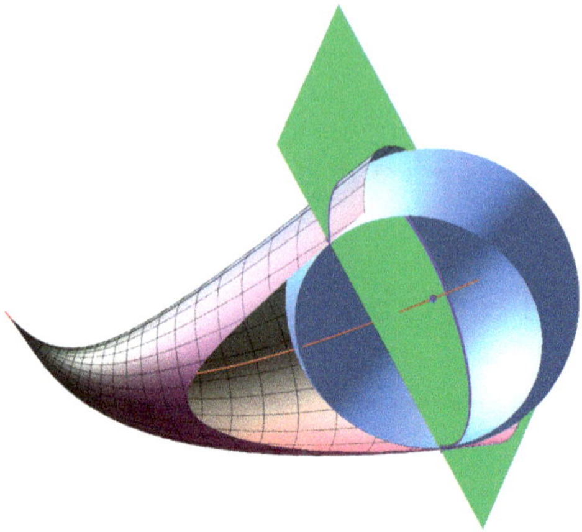

FIGURE 8.1 A piece of a canal surface and a corresponding spine curve segment (in red), with, for a given parameter t_0, the plane $\dot{\Sigma}(t_0)$ and the sphere $\Sigma(t_0)$ with center $\mathbf{c}(t_0)$ and radius $r(t_0)$, intersecting in the characteristic circle $k(t_0)$.

where $\dot{F}$ represents the derivative with respect to the parameter t; in this sense, the above equations define the envelope $\mathcal{E}$. For a fixed t_0, $\{F(\boldsymbol{x}, t_0) = 0, \dot{F}(\boldsymbol{x}, t_0) = 0\}$ represents a space curve $\mathcal{C}_{t_0}$, called a *characteristic* curve, contained in $\mathcal{E}$. Notice that $\mathcal{E}$ can be seen as the union of all its characteristic curves.

A *canal surface* $S = S_{\mathbf{c},r} \subset \mathbb{R}^3$ is the envelope of a 1-parameter family $\Sigma = \Sigma_{\mathbf{c},r}$ of spheres, centered at a *spine curve* $\mathbf{c}$ and with *radius function* r. The defining equations for S are

$$\Sigma_{\mathbf{c},r}(t) : \|\boldsymbol{x} - \mathbf{c}(t)\|^2 - r^2(t) = 0, \tag{8.1a}$$

$$\dot{\Sigma}_{\mathbf{c},r}(t) : \langle \boldsymbol{x} - \mathbf{c}(t), \dot{\mathbf{c}}(t) \rangle + r(t)\dot{r}(t) = 0, \tag{8.1b}$$

where t is the parameter of the family. For a fixed t, the first equation is a sphere and the second is a plane; thus the characteristic curve is a circle $k(t) = k_{\mathbf{c},r}(t) := \Sigma_{\mathbf{c},r}(t) \cap \dot{\Sigma}_{\mathbf{c},r}(t)$. Fig. 8.1 illustrates all these notions; this figure and other figures in this chapter are taken from [19], whose ideas we will be following here.

It follows from Eq. (8.1) that $k(t)$ is nonempty (over the real numbers) precisely when $\|\dot{\mathbf{c}}(t)\|^2 \geq \dot{r}(t)^2$, degenerating to a single point if equality holds at t, and to a curve plus the 1-parameter family of tangent planes $\dot{\Sigma}(t)$ if equality holds in a nonempty interval [104, §2]. To exclude these degenerate cases, we will assume that $\|\dot{\mathbf{c}}(t)\|^2 > \dot{r}(t)^2$ holds for all parameters t for which $\mathbf{c}(t)$ and $r(t)$ are defined, therefore guaranteeing that Eq. (8.1) defines a real surface S. For these statements and other general results on canal surfaces that we recall in this section, we refer the interested reader to [47, 48, 79, 104]. To this list we can also add [130], where we can find an algorithm to recognize whether or not a given implicit surface is a rational canal surface; the algorithm in [130] also allows us to compute the spine curve and the radius function.

Remark 15. *The family Σ of spheres can be identified with a curve in the 4-dimensional Minkowski space $\mathbb{R}^{3,1}$, where the point $\big(\mathbf{c}(t); r(t)\big)$ corresponds to a sphere $\Sigma(t)$ with center $\mathbf{c}(t)$ and radius $r(t)$. The sign of the radius gives us the* orientation *of the sphere (toward the center when $r(t) > 0$, and outwards when $r(t) < 0$), which is inherited by the canal surface. Thus an* oriented *canal surface corresponds to a curve $\big(\mathbf{c}; r\big)$ in $\mathbb{R}^{3,1}$.*

The canal surface S admits parametrizations of the form

$$\boldsymbol{x}(t,s) = \boldsymbol{x}_{\mathbf{c},r}(t,s) := \mathbf{c}(t) + r(t)\boldsymbol{N}(t,s), \;\; \boldsymbol{N}(t,s) = \frac{\boldsymbol{x}_t \times \boldsymbol{x}_s}{\|\boldsymbol{x}_t \times \boldsymbol{x}_s\|}. \tag{8.2}$$

Note, and we will use this observation afterwards, that if we replace r by $-r$, Eq. (8.1) is unaffected and also Eq. (8.2); this last fact follows because the change in sign of the radius results in a change of orientation of the canal surface, i.e., a reversal of the direction of the unit normal vector field described by $\boldsymbol{N}(t,s)$, so the sign of the second term of the parametrization in Eq. (8.2) remains positive.

Replacing $\boldsymbol{x}$ in Eq. (8.1) by Eq. (8.2) and factoring out $r(t)$ yields

$$\|\boldsymbol{N}(t,s)\|^2 = 1, \tag{8.3a}$$

$$\langle \boldsymbol{N}(t,s), \dot{\mathbf{c}}(t)\rangle + \dot{r}(t) = 0. \tag{8.3b}$$

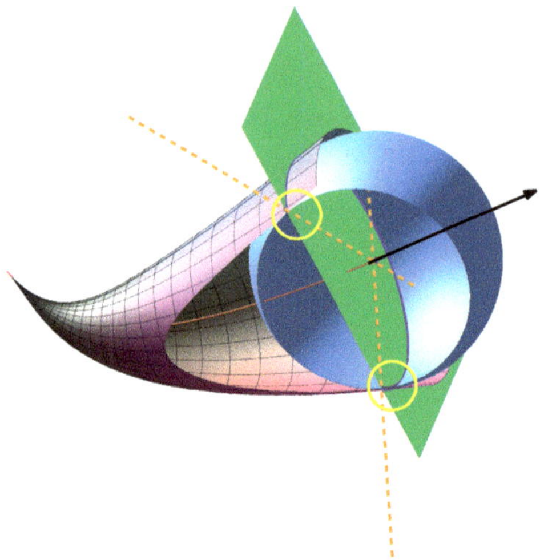

FIGURE 8.2 Surface normals (in orange) at two points (circled in yellow) of a characteristic circle intersecting the spine curve at the center of the characteristic circle; vector $\dot{\mathbf{c}}(t)$ (in black) at the center of the characteristic circle normal to the plane containing the characteristic circle.

Also, for each characteristic circle $k(t)$ the normal lines to S along the circle intersect at the point $\mathbf{c}(t)$ on the spine curve, and the derivative $\dot{\mathbf{c}}(t)$ at this point is normal to the plane containing the characteristic circle; see Fig. 8.2. In the case when $r(t)$ is a constant we have a special type of canal surface, called a *pipe*, or *tubular*, surface. Furthermore, if $\mathbf{c}$ is a line, then S is a surface of revolution.

We will be assuming that $\mathbf{c}$ and r are real and rational, with $\mathbf{c}$ properly parametrized. If $\mathbf{c}, r$ are rational, finding a rational parametrization of type (8.2) is possible, and reduces to finding a rational parametrization of $\boldsymbol{N}$. This was first accomplished in [104], and minimal degree rational parametrizations were obtained in [79]. Thus, if $\mathbf{c}, r$ are rational, then S is also rational. Let us also mention [26] that in fact a canal surface is rational if and only if $\mathbf{c}, r^2$ are rational, so we can also have rational canal surfaces where $r(t)$ is the square-root of a rational function.

One can wonder if $\mathbf{c}$ and r are unique for a given canal surface S, or if there are surfaces that are canal surfaces in at least two different ways. This question was answered by Maxwell [88], who characterized the surfaces that are the envelope of (exactly two) different 1-parameter families of spheres centered at distinct spine curves. These surfaces are the *Dupin cyclides*, which had been introduced years before, and named *cyclides*, by Charles Pierre Dupin as surfaces different from the sphere whose lines of curvature (see the previous chapter, when we introduced principal curvatures) are circles. There are other alternative definitions of Dupin cyclides, related to their rich mathematical properties; a good account of definitions and properties appears in [40].

Maxwell [88] also showed that the spine curves of Dupin cyclides must be conics lying in perpendicular planes and passing through each other's foci, yielding three different cases corresponding to the nature of the spine curves. For each of these cases (Type I, Type II, Type III) one can provide [51, §23.2] a canonical description of the spine curves $\mathbf{c}_1, \mathbf{c}_2$, radius functions r_1, r_2 and implicit forms, shown in Table 8.1. This form depends on certain parameters a, b, c, ξ, g satisfying that $\xi^2 = a^2 - b^2$. Notice that $a, c \neq 0$ in Type I, $a, b \neq 0$ and $\xi > 0$ in Type II, and $g \neq 0$ in Type III. Note also that Type II degenerates to Type I in the limit $\xi \to 0$. We say that a Dupin cyclide is in *canonical form*, if it is the zeroset of F^{I}, F^{II} or F^{III} in Table 8.1 for certain parameters a, b, c, ξ, g satisfying the above constraints. The implicit forms in Table 8.1 show that the Dupin cyclides of Types I and II are quartics, while the Dupin cyclides of Type III are cubics. Furthermore, the Dupin cyclides of Type I are tori, and these surfaces are surfaces of revolution, since the characteristic circles are rotationally invariant about the axis traced out by $\mathbf{c}_2$.

8.2 COMPUTATION OF SYMMETRIES

Here we consider the computation of the symmetries of a canal surface S that is not a Dupin cyclide, so S has a unique spine curve $\mathbf{c}$; Dupin cyclides will be addressed in the next section. We will see that such symmetries are, essentially, the symmetries of

TABLE 8.1 Parametric and corresponding implicit canonical forms for Dupin cyclides.

I: Quartic	spine curve: circle/line	radius function
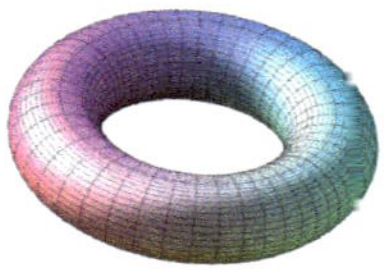	$\mathbf{c}_1^{\mathrm{I}}(t) = a\left(\frac{1-t^2}{1+t^2}, \frac{2t}{1+t^2}, 0\right)$ $\mathbf{c}_2^{\mathrm{I}}(t) = a\left(0, 0, \frac{2t}{1-t^2}\right)$	$r_1^{\mathrm{I}}(t) = c$ $r_2^{\mathrm{I}}(t) = c - a\frac{1+t^2}{1-t^2}$
$0 = F^{\mathrm{I}} := (x^2+y^2+z^2+a^2-c^2)^2 - 4a^2(x^2+y^2)$		
II: Quartic	spine curve: ellipse/hyperbola	radius function
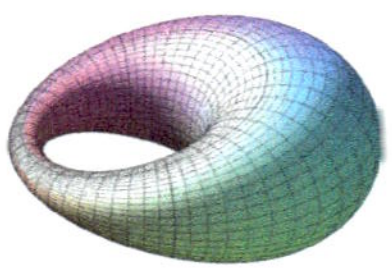	$\mathbf{c}_1^{\mathrm{II}}(t) = \left(\frac{1-t^2}{1+t^2}a, \frac{2t}{1+t^2}b, 0\right)$ $\mathbf{c}_2^{\mathrm{II}}(t) = \left(\frac{1+t^2}{1-t^2}\xi, 0, \frac{2t}{1-t^2}b\right)$	$r_1^{\mathrm{II}}(t) = c - \xi\frac{1-t^2}{1+t^2}$ $r_2^{\mathrm{II}}(t) = c - a\frac{1+t^2}{1-t^2}$
$0 = F^{\mathrm{II}} := (x^2+y^2+z^2+a^2-\xi^2-c^2)^2 - 4(ax-c\xi)^2 - 4y^2(a^2-\xi^2)$		
III: Cubic	spine curve: parabola/parabola	radius function
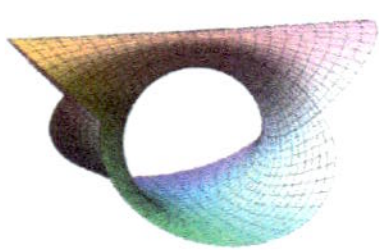	$\mathbf{c}_1^{\mathrm{III}}(t) = g\left(t^2-\frac{1}{2}, 2t, 0\right)$ $\mathbf{c}_2^{\mathrm{III}}(t) = g\left(\frac{1}{2}-t^2, 0, 2t\right)$	$r_1^{\mathrm{III}}(t) = c + g\left(t^2+\frac{1}{2}\right)$ $r_2^{\mathrm{III}}(t) = c - g\left(t^2+\frac{1}{2}\right)$
$0 = F^{\mathrm{III}} := (x+c)(x^2+y^2+z^2) + (y^2-z^2)g - (g^2+c^2)x + (g^2-c^2)c$		

the spine curve that are "compatible" with the radius function, therefore reducing the problem to symmetries of rational space curves. Also, we assume that $\mathbf{c}$ is not a line, since otherwise S is a surface of revolution, whose symmetries have been addressed in Chapter 6. We also assume that $\mathbf{c}, r$ are rational, in which case S is also rational. Recall that every rational surface is irreducible, a fact that we will need later.

Some preliminaries are required. Since $\mathbf{c}$ is not a line, for all but finitely many parameters t, the *unit tangent*, *principal normal*, and *bi-normal* vectors of $\mathbf{c}$ at t,

$$\boldsymbol{t}_{\mathbf{c}} = \boldsymbol{t}_{\mathbf{c}}(t) := \frac{\dot{\mathbf{c}}(t)}{\|\dot{\mathbf{c}}(t)\|}, \quad \boldsymbol{n}_{\mathbf{c}} = \boldsymbol{n}_{\mathbf{c}}(t) := \frac{\dot{\boldsymbol{t}}_{\mathbf{c}}(t)}{\|\dot{\boldsymbol{t}}_{\mathbf{c}}(t)\|},$$
$$\boldsymbol{b}_{\mathbf{c}} = \boldsymbol{b}_{\mathbf{c}}(t) := \boldsymbol{t}_{\mathbf{c}}(t) \times \boldsymbol{n}_{\mathbf{c}}(t)$$

are well defined and together form the *Frenet frame* $(\boldsymbol{t}_{\mathbf{c}}, \boldsymbol{n}_{\mathbf{c}}, \boldsymbol{b}_{\mathbf{c}})$ of the curve at t. We have the following result, that we will use later.

Lemma 155. *Let $f(\boldsymbol{x}) = A\boldsymbol{x} + \boldsymbol{b}$ be an isometry of $\mathbb{R}^3$ and $\mathbf{c}$ a parametrized curve with Frenet frame $(\boldsymbol{t}_{\mathbf{c}}, \boldsymbol{n}_{\mathbf{c}}, \boldsymbol{b}_{\mathbf{c}})$. Then the curve $f \circ \mathbf{c}$ has Frenet frame*

$$(\boldsymbol{t}_{f\circ\mathbf{c}}, \boldsymbol{n}_{f\circ\mathbf{c}}, \boldsymbol{b}_{f\circ\mathbf{c}}) = (A\boldsymbol{t}_{\mathbf{c}}, A\boldsymbol{n}_{\mathbf{c}}, \det(A)A\boldsymbol{b}_{\mathbf{c}}).$$

Proof. One has

$$\boldsymbol{t}_{f\circ\mathbf{c}} = \frac{\mathrm{d}f\circ\mathbf{c}}{\mathrm{d}t} \Big/ \left\|\frac{\mathrm{d}f\circ\mathbf{c}}{\mathrm{d}t}\right\| = A\frac{\mathrm{d}\mathbf{c}}{\mathrm{d}t} \Big/ \left\|A\frac{\mathrm{d}\mathbf{c}}{\mathrm{d}t}\right\| = A\boldsymbol{t}_{\mathbf{c}},$$
$$\boldsymbol{n}_{f\circ\mathbf{c}} = \frac{\mathrm{d}\boldsymbol{t}_{f\circ\mathbf{c}}}{\mathrm{d}t} \Big/ \left\|\frac{\mathrm{d}\boldsymbol{t}_{f\circ\mathbf{c}}}{\mathrm{d}t}\right\| = A\frac{\mathrm{d}\boldsymbol{t}_{\mathbf{c}}}{\mathrm{d}t} \Big/ \left\|A\frac{\mathrm{d}\boldsymbol{t}_{\mathbf{c}}}{\mathrm{d}t}\right\| = A\boldsymbol{n}_{\mathbf{c}},$$
$$\boldsymbol{b}_{f\circ\mathbf{c}} = (A\boldsymbol{t}_{\mathbf{c}}) \times (A\boldsymbol{n}_{\mathbf{c}}) = \det(A)A(\boldsymbol{t}_{\mathbf{c}} \times \boldsymbol{n}_{\mathbf{c}}) = \det(A)A\boldsymbol{b}_{\mathbf{c}},$$

where we used the identity for the cross product in Eq. (3.9). □

Using the Frenet frame of $\mathbf{c}$ we can write the normal vector $\boldsymbol{N}$ in the following (in general nonrational) way; see Eq. (3.12) in

[47]:

$$\boldsymbol{N}_{\mathbf{c},r}(t,s) = \frac{\dot{r}(t)}{\|\dot{\mathbf{c}}(t)\|}\boldsymbol{t}_{\mathbf{c}} + \sqrt{1-\left(\frac{\dot{r}(t)}{\|\dot{\mathbf{c}}(t)\|}\right)^2}\left(\frac{1-s^2}{1+s^2}\boldsymbol{n}_{\mathbf{c}} + \frac{2s}{1+s^2}\boldsymbol{b}_{\mathbf{c}}\right). \tag{8.4}$$

The following result on real rational functions will also be necessary.

Lemma 156. *Let g, h be two real rational functions. If $|g(t)| = |h(t)|$ holds for every $t \in \mathbb{R}$ for which both sides are defined, then either $g = h$ or $g = -h$.*

Proof. Since g, h are real rational functions, there exists an open interval $I \subset \mathbb{R}$ where g, h are both defined and have constant sign. Therefore, for any $t \in I$, $|g(t)| - |h(t)|$ is defined, and $g(t) - h(t) = 0$ or $g(t) + h(t) = 0$ holds identically for $t \in I$. Assume that $g(t) - h(t) = 0$ holds identically for $t \in I$; the second possibility is analogous. Since the restriction of $g - h$ to I is rational and well defined, it is analytic on its domain I. Therefore, if this restriction is zero in I, the Identity Theorem [74] implies that $g = h$. □

Now let us address our problem. We start with a series of lemmas that are also applicable to Dupin cyclides, and that will be also used in the next section. In particular, the following lemma shows that if $f(\boldsymbol{x}) = A\boldsymbol{x} + \boldsymbol{b}$ is an symmetry of S, then $\tilde{\mathbf{c}} := f \circ \mathbf{c}$ is also a spine curve of S; the subsequent lemmas describe its relation to $\mathbf{c}$.

Lemma 157. *Let f be a symmetry of $S_{\mathbf{c},r}$. Then $S_{f\circ\mathbf{c},\tilde{r}} = S_{\mathbf{c},r}$, where either $\tilde{r} = r$ or $\tilde{r} = -r$. In particular, $f \circ \mathbf{c}$ is also a spine curve of $S_{\mathbf{c},r}$.*

Proof. Let $\Sigma_{\mathbf{c},r}$ be the 1-parameter family of spheres (8.1a) corresponding to the pair $(\mathbf{c}, r)$. For every t, the isometry f maps

$$\Sigma_{\mathbf{c},r}(t) \xrightarrow{f} \Sigma_{f\circ\mathbf{c},\tilde{r}}(t), \qquad \dot{\Sigma}_{\mathbf{c},r}(t) \xrightarrow{f} \dot{\Sigma}_{f\circ\mathbf{c},\tilde{r}}(t),$$

where $\tilde{r}^2 = r^2$, so by Lemma 156 we have either $\tilde{r} = r$ or $\tilde{r} = -r$. The envelope of the spheres $\Sigma_{f\circ\mathbf{c},\tilde{r}}(t)$ defines a canal surface $S_{f\circ\mathbf{c},\tilde{r}}$, and f maps characteristic circles of $S_{\mathbf{c},r}$ to characteristic circles of $S_{f\circ\mathbf{c},\tilde{r}}$,

$$k_{\mathbf{c},r}(t) \xrightarrow{f} k_{f\circ\mathbf{c},\tilde{r}}(t).$$

Moreover, since f is a symmetry of $S_{\mathbf{c},r}$, the latter characteristic circles are contained in $S_{\mathbf{c},r}$. Since $S_{\mathbf{c},r}$ is irreducible, it follows that $S_{\mathbf{c},r} = S_{f\circ\mathbf{c},\tilde{r}}$. In particular f maps each spine curve of $S_{\mathbf{c},r}$ to a spine curve of $S_{\mathbf{c},r}$. □

Lemma 158. *The spine curves* $\mathbf{c}$ *and* $\tilde{\mathbf{c}}$ *have identical speed.*

Proof. Since A is orthogonal,

$$\|\dot{\tilde{\mathbf{c}}}(t)\| = \left\|\frac{\mathrm{d}}{\mathrm{d}t}(f\circ\mathbf{c})(t)\right\| = \|A\dot{\mathbf{c}}(t)\| = \|\dot{\mathbf{c}}(t)\|. \tag{8.5}$$

□

Lemma 159. *The spine curves* $\mathbf{c}$ *and* $\tilde{\mathbf{c}}$*, together with the radius function* r*, have surface normal parametrizations* (8.4) *related by*

$$A\boldsymbol{N}_{\mathbf{c},r}(t,s) = \boldsymbol{N}_{\tilde{\mathbf{c}},r}\big(t, \det(A)s\big). \tag{8.6}$$

Proof. The result follows after multiplying Eq. (8.4) by A, using Lemmas 155 and 158. □

Now we can state the characterization theorem for the existence of symmetries of canal surfaces with a single spine curve.

Theorem 160. *Let* $S_{\mathbf{c},r}$ *be a canal surface, not a Dupin cyclide, with non-linear spine curve* $\mathbf{c}$*. The isometry* $f(\boldsymbol{x}) = A\boldsymbol{x} + \boldsymbol{b}$ *is a symmetry of* $S_{\mathbf{c},r}$ *if and only if there exists a Möbius transformation* φ *such that*

C1: the spine curve satisfies that $f\circ\mathbf{c} = \mathbf{c}\circ\varphi$*;*

C2: the radius function satisfies that $r^2 = (r\circ\varphi)^2$*.*

Proof. ($\Rightarrow$) By Lemma 157, f must be a symmetry of the spine curve $\mathbf{c}$. By Theorem 45 in Chapter 3 this is equivalent to the existence of a Möbius transformation φ for which $f \circ \mathbf{c} = \mathbf{c} \circ \varphi$, establishing C1. Using Condition C1, then Lemma 157 with $\tilde{r} = \pm r$, and finally that φ is a birational map on the real line, one obtains

$$S_{\mathbf{c}\circ\varphi,\tilde{r}} = S_{f\circ\mathbf{c},\tilde{r}} = S_{\mathbf{c},r} = S_{\mathbf{c}\circ\varphi,r\circ\varphi}$$

and deduces $r^2 = \tilde{r}^2 = (r \circ \varphi)^2$, establishing C2.

($\Leftarrow$) Let $\boldsymbol{x}_{\mathbf{c},r}(t,s)$ be the parametrization (8.2) of $S_{\mathbf{c},r}$ with normals $\boldsymbol{N}_{\mathbf{c},r}(t,s)$ as in Eq. (8.4). Since the radius function r and $\tilde{r} := r \circ \varphi$ are real rational functions satisfying that $r^2 = \tilde{r}^2$, Lemma 156 implies $r = \pm\tilde{r}$. As a change of sign of the radius function results in a change of orientation of the canal surface and leaves the geometric shape unchanged, we may safely assume that $r = \tilde{r}$. Then using Lemma 159 with $\tilde{\mathbf{c}} := \mathbf{c}\circ\varphi = f\circ\mathbf{c}$, the symmetry f maps any point $\boldsymbol{x}_{\mathbf{c},r}(t,s)$ on $S_{\mathbf{c},r}$ to

$$\begin{aligned} f \circ \boldsymbol{x}_{\mathbf{c},r}(t,s) &= (f \circ \mathbf{c})(t) + r(t)A\boldsymbol{N}_{\mathbf{c},r}(t,s) \\ &= \tilde{\mathbf{c}}(t) + \tilde{r}(t)\boldsymbol{N}_{\tilde{\mathbf{c}},\tilde{r}}\big(t, \det(A)s\big) = \boldsymbol{x}_{\tilde{\mathbf{c}},\tilde{r}}\big(t, \det(A)s\big). \end{aligned}$$

It follows that $f \circ \boldsymbol{x}_{\mathbf{c},r}(t,s)$ is a point on the canal surface with spine curve $\tilde{\mathbf{c}}$, radius function $\tilde{r}$ and corresponding normal $\boldsymbol{N}_{\tilde{\mathbf{c}},\tilde{r}}$ as in Eq. (8.4), which is just a reparametrization of $S_{\mathbf{c},r}$. Therefore $f(S_{\mathbf{c},r}) = S_{\mathbf{c},r}$, and f is a symmetry of $S_{\mathbf{c},r}$. □

In the specific case of a pipe surface, Theorem 160 takes the following form.

Corollary 161. *Let S be a pipe surface, not a Dupin cyclide, with a non-linear spine curve* $\mathbf{c}$. *An isometry is a symmetry of S if and only if it is a symmetry of* $\mathbf{c}$.

In order to derive an algorithm from Theorem 160 to compute the symmetries of S, we invoke once again the notion of a Möbius-like polynomial (see Eq. (3.13) in Chapter 3), $F(t,\omega) = \omega(\gamma t + \delta) - (\alpha t + \beta)$, where $\varphi(t) = \frac{\alpha t+\beta}{\gamma t+\delta}$ and $\omega = \varphi(t)$. Writing $\omega = \varphi(t)$, Condition C2 of Theorem 160 takes the form $r^2(t) - r^2(\omega) = 0$.

Writing $r(t) = A(t)/B(t)$, with $A(t), B(t)$ coprime, and clearing denominators yields the polynomial equation

$$R(t, \omega) = A^2(t)B^2(\omega) - A^2(u)B^2(\omega) = 0. \qquad (8.7)$$

The following result shows how Condition C2 in Theorem 160 can be tested by checking for Möbius-like factors of R.

Proposition 162. *Let $r = A/B$ be a real rational function, with corresponding bivariate polynomial R as in (8.7), and let φ be a Möbius transformation with corresponding Möbius-like polynomial F. Then Condition C2 holds precisely when F divides R.*

Proof. The equation $r^2(t) - r^2\big(\varphi(t)\big) = 0$ holds identically iff $R\big(t, \varphi(t)\big) = 0$ holds identically. In that case, the zeroset of R contains the graph of φ, which is equal to the zeroset of F. Since F is irreducible, it follows that this happens precisely when F divides R. □

Combining Theorem 160 and Proposition 162, we observe that we can determine the symmetries of a canal surface by:

(i) testing Condition C2 on the radius function, by finding the Möbius-like factors F of R and corresponding tentative Möbius transformations φ;

(ii) testing Condition C1 on the spine curve for each such φ, by determining whether φ corresponds to a symmetry f of the spine curve $\mathbf{c}$.

After computing φ, we can compute the symmetry f as in the case of space rational curves. Thus we arrive at Algorithm `SymCanal` for computing the symmetries of a canal surface with a unique non-linear spine curve. Notice that in this algorithm we distinguish pipe surfaces as a special case, since in that case R is identically zero.

Example 19. *Consider the crunode spine curve and radius function*

$$\mathbf{c}(t) = \left(\frac{t}{t^4+1}, \frac{t^2}{t^4+1}, \frac{t^3}{t^4+1} \right), \qquad r(t) = \frac{t^2}{t^4+1},$$

Algorithm 5 `SymCanal`

Require: A real, rational, properly parametrized, non-linear spine curve $\mathbf{c}$, and a real, rational radius function r, defining a parametrization (8.2) of a canal surface S that is not a Dupin cyclide.

Ensure: The symmetries of S.

if r is constant **then**
  return the symmetries of the spine curve
else
  find the Möbius-like factors of R and associated Möbius transformations φ
  for each Möbius transformation φ **do**
    find the isometry f associated with φ using the method for space rational curves
    if $f \circ \mathbf{c} = \mathbf{c} \circ \varphi$, return the isometry f
  end for
end if

with corresponding canal surface shown in Fig. 8.3, left. Then

$$R(t, \omega) = (\omega - t)(\omega + t)(\omega t - 1)(\omega t + 1)(\omega^2 t^2 + 1)(\omega^2 + t^2),$$

whose Möbius-like factors correspond to Möbius transformations

$$\varphi_1(t) = t, \qquad \varphi_2(t) = -t, \qquad \varphi_3(t) = 1/t, \qquad \varphi_4(t) = -1/t.$$

Suppose Condition C1 holds with $\varphi = \varphi_1, \varphi_2$*, i.e.,*

$$A\mathbf{c}(t) + \boldsymbol{b} = f \circ \mathbf{c}(t) = \mathbf{c} \circ \varphi_i(t) = \mathbf{c}\big((-1)^{i+1} t\big), \qquad i = 1, 2. \quad (8.8)$$

Substituting $t = 0$ *we immediately determine* $\boldsymbol{b} = 0$*. Taking, for* $n = 1, 2, 3$*, the* n*-th derivative of* (8.8) *and evaluating at* $t = 0$*, we get* A *from its action on* $\{\mathbf{c}'(0), \mathbf{c}''(0), \mathbf{c}'''(0)\}$*, and*

$$f_i(\boldsymbol{x}) = A_i \boldsymbol{x}, \qquad A_i = \begin{bmatrix} (-1)^{i+1} & 0 & 0 \\ 0 & 1 & 0 \\ 0 & 0 & (-1)^{i+1} \end{bmatrix}, \qquad i = 1, 2,$$

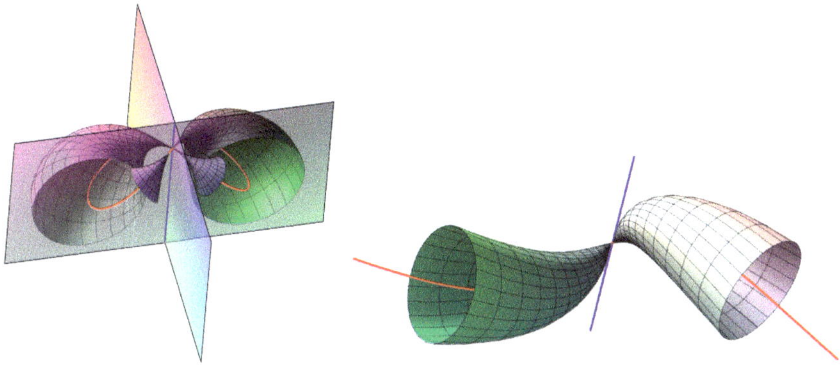

FIGURE 8.3 Canal surfaces together with their spine curves and symmetry elements for Example 19 (left) and Example 22 (right).

corresponding to the trivial symmetry and the half-turn around the y-axis. One directly verifies that $f_i \circ \mathbf{c}(t) = \mathbf{c} \circ \varphi_i(t)$ *for* $i = 1, 2$, *confirming that Condition C1 holds for* f_1, f_2; *so* f_1, f_2 *are symmetries of the canal surface defined by* $(\mathbf{c}, r)$.

Next suppose Condition C1 holds with $\varphi = \varphi_3, \varphi_4$. *With the above procedure, we obtain isometries*

$$f_i(\boldsymbol{x}) = A_i\boldsymbol{x}, \qquad A_i = \begin{bmatrix} 0 & 0 & (-1)^{i+1} \\ 0 & 1 & 0 \\ (-1)^{i+1} & 0 & 0 \end{bmatrix}, \qquad i = 3, 4,$$

which are reflections in the planes $x + (-1)^i z = 0$. *Verifying* $f_i \circ \mathbf{c}(t) = \mathbf{c} \circ \varphi_i(t)$ *for* $i = 3, 4$, *we confirm that* f_3 *and* f_4 *are also symmetries of the canal surface defined by* $(\mathbf{c}, r)$.

8.3 SYMMETRIES OF DUPIN CYCLIDES

In this section we consider the remaining case of a Dupin cyclide S with two distinct spine curves $\mathbf{c}_1, \mathbf{c}_2$ and corresponding radius functions r_1, r_2. First we provide a characterization theorem for an isometry to be a symmetry of a Dupin cyclide. Based on this theorem, we provide a complete classification of the symmetries of the three types of Dupin cyclides, together with the symmetry group in each case; this classification is classical, but it is interesting to see how it follows easily from the algorithms. Finally, based on this classification, we present an algorithm for computing the symmetries of a Dupin cyclide represented by pairs $(\mathbf{c}_i, r_i)$, with $i = 1, 2$, not necessarily in canonical form.

8.3.1 Characterization

Using Lemmas 157–159, we establish the following characterization theorem for the symmetries of Dupin cyclides.

Theorem 163. *Let S be a Dupin cyclide with non-linear spine curves $\mathbf{c}_1, \mathbf{c}_2$ and radius functions r_1, r_2. The isometry $f(\boldsymbol{x}) = A\boldsymbol{x} + \boldsymbol{b}$ is a symmetry of S if and only if there exist Möbius transformations φ_1, φ_2 such that either*

A1: the spine curves satisfy $f \circ \mathbf{c}_1 = \mathbf{c}_1 \circ \varphi_1$ and $f \circ \mathbf{c}_2 = \mathbf{c}_2 \circ \varphi_2$;

A2: the radius functions satisfy $r_1^2 = (r_1 \circ \varphi_1)^2$ and $r_2^2 = (r_2 \circ \varphi_2)^2$,

or

B1: the spine curves satisfy $f \circ \mathbf{c}_1 = \mathbf{c}_2 \circ \varphi_1$ and $f \circ \mathbf{c}_2 = \mathbf{c}_1 \circ \varphi_2$;

B2: the radius functions satisfy $r_1^2 = (r_2 \circ \varphi_1)^2$ and $r_2^2 = (r_1 \circ \varphi_2)^2$.

Proof. ($\Rightarrow$) By Lemma 157, for $i = 1, 2$ we have that $f \circ \mathbf{c}_i$ must also be a spine curve of S. Suppose f maps one of the spine curves of S, say $\mathbf{c}_1$, to itself. Since f is a bijection, it follows that f also maps $\mathbf{c}_2$ to itself. Thus f is a symmetry of both $\mathbf{c}_1$ and $\mathbf{c}_2$, and by Theorem 45 in Chapter 3 this is equivalent to the existence of Möbius transformations φ_1 and φ_2 for which $f \circ \mathbf{c}_i = \mathbf{c}_i \circ \varphi_i$, $i = 1, 2$, establishing A1. Then, for $i = 1, 2$, A2 is established as in the implication "$\Longrightarrow$" of Theorem 160.

Now let C_1, C_2 be the curves defined by $\mathbf{c}_1, \mathbf{c}_2$ and suppose that f maps C_1 to C_2. Then, since f is a bijection, it also maps C_2 to C_1 by Lemma 157. In particular f^2 is a symmetry of both C_1 and C_2. By Theorem 45 there exist Möbius transformations φ_1, φ_2 such that $\boldsymbol{f} \circ \mathbf{c}_1 = \mathbf{c}_2 \circ \varphi_1$ and $\boldsymbol{f} \circ \mathbf{c}_2 = \mathbf{c}_1 \circ \varphi_2$, establishing B1. Using first Condition B1, then Lemma 157 with $r = r_1$ and $\tilde{r} = \tilde{r}_1 = \pm r_1$, next that $(\mathbf{c}_1, r_1)$ and $(\mathbf{c}_2, r_2)$ both define S, and finally that φ_1 is a birational map on the real line, one obtains

$$S_{\mathbf{c}_2\circ\varphi_1,\tilde{r}_1} = S_{f\circ\mathbf{c}_1,\tilde{r}_1} = S_{\mathbf{c}_1,r_1} = S_{\mathbf{c}_2,r_2} = S_{\mathbf{c}_2\circ\varphi_1,r_2\circ\varphi_1}$$

and deduces $r_1^2 = \tilde{r}_1^2 = (r_2 \circ \varphi_1)^2$, establishing B2.

($\Leftarrow$) Let $i, j \in \{1, 2\}$ and suppose $i = j$ (resp. $i \neq j$). Let $\boldsymbol{x}_{\mathbf{c}_i,r_i}(t, s)$ be the parametrization (8.2) of S with normals $\boldsymbol{N}_{\mathbf{c}_i,r_i}(t, s)$ as in Eq. (8.4). Since the radius function r_i and $\tilde{r}_j := r_j \circ \varphi_i$ are real rational functions satisfying $r_i^2 = \tilde{r}_j^2$ by A2 (resp. B2), Lemma 156 implies $r_i = \pm\tilde{r}_j$. As a change of sign of the radius function results in a change of orientation of the canal surface and leaves the geometric shape unchanged, we may safely assume that $r_i = \tilde{r}_j$. Then using Lemma 159 with $\mathbf{c} = \mathbf{c}_i$, $\tilde{\mathbf{c}} = \tilde{\mathbf{c}}_j := \mathbf{c}_j \circ \varphi_i = f \circ \mathbf{c}_i$, $r = r_i = \tilde{r}_j$, we get $A\boldsymbol{N}_{\mathbf{c}_i,r_i}(t, s) = \boldsymbol{N}_{\tilde{\mathbf{c}}_j,\tilde{r}_j}\big(t, \det(A)s\big)$, and the proof proceeds as in the implication "$\Longleftarrow$" of Theorem 160. □

Remark 16. *Notice that φ_1 is typically not equal to φ_2. Case A of Theorem 163 implies that f is a symmetry of each spine curve of S; in particular, in this case f maps each spine curve to itself. Case B implies that the spine curves $\mathcal{C}_1$ and $\mathcal{C}_2$ are mapped to each other by an isometry f that is not a symmetry of either $\mathcal{C}_1$ or $\mathcal{C}_2$. Nevertheless, f^2 is a symmetry of both $\mathcal{C}_1$ and $\mathcal{C}_2$ with associated Möbius transformations $\varphi_2 \circ \varphi_1$ and $\varphi_1 \circ \varphi_2$.*

Analogous to the method for canal surfaces with a unique spine curve, we can use Conditions A2 and B2 to compute the symmetries of the surface. Again introducing the auxiliary variable ω satisfying $\omega - \varphi(t) = 0$, Conditions A2 and B2 of Theorem 163 take the form $r_i^2(t) - r_j^2(\omega) = 0$, with $(i, j) = (1, 1), (2, 2)$ for Condition A2 and $(i, j) = (1, 2), (2, 1)$ for Condition B2. Clearing denominators yields corresponding polynomial conditions

$$R_{ij}(t, \omega) := A_i^2(t)B_j^2(\omega) - A_j^2(\omega)B_i^2(t) = 0, \qquad i, j = 1, 2, \quad (8.9)$$

where $r_1 = A_1/B_1$ and $r_2 = A_2/B_2$, with (A_1, B_1) and (A_2, B_2) pairs of coprime univariate polynomials. The following proposition, which has a proof analogous to that of Proposition 162, shows how the tentative Möbius transformations can be determined in the form of Möbius-like factors of the polynomials R_{ij}.

Proposition 164. *For $i, j = 1, 2$, let $r_i = A_i/B_i$ be real rational functions, with corresponding bivariate polynomials R_{ij} as in (8.9), and let φ_1, φ_2 be Möbius transformations with corresponding Möbius-like polynomials F_1, F_2. Then*

A2 holds precisely when F_1 divides R_{11} and F_2 divides R_{22};

B2 holds precisely when F_1 divides R_{12} and F_2 divides R_{21}.

Example 20. *In the proof of Theorem 167, it is derived that the Möbius transformations $\varphi_\pm(t) := \pm t$ satisfy Condition B2 in Theorem 163 for a Dupin cyclide of Type III with $c = 0$. Let us determine the associated symmetries.*

Let $f(\boldsymbol{x}) = A\boldsymbol{x} + \boldsymbol{b}$, and suppose that $f \circ \mathbf{c}_1^{\mathrm{III}} = \mathbf{c}_2^{\mathrm{III}} \circ \varphi_\pm$. Then

$$A\begin{bmatrix} t^2 - \frac{1}{2} \\ 2t \\ 0 \end{bmatrix} + \boldsymbol{b} = A\mathbf{c}_1^{\mathrm{III}}(t) + \boldsymbol{b} = f \circ \mathbf{c}_1^{\mathrm{III}}(t) = \mathbf{c}_2^{\mathrm{III}}(\pm t) = \begin{bmatrix} \frac{1}{2} - t^2 \\ 0 \\ \pm 2t \end{bmatrix}. \tag{8.10}$$

Differentiating once and twice, evaluating at $t = 0$, and taking the cross product,

$$A\begin{bmatrix} 0 \\ 1 \\ 0 \end{bmatrix} = \begin{bmatrix} 0 \\ 0 \\ \pm 1 \end{bmatrix}, \quad A\begin{bmatrix} 1 \\ 0 \\ 0 \end{bmatrix} = \begin{bmatrix} -1 \\ 0 \\ 0 \end{bmatrix}, \quad A\begin{bmatrix} 0 \\ 0 \\ 1 \end{bmatrix} = \begin{bmatrix} 0 \\ \pm\det A \\ 0 \end{bmatrix}. \tag{8.11}$$

Applying the rules of matrix block multiplication to (8.11), we obtain

$$A = \begin{bmatrix} -1 & 0 & 0 \\ 0 & 0 & \pm\det A \\ 0 & \pm 1 & 0 \end{bmatrix}.$$

Substituting this A and $t = 0$ in (8.10) yields $\boldsymbol{b} = 0$, and it follows that $f(x, y, z) = (-x, \pm\det Az, \pm y)$. Moreover, $f \circ \mathbf{c}_2^{\mathrm{III}} = \mathbf{c}_1^{\mathrm{III}} \circ \varphi_{\pm\det A}$, so that Conditions B1 and B2 are satisfied for the Cases (i)–(l) in Table 8.2.

TABLE 8.2 For the Types II and III of Dupin cyclides in Table 8.1, the table lists the discrete symmetries f and Möbius transformations (φ_1, φ_2) associated with f via Case A and Case B in Theorem 163. Here $\Pi_0 : x = 0$ is the plane through $O := (0, 0, 0)$, the barycenter of the ellipse and hyperbola and average of the foci (which are, as a set, also the vertices) of the parabolas, and perpendicular to the planes Π_1, Π_2 containing the spine curves $\mathbf{c}_1, \mathbf{c}_2$; $\Pi_3 : z - y = 0, \Pi_4 : z + y = 0$ are the bisector planes of Π_1, Π_2.

Case A	(φ_1, φ_2)	$f(x, y, z)$	description of the symmetry	Type
(a)	$(+t, +t)$	$(+x, +y, +z)$	trivial symmetry	II, III
(b)	$(+t, -t)$	$(+x, +y, -z)$	reflection in the plane Π_1	II, III
(c)	$(-t, +t)$	$(+x, -y, +z)$	reflection in the plane Π_2	II, III
(d)	$(-t, -t)$	$(+x, -y, -z)$	half-turn about the line $\Pi_1 \cap \Pi_2$	II, III
(e)	$(+\frac{1}{t}, -\frac{1}{t})$	$(-x, +y, +z)$	reflection in the plane Π_0	II, $c = 0$
(f)	$(+\frac{1}{t}, +\frac{1}{t})$	$(-x, +y, -z)$	half-turn about the line $\Pi_0 \cap \Pi_1$	II, $c = 0$
(g)	$(-\frac{1}{t}, -\frac{1}{t})$	$(-x, -y, +z)$	half-turn about the line $\Pi_0 \cap \Pi_2$	II, $c = 0$
(h)	$(-\frac{1}{t}, +\frac{1}{t})$	$(-x, -y, -z)$	central symmetry about O	II, $c = 0$
Case B	(φ_1, φ_2)	$f(x, y, z)$	description of the symmetry	Type
(i)	$(+t, +t)$	$(-x, +z, +y)$	half-turn about the line $\Pi_0 \cap \Pi_3$	III, $c = 0$
(j)	$(-t, -t)$	$(-x, -z, -y)$	half-turn about the line $\Pi_0 \cap \Pi_4$	III, $c = 0$
(k)	$(+t, -t)$	$(-x, -z, +y)$	composition of a reflection in the plane Π_0 and a quarter-turn about the line $\Pi_1 \cap \Pi_2$	III, $c = 0$
(l)	$(-t, +t)$	$(-x, +z, -y)$	composition of a reflection in the plane Π_0 and a quarter-turn about the line $\Pi_1 \cap \Pi_2$	III, $c = 0$

TABLE 8.3 The symmetries and symmetry groups for the Dupin cyclides of Types I, II, III in Table 8.1 with parameters a, b, c, ξ, g. Here $\mathbb{Z}_2 \times \mathbb{S}^1$ is identified with the subgroup of rotations about the z-axis and reflections in planes containing the z-axis.

Type	graphic	parameters	symmetries	group
I: Quartic		all	$\{(a),(b)\} \times \mathbb{Z}_2 \times \mathbb{S}^1$	$\mathbb{Z}_2^2 \times \mathbb{S}^1$
II: Quartic		$c \neq 0$	(a)–(d)	$\mathbb{Z}_2^2$
		$c = 0$	(a)–(h)	$\mathbb{Z}_2^3$
III: Cubic		$c \neq 0$	(a)–(d)	$\mathbb{Z}_2^2$
		$c = 0$	(a)–(d), (i)–(l)	D_4

8.3.2 Classification

In this subsection we apply the preceding results and ideas to classify the symmetries of Dupin cyclides, providing the symmetry groups in each case. For this purpose, we assume each Dupin cyclide to be in canonical form, i.e., with spine curves $\mathbf{c}_1^\alpha, \mathbf{c}_2^\alpha$ and radius functions r_1^α, r_2^α for one of the Types $\alpha = \text{I}, \text{II}$ or III as in Table 8.1, and corresponding implicit equation F^α. The results are summarized in Tables 8.2 and 8.3.

8.3.2.1 Type I

Let us address first the symmetries of the Dupin cyclides of Type I, i.e., tori. This can be deduced from results on symmetries of surfaces of revolution. However, we will show that the previous results can also be used to easily derive these symmetries.

Even though one of the spine curves of the Dupin cyclide of Type I is a line, the following theorem shows that Theorem 163 extends to these surfaces as well and explicitly describes the symmetries and associated Möbius transformations.

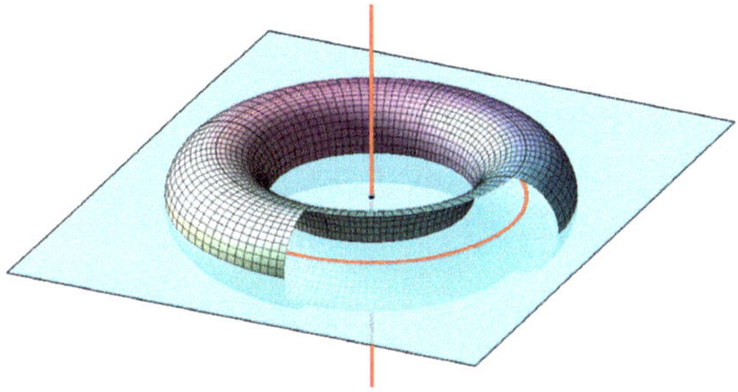

FIGURE 8.4 The Dupin cyclide S of Type I with parameters $a = 2$ and $c = 1$, together with its spine curves, the line $\mathbf{c}_2^{\mathrm{I}}$ passing through the barycenter of the circle $\mathbf{c}_1^{\mathrm{I}}$. The axis of the rotational symmetry coincides with the spine curve $\mathbf{c}_2^{\mathrm{I}}$, and the plane of $\mathbf{c}_1^{\mathrm{I}}$ corresponds to a mirror symmetry of S.

Theorem 165. *The isometry $f(\boldsymbol{x}) = A\boldsymbol{x} + \boldsymbol{b}$ is a symmetry of the Dupin cyclide S of Type I in Table 8.1 if and only if there exist Möbius transformations φ_1, φ_2 such that Conditions A1 and A2 in Theorem 163 hold. Moreover, with signs $\varepsilon_1, \varepsilon_2 \in \{-1, 1\} \simeq \mathbb{Z}_2$ and angle $\theta \in [0, 2\pi) \simeq \mathbb{S}^1$, these Möbius transformations take the form*

$$\varphi_1(t) = -\left(\frac{\cos(\theta/2)t + \sin(\theta/2)}{\sin(\theta/2)t - \cos(\theta/2)}\right)^{\varepsilon_2}, \qquad \varphi_2(t) = \varepsilon_1 t, \tag{8.12}$$

with associated symmetries

$$f(\boldsymbol{x}) = A\boldsymbol{x}, \qquad A = \begin{bmatrix} \varepsilon_2 \cos\theta & -\varepsilon_2 \sin\theta & 0 \\ \sin\theta & \cos\theta & 0 \\ 0 & 0 & \varepsilon_1 \end{bmatrix} \tag{8.13}$$

forming a symmetry group isomorphic to $\mathbb{Z}_2^2 \times \mathbb{S}^1$.

Proof. We first determine the isometries f and associated Möbius transformations φ_1, φ_2 satisfying Conditions A1, A2. Substituting r_2^{I} into (8.9) yields

$$R_{22}^{\mathrm{I}}(t, \omega) = 4a(\omega + t)(u - t)(t^2\omega^2 a + t^2\omega^2 c - t^2 c - \omega^2 c - a + c).$$

Since $a, c \neq 0$, the right-most factor in R^{I}_{22} does not split, and the only Möbius-like factors are $\omega - t$ and $\omega + t$, corresponding to Möbius transformations $\varphi_2 = \pm t$ satisfying Condition A2 by Proposition 164. The relation $f \circ \mathbf{c}_2 = \mathbf{c}_2 \circ \varphi_2$ implies that f satisfies $f(0,0,z) = (0,0,\pm z)$ for any point $(0,0,z)$ on the line $\mathbf{c}^{\mathrm{I}}_2$, i.e.,

$$f(x,y,z) = \left[\begin{array}{c|c} \widehat{A} & 0 \\ \hline 0 & \pm 1 \end{array}\right] \begin{bmatrix} x \\ y \\ z \end{bmatrix}, \qquad \widehat{A}\widehat{A}^{\mathrm{T}} = I.$$

On the other hand, since r^{I}_1 is constant, Condition A2 does not determine (or even restrict) φ_1. However, it is well known that any orthogonal matrix $\widehat{A} \in \mathbb{R}^{2,2}$ maps the circle to itself. This can be shown explicitly using the trigonometric reparametrization $\mathbf{c}^{\mathrm{I}}_1\big(\tan(\phi/2)\big) = a(\cos\phi, \sin\phi, 0)$ and the representation

$$\widehat{A} \in \left\{ \begin{bmatrix} \varepsilon_2 \cos\theta & -\varepsilon_2 \sin\theta \\ \sin\theta & \cos\theta \end{bmatrix} : \varepsilon_2 \in \{\pm 1\},\ \theta \in [0, 2\pi) \right\}.$$

Thus f necessarily takes the form (8.13) and corresponds to the reparametrizations $\phi \longmapsto \phi + \theta$ and $\phi \longmapsto \pi - \phi - \theta$, or equivalently to the Möbius transformations (8.12). A direct calculation shows that such φ_1, φ_2 and f satisfy Conditions A1 and A2.

($\Rightarrow$) Using in addition that the circle $\mathbf{c}^{\mathrm{I}}_1$ and line $\mathbf{c}^{\mathrm{I}}_2$ are not related by an isometry, this is established as in Theorem 163.

($\Leftarrow$) Using the above explicit form of f, a straightforward calculation shows that these indeed are symmetries of S. □

Remark 17. *Notice that the matrix A in* (8.13) *can be decomposed as a product*

$$\begin{bmatrix} \varepsilon_2 \cos\theta & -\varepsilon_2 \sin\theta & 0 \\ \sin\theta & \cos\theta & 0 \\ 0 & 0 & \varepsilon_1 \end{bmatrix} = \begin{bmatrix} \varepsilon_2 & 0 & 0 \\ 0 & 1 & 0 \\ 0 & 0 & 1 \end{bmatrix} \begin{bmatrix} \cos\theta & -\sin\theta & 0 \\ \sin\theta & \cos\theta & 0 \\ 0 & 0 & 1 \end{bmatrix} \begin{bmatrix} 1 & 0 & 0 \\ 0 & 1 & 0 \\ 0 & 0 & \varepsilon_1 \end{bmatrix}.$$

This shows that the symmetries of cyclides of Type I are the rotations about the line $\mathbf{c}^{\mathrm{I}}_2$, together with the composition of these rotations with the reflection in the plane containing the circle $\mathbf{c}^{\mathrm{I}}_1$ and/or with the reflection in any plane containing the line $\mathbf{c}^{\mathrm{I}}_2$.

8.3.2.2 Type II

Now we apply Theorem 163 to derive the symmetries of Dupin cyclides of Type II. Every such Dupin cyclide has an ellipse and hyperbola as its spine curves, which are not related by an isometry. It follows that for these Dupin cyclides Case B in Theorem 163 cannot happen. We distinguish cases according to whether the parameter c vanishes; see Fig. 8.5.

Theorem 166. *For any Dupin cyclide S of Type II in Table 8.1, Conditions A1 and A2 in Theorem 163 are satisfied if and only if:*

- *$c \neq 0$ and (φ_1, φ_2) and f are given by Cases (a)–(d) in Table 8.2, forming a symmetry group isomorphic to $\mathbb{Z}_2^2$, the Klein four group;*
- *$c = 0$ and (φ_1, φ_2) and f are given by Cases (a)–(h) in Table 8.2, forming a symmetry group isomorphic to $\mathbb{Z}_2^3$, the elementary abelian group of order 8.*

Proof. Inserting the radius functions $r_1^{\mathrm{II}}, r_2^{\mathrm{II}}$ from Table 8.1 into (8.9) yields

$$R_{11}^{\mathrm{II}}(t,\omega) = -4\xi(\omega - t)(\omega + t)(c\omega^2 t^2 + \xi\omega^2 t^2 + c\omega^2 + ct^2 - \xi + c),$$
$$R_{22}^{\mathrm{II}}(t,\omega) = +4a(\omega - t)(\omega + t)(a\omega^2 t^2 + c\omega^2 t^2 - c\omega^2 - ct^2 - a + c),$$

each of which has the Möbius-like factors $F_1(t,\omega) = \omega - t$ and $F_2(t,\omega) = \omega + t$. In each case the remaining factor has degree two, and whether it splits into two additional Möbius-like factors

$$F_3(t,\omega) = \omega(\gamma t + \delta) - (\alpha t + \beta), \qquad F_4(t,\omega) = \omega(\gamma' t + \delta') - (\alpha' t + \beta')$$

depends on the parameters a, c, f.

In particular for the polynomial R_{11}^{II}, if the remaining factor satisfies

$$c\omega^2 t^2 + f\omega^2 t^2 + c\omega^2 + ct^2 - \xi + c = \omega^2 \cdot [(c+\xi)t^2 + c] + (ct^2 + c - \xi)$$
$$= F_3(t,\omega)F_4(t,\omega),$$

comparing the coefficients of ω on each side yields

$$(\gamma t + \delta)(\alpha' t + \beta') = -(\gamma' t + \delta')(\alpha t + \beta),$$

implying that the Möbius transformations corresponding to $F_3(t,\omega)$ and $F_4(t,\omega)$ are opposite. Hence, after an appropriate scaling of F_3 or F_4, the remaining factor satisfies that

$$\omega^2 \cdot [(c+\xi)t^2 + c] + (ct^2 + c - \xi) = \omega^2 \cdot (\gamma t + \delta)^2 - (\alpha t + \beta)^2.$$

Comparing the coefficients of t and the coefficients of $\omega^2 t$ yields $c = 0$, in which case

$$R_{11}^{\mathrm{II}}(t,\omega)|_{c=0} = -4\xi^2(\omega - t)(\omega + t)(\omega t - 1)(\omega t + 1).$$

A similar argument shows that the remaining factor of R_{22}^{II} only factors when $c = 0$, in which case

$$R_{22}^{\mathrm{II}}(t,\omega)|_{c=0} = 4a^2(\omega - t)(\omega + t)(\omega t - 1)(\omega t + 1).$$

Thus R_{11}^{II} and R_{22}^{II} each determine tentative Möbius transformations $\varphi_1(t) = t, \varphi_2(t) = -t$, and two additional Möbius transformations $\varphi_3(t) = 1/t, \varphi_4(t) = -1/t$ if and only if $c = 0$.

When $c \neq 0$, we therefore only get the Möbius transformations $\varphi_1(t) = t, \varphi_2(t) = -t$, for both $\mathbf{c}_1^{\mathrm{II}}$ and $\mathbf{c}_2^{\mathrm{II}}$, which combine in pairs associated with four potential symmetries. The nature of these symmetries can be determined as before, yielding Cases (a)–(d) in Table 8.2. When $c = 0$, we get the Möbius transformations $\varphi_1, \varphi_2, \varphi_3, \varphi_4$ for both $\mathbf{c}_1^{\mathrm{II}}$ and $\mathbf{c}_2^{\mathrm{II}}$, which combine in pairs associated with 16 potential symmetries. Proceeding as in Example 20, one finds that only 8 of these correspond to a symmetry of S, namely Cases (a)–(h) in Table 8.2.

Noting from the explicit representations in Table 8.2 that the symmetries (b)–(h) have order 2, the second part follows from comparing to a list of groups of small order [109, p. 85]. □

8.3.2.3 Type III

Next we apply Theorem 163 to derive the symmetries of Dupin cyclides of Type III. Every such Dupin cyclide has parabolas as its spine curves, which might be related by an isometry. Hence,

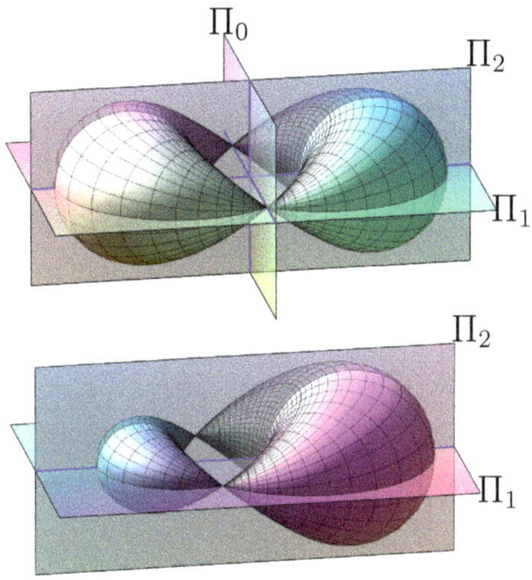

FIGURE 8.5 The Dupin cyclide of Type II in Table 8.1 with parameters $a = 5, b = 4, c = 0, \xi = 3$ (left) and parameters $a = 5, b = 4, c = -1, \xi = 3$ (right)

for such Dupin cyclides, it is necessary to analyze Case B of Theorem 163 as well. We distinguish cases according to whether the parameter c vanishes; see Fig. 8.6.

Theorem 167. *For any Dupin cyclide S of Type III in Table 8.1:*

- *Conditions A1 and A2 in Theorem 163 are satisfied if and only if (φ_1, φ_2) and f are given by Cases (a)–(d) in Table 8.2.*

- *Conditions B1 and B2 in Theorem 163 are satisfied if and only if $c = 0$ and (φ_1, φ_2) and f are given by Cases (i)–(l) in Table 8.2.*

In particular:

- *If $c \neq 0$, the symmetries of S are (a)–(d), forming a group isomorphic to $\mathbb{Z}_2^2$, the Klein four group.*

- *If $c = 0$, the symmetries of S are (a)–(d) and (i)–(l), forming a group isomorphic to D_4, the dihedral group of order eight.*

Proof. Case A: Inserting the radius functions $r_1^{\mathrm{III}}, r_2^{\mathrm{III}}$ from Table 8.1 into (8.9) yields

$$R_{11}^{\mathrm{III}}(t,\omega) = -16g(\omega+t)(u-t)\big(g(\omega^2+t^2+1)+2c\big),$$
$$R_{22}^{\mathrm{III}}(t,\omega) = -16g(\omega+t)(u-t)\big(g(\omega^2+t^2+1)-2c\big),$$

each of which has the Möbius-like factors $F_1(t,\omega) = \omega - t$ and $F_2(t,\omega) = \omega+t$. As the remaining factor is irreducible in each case, we find that the tentative Möbius transformations are $\varphi_1(t) = t$ and $\varphi_2(t) = -t$ for both $\mathbf{c}_1^{\mathrm{III}}$ and $\mathbf{c}_2^{\mathrm{III}}$. These again combine to four potential symmetries, corresponding exactly to the Cases (a)–(d) in Table 8.2.

Case B: One computes

$$R_{12}^{\mathrm{III}}(t,\omega) = (r_1^{\mathrm{III}})^2(t)-(r_2^{\mathrm{III}})^2(\omega) = -16g(g\omega^2-gt^2-2c)(\omega^2+t^2+1).$$

The factor ω^2+t^2+1 is irreducible. If $c \neq 0$, then the factor $-g\omega^2+gt^2+2c$ defines a hyperbola, since $g \neq 0$ and is therefore also irreducible. It follows that R_{12}^{III} does not have Möbius-like factors, so that there are no symmetries corresponding to Case B of Theorem 163 when $c \neq 0$. However, if $c = 0$, then R_{12}^{III} has the Möbius-like factors $F_{\pm}(t,\omega) = \omega \mp t$, corresponding to the Möbius transformations $\varphi_{\pm}(t) = \pm t$. From Example 20 it follows that Conditions B1 and B2 are satisfied for the Cases (i)–(l) in Table 8.2.

Noting from the explicit representations in Table 8.2 that the symmetries (b), (c), (d), (i), (j) have order 2 and (k), (l) have order 4, the second part follows from comparing to a list of groups of small order [109, p. 85]. □

While it is well known that any cyclide of Type II or III is symmetric with respect to the planes containing each of the spine curves, and therefore also with respect to the intersection line of these two planes, the preceding results show that when $c \neq 0$, such a cyclide cannot have any other symmetry. In fact, we have proven that cyclides of Type II and III have either 4 or 8 symmetries; in this last case, we say that it is a *super-symmetric cyclide* (see Fig. 8.5, left, and Fig. 8.6, left).

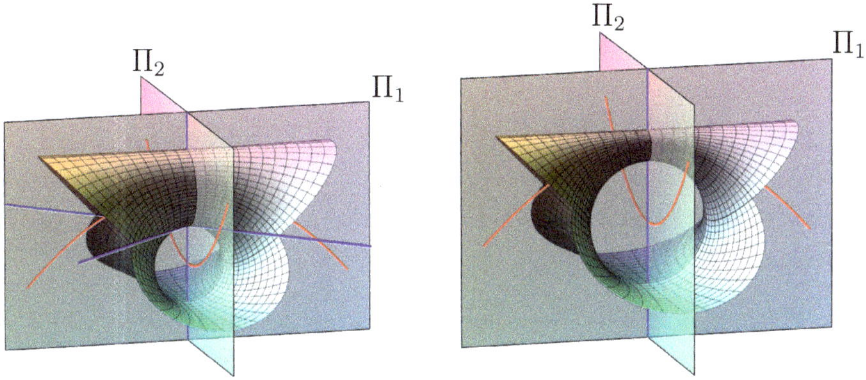

FIGURE 8.6 The Dupin cyclide of Type III in Table 8.1 with parameters $c = 0, g = 1$ (left) and parameters $c = 0.3, g = 1$ (right).

8.3.3 Algorithm

We end with an algorithm for computing the symmetries of a Dupin cyclide S defined by spine curves $\mathbf{c}_1, \mathbf{c}_2$ and corresponding radius functions r_1, r_2, not necessarily given in canonical form. Whether the Dupin cyclide is of Type I, II, or III follows from the nature of the conics; this is easily determined, for instance by implicitization or by computing the curvature, which is independent of position and orientation. Moreover, we have the following result.

Lemma 168. *A Dupin cyclide $S = S_{\mathbf{c}_1,r_1} = S_{\mathbf{c}_2,r_2}$ is supersymmetric if and only if one of the following cases holds:*

- *S is of Type II, and the radius function corresponding to the ellipse has minimum $r_{\min}$ and maximum $r_{\max}$ satisfying $r_{\min} + r_{\max} = 0$.*
- *S is of Type III, and $r_1 + r_2 = 0$ holds identically.*

Proof. Since the conditions $r_{\min} + r_{\max} = 0$ and $r_1 + r_2 = 0$ remain valid under reparametrization, we can assume that $(\mathbf{c}_1, r_1)$ and $(\mathbf{c}_2, r_2)$ take the canonical forms in Table 8.1. In the first case of the lemma,

$$r_{\min} = r_1^{\mathrm{II}}(0) = c - \xi, \qquad r_{\max} = \lim_{t \to \pm\infty} r_1^{\mathrm{II}}(t) = c + \xi,$$

while in the second case

$$r_1 = r_1^{\text{III}} = c + g\left(t^2 + \frac{1}{2}\right), \qquad r_2 = r_2^{\text{III}} = c - g\left(t^2 + \frac{1}{2}\right).$$

In either case the sum is zero precisely when $c = 0$, which proves the lemma. □

We can now sketch a method for determining the symmetries of S.

1. *Determine the Type of S from the nature of its spine curves.*

2. *Determine the invariants of the conics referenced in Table 8.2.*
   Find planes Π_1, Π_2 containing $\mathbf{c}_1, \mathbf{c}_2$. Next determine O, i.e., for

   - Type I, the barycenter of the circle;
   - Type II, the barycenter of the ellipse/hyperbola;
   - Type III, the average of the vertex and focal point, for each parabola.

   Let Π_0 be the plane passing through O and perpendicular to Π_1, Π_2. For Type III, we also determine the bisector planes Π_3, Π_4 of Π_1, Π_2.

3. *For Types II and III, determine if S is super-symmetric using Lemma 168.*

4. *Look up the symmetry groups and symmetries of S in Tables 8.2 and 8.3.*

8.4 APPLICATIONS TO BUILD SYMMETRIC BLENDS

The most popular application of canal surfaces is *blending*, an operation in Solid Modeling which consists of connecting two different surface patches with a good continuity level. In this sense, canal surfaces are useful when the endings of the patches to be connected are circular. Because of the nature of the solid to be modeled, we may require the blend to satisfy certain symmetry conditions. As an application of the results in the previous sections, here we will see how to construct patches of canal surfaces with a prescribed planar, axial or central symmetry, and how to compute blendings of two non-intersecting canal surfaces, under certain conditions, with a prescribed symmetry. In either case we apply Theorem 160, choosing a symmetric rational spine curve (Condition C1) and a rational radius function that respects the symmetries of the spine curve (Condition C2).

8.4.1 Building symmetric (patches of) canal surfaces

In order to build a patch of a rational canal surface $S_{\mathbf{c},r}$ with a prescribed symmetry f, we first need to find a rational spine curve $\mathbf{c}$ invariant under f. For this purpose, we can take advantage of Bézier curves, and its properties (see Section 3.3.1). Thus, let

$$\mathbf{c}(t) = \sum_{i=0}^{n} \boldsymbol{b}_i B_{i,n}(t), \qquad t \in [0,1], \tag{8.14}$$

define a Bézier curve with (space) control points $\boldsymbol{b}_0, \ldots, \boldsymbol{b}_n$, where the $B_{i,n}(t)$ are the Bernstein polynomials, and let $\mathcal{P}$ be the control polygon of the curve. If $\mathcal{P}$ is invariant under f, we know from Section 3.3.1 that $\mathbf{c}$ is also invariant under f. Hence, by Condition C1 of Theorem 160 there exists a Möbius transformation φ such that

$$\sum_{i=0}^{n} f(\boldsymbol{b}_i) B_{i,n}(t) = f \circ \mathbf{c}(t) = \mathbf{c} \circ \varphi(t) = \sum_{i=0}^{n} \boldsymbol{b}_i B_{i,n}(\varphi(t)).$$

Using that $B_{i,n}(1-t) = B_{n-i,n}(t)$, one obtains two cases:

$$\begin{aligned} &\text{Case 1}: \quad \varphi(t) = t, \qquad f(\boldsymbol{b}_i) = \boldsymbol{b}_i, \qquad i = 0, \ldots, n; \\ &\text{Case 2}: \quad \varphi(t) = 1 - t, \quad f(\boldsymbol{b}_i) = \boldsymbol{b}_{n-i}, \quad i = 0, \ldots, n. \end{aligned}$$

In the first case, Condition C2 of Theorem 160 holds trivially. In the second case, we require a radius function $r(t)$ satisfying $r^2(t) = (r \circ \varphi)^2(t) = r^2(1-t)$, which is achieved using the following result.

Theorem 169. *A degree n polynomial r satisfies* $r^2(t) = r^2(1-t)$ *if and only if*

$$r(t) = \sum_{i=0}^{n} a_i B_{i,n}(t), \qquad a_i = (-1)^n a_{n-i}, \qquad i = 0, \ldots, n. \quad (8.15)$$

Proof. (⇒) By Lemma 156, $r(t) = \pm r(1-t)$. Expressing r in the Bernstein basis and using that $B_{i,n}(1-t) = B_{n-i,n}(t)$, it follows

$$0 = \sum_{i=0}^{n} a_i B_{i,n}(t) \mp \sum_{i=0}^{n} a_i B_{n-i,n}(t) = \sum_{i=0}^{n} (a_i \mp a_{n-i}) B_{i,n}(t),$$

implying either $a_i = a_{n-i}$ for $i = 0, \ldots, n$, or $a_i = -a_{n-i}$ for $i = 0, \ldots, n$. Let Δ^k denote the k-th forward difference operator recursively defined by

$$\Delta^0 a_i = a_i, \quad \Delta^k a_i = \Delta^{k-1} a_{i+1} - \Delta^{k-1} a_i, \quad i = 0, \ldots, n-k, \quad k \geq 1. \quad (8.16)$$

It follows by induction that

$$\frac{\mathrm{d}^n r}{\mathrm{d}t^n} = n! \Delta^n a_0 = n! \sum_{j=0}^{n} (-1)^j \binom{n}{j} a_{n-j},$$

which is zero when n is odd and $a_i = a_{n-i}$, or when n is even and $a_i = -a_{n-i}$. Since r has degree n, it takes the form (8.15).

(⇐) By hypothesis, the coefficients a_i in the Bernstein form (8.15) of r satisfy $a_i = (-1)^n a_{n-i}$. Using that $B_{i,n}(1-t) = B_{n-i,n}(t)$, it follows that $r(1-t) = (-1)^n \cdot r(t)$. □

Thus we can apply Theorem 160 to construct a canal surface $S_{\mathbf{c},r}$ with the prescribed symmetry f.

Example 21. *The case of a symmetry f with respect to a plane* Π *is especially simple, as it suffices to assume that* $\mathbf{c}$ *is contained in* Π*. In this case the restriction* $f|_\Pi$ *is the identity, and Conditions C1 and C2 in Theorem 160 are trivially satisfied taking* $\varphi(t) = t$*.*

Example 22. *Let $n = 3$, and consider the Bézier curve $\mathbf{c}$ as in (8.14) with*

$$\boldsymbol{b}_0 = \begin{bmatrix} -1 \\ 1 \\ -1 \end{bmatrix}, \qquad \boldsymbol{b}_1 = \begin{bmatrix} -\frac{1}{3} \\ -\frac{1}{3} \\ 1 \end{bmatrix}, \qquad \boldsymbol{b}_2 = \begin{bmatrix} \frac{1}{3} \\ -\frac{1}{3} \\ -1 \end{bmatrix}, \qquad \boldsymbol{b}_3 = \begin{bmatrix} 1 \\ 1 \\ 1 \end{bmatrix}.$$

This is just a reparametrization of the twisted cubic curve, because

$$\mathbf{c}\big((s+1)/2\big) = (s, s^2, s^3), \qquad s \in [-1, 1].$$

Clearly the half-turn $f(x, y, z) = (-x, y, -z)$ is a symmetry of the control polygon of $\mathbf{c}$, since $f(\boldsymbol{b}_i) = \boldsymbol{b}_{3-i}$ for $i = 0, 1, 2, 3$. Hence Condition C1 holds for the curve $\mathbf{c}$ with the symmetry $\boldsymbol{f}$ and Möbius transformation $\varphi(t) = 1 - t$. Taking the degree to be one in (8.15) — the simplest case not yielding a pipe surface — and $a_1 = -a_0 = 1/2$, we obtain the radius function $r(t) = t - 1/2$, which satisfies Condition C2. The resulting symmetric canal surface $S_{\mathbf{c},r}$ is shown in Fig. 8.3, right.

Notice that since the parametrization (8.14) is polynomial, the surface S produced this way is unbounded; bounded surfaces can be obtained from bounded spine curves, for instance represented by NURBS.

8.4.2 Building symmetric blends

Suppose that for two canal surfaces $S_i = S_{\mathbf{c}_i, r_i}$, with $i = 1, 2$, we wish to construct a symmetric canal surface blend S connecting characteristic circles $k_i(t_i) \subset S_i$, i.e., $S \cap S_i = k_i(t_i)$; see Fig. 8.7. Now when we connect two surfaces, we may require different levels of continuity: G^0 means that both patches match along the common curve, G^1 means that the tangent planes of both surfaces match, and G^2 implies that the curvatures match. Taking into account that, from Remark 15, each S_i can be identified with a rational parametric curve $\big(\mathbf{c}_i(t); r_i(t)\big)$ in four-dimensional Minkowski space $\mathbb{R}^{3,1}$, in [48] it is shown that the level of continuity is achieved whenever the corresponding curves in $\mathbb{R}^{3,1}$ meet with the same level of continuity. Let us see how to apply this to two different subproblems, by enforcing G^N-continuity on the curves in $\mathbb{R}^{3,1}$, thus resulting in a G^N-continuous canal surface blend.

8.4.2.1 Symmetry with respect to the plane of coplanar spine curves

Suppose f is a reflection in a plane Π containing both spline curves $\mathbf{c}_1, \mathbf{c}_2$. We wish to construct a canal surface $S = S_{\mathbf{c},r}$ that is symmetric with respect to Π, i.e., invariant under f. This situation arises, for instance, when S_1, S_2 are surfaces of revolution with intersecting axes, which we wish to blend with a patch symmetric with respect to a plane containing the axes.

As explained in Example 21, a patch is symmetric with respect to Π whenever its spine curve $\mathbf{c}$ is contained in Π. Thus, we consider a Bézier curve $\mathbf{c}$ as in (8.14) of degree $n = 2N+1$, in which case the C^N-continuity conditions form the Hermite interpolation problem

$$\begin{aligned} \frac{n!}{(n-k)!}\Delta^k \boldsymbol{b}_0 &= \mathbf{c}^{(k)}(0) = \mathbf{c}_1^{(k)}(t_1) \\ \frac{n!}{(n-k)!}\Delta^k \boldsymbol{b}_{n-k} &= \mathbf{c}^{(k)}(1) = \mathbf{c}_2^{(k)}(t_2) \end{aligned}, \qquad k = 0, \dots, N, \tag{8.17}$$

where Δ^k is the k-th forward difference operator as in (8.16). Starting from $\boldsymbol{b}_0 := \mathbf{c}_1(t_1)$ and $\boldsymbol{b}_n = \mathbf{c}_2(t_2)$, the condition for C^0-continuity, the control points $\boldsymbol{b}_k, \boldsymbol{b}_{n-k}$ of $\mathbf{c}$ are determined from (8.17) for $k = 1, 2, \dots, N$. Notice in particular that no symmetry conditions are imposed on the $\boldsymbol{b}_i$.

Additionally, as explained in Example 21, any choice of the radius function r provides a canal surface symmetric with respect to Π. However, we need to impose continuity conditions on r as well, resulting in the Hermite interpolation problem

$$r^{(k)}(0) = r_1^{(k)}(t_1), \qquad r^{(k)}(1) = r_2^{(k)}(t_2), \qquad k = 0, \dots, N. \tag{8.18}$$

A polynomial r satisfying these conditions can be found as above, by expressing it in the Bernstein basis.

Example 23. *Let us consider the two cylinders S_1, S_2 in Fig. 8.7 with corresponding radii $r_1 = 1/2$, $r_2 = 1/4$ and axes (spine curves)*

$$\mathbf{c}_1(t) = (0, 0, -2t + 1), \ \ \mathbf{c}_2(t) = (0, 2t - 1, 0)$$

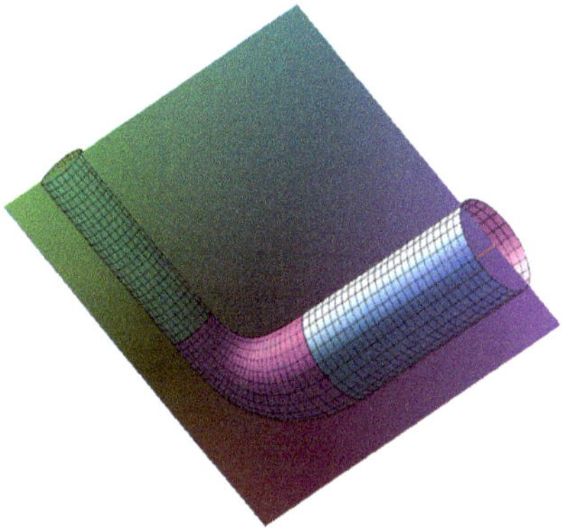

FIGURE 8.7 The two cylinders S_1 (right) and S_2 (left) from Example 23, blended by the symmetric canal surface patch S (middle).

intersecting in the origin. Applying the results in this section, we let $t_1 = 0$, $t_2 = 1$ *and connect* $\mathbf{c}_1((-\infty, t_1])$, $\mathbf{c}_2([t_2, \infty))$ *smoothly by a canal surface patch* S*, shown in Fig. 8.7, with cubic Bézier spine curve* $\mathbf{c}$ *as in* (8.14) *and radius function* $r(t) = \sum_{i=0}^{3} a_i B_{i,3}(t)$. *From* (8.17) *we determine*

$$\boldsymbol{b}_0 = \mathbf{c}(0) = \mathbf{c}_1(0) = (0,0,1), \qquad \boldsymbol{b}_3 = c(1) = \mathbf{c}_2(1) = (0,1,0),$$

$$\boldsymbol{b}_1 = \boldsymbol{b}_0 + \frac{1}{3}\mathbf{c}_1'(0) = (0,0,1/3), \qquad \boldsymbol{b}_2 = \boldsymbol{b}_3 - \frac{1}{3}\mathbf{c}_2'(1) = (0,1/3,0),$$

which reduces to a quadratic spine curve $\mathbf{c}(t) = \big(0, t^2, (t-1)^2\big)$, *while* (8.18) *yields*

$$a_0 = r(0) = r_1(0) = 1/2, \qquad a_3 = r(1) = r_2(1) = 1/4,$$

$$a_1 = a_0 + \frac{1}{3}r_1'(0) = 1/2, \qquad a_2 = a_3 - \frac{1}{3}r_2'(1) = 1/4,$$

so that $r(t) = (2-3t^2+2t^3)/4$. *Observe that the patch is symmetric with respect to the plane spanned by the axes of the cylinders.*

8.4.2.2 A more general case

More generally, suppose that we are given two canal surfaces S_1, S_2 and a nontrivial isometry f such that, for some integer N

and parameters t_1, t_2,

$$f \circ \mathbf{c}_1^{(k)}(t_1) = \mathbf{c}_2^{(k)}(t_2), \qquad k = 0, \ldots, N; \tag{8.19}$$

$$r_1^{(k)}(t_1) = \pm(-1)^k r_2^{(k)}(t_2), \qquad k = 0, \ldots, N. \tag{8.20}$$

For instance, Conditions (8.19) and (8.20) hold for any integer N and certain parameters t_1, t_2 when S_1, S_2 each have a unique spine curve and are related by an isometry f. Under these conditions, there exists a symmetric patch (invariant under f) with degree of continuity equal to N.

As before, we first choose a Bézier curve $\mathbf{c}$ of degree $n := 2N+1$ that satisfies the continuity conditions (8.17). Using (8.19) it follows that

$$\Delta^k \mathbf{b}_{n-k} = \frac{(n-k)!}{n!}\mathbf{c}_2^{(k)}(t_2) = \frac{(n-k)!}{n!} f \circ \mathbf{c}_1^{(k)}(t_1) = f\big(\Delta^k \boldsymbol{b}_0\big),$$

implying that the Bézier curve $\mathbf{c}$ is invariant under f. Hence, as shown in Section 8.4.1, $f \circ \mathbf{c} = \mathbf{c} \circ \varphi$ with $\varphi(t) = 1-t$. By Theorem 160, a canal surface with spine curve $\mathbf{c}$ is invariant under f when, in addition, its radius function r satisfies $r^2(t) = r^2(1-t)$, taking the form (8.15). Together with our assumption (8.20), the continuity conditions (8.18) become

$$r^{(k)}(0) = r_1^{(k)}(t_1) = \pm(-1)^k r_2^{(k)}(t_2) = \pm(-1)^k r^{(k)}(1), \quad k=0,\ldots,N,$$

which is compatible with our invariance condition $r^2(t) = r^2(1-t)$.

Example 24. *Consider the symmetric canal surface S from Example 22. We wish to replace a piece surrounding the singularity by a regular canal surface patch, without breaking the symmetry. For this purpose, we do not modify the spine curve, but replace, in the interval $[t_1, t_2] = [0, 1]$, the linear radius function $r_1(t) = t - 1/2$ by a quadratic radius function $r(t) = \sum_{i=0}^{2} a_i B_{i,2}(t)$. From (8.18), with $r_2 = -r_1$, imposing smoothness yields*

$$a_0 = r(0) = r_1(0) = -\frac{1}{2} = r_2(1) = r(1) = a_2,$$

$$0 = a_0 + \frac{1}{2}r_1'(0) = a_1 = a_2 - \frac{1}{2}r_2'(1) = 0.$$

Since the latter equation is consistent, we obtain a quadratic solution $r(t) = t^2 - t + 1/2$. The resulting smooth piecewise canal surface is shown in Fig. 8.8.

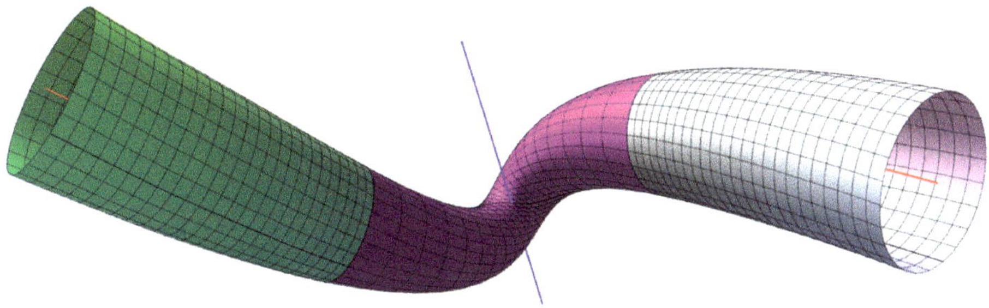

FIGURE 8.8 The symmetric canal surface from Example 22, with its middle singular part replaced by a regular canal surface patch with the same symmetry.

8.5 FINAL REMARKS AND OPEN PROBLEMS

Canal surfaces can be seen as the result of sweeping a circle of variable radius along a directrix curve, with the property that the center of the circle lies on the directrix, and the plane containing the circle is normal to the directrix. If we drop this last condition, so that the circle lies in a plane of changing direction, not necessarily normal to the directrix, we have a generalization of canal surfaces, called *ringed surfaces* [26, 27]. Like canal surfaces, if the directrix, the radius function and the normal vector field, i.e., the field providing the directions of the planes containing the circles, are rational, the ringed surface is also rational. Furthermore, ringed surfaces are used for blending purposes too, and in some cases, e.g. blending of oblique cones, ringed surfaces are preferred over canal surfaces [27]. Thus, we have the following, potential research question.

Open Problem 24. *Generalize the results in this chapter to the case of rational ringed surfaces*, including algorithms to generate blends with prescribed symmetries.

CHAPTER 9

Rational surfaces (III): translational and minimal

In this chapter we address *translational surfaces*, which are, essentially, rational surfaces obtained by sliding a rational curve along another rational curve; these two curves generate the surface and are contained in it, so translational surfaces give a simple solution to the problem of interpolating a surface containing two given curves. For these surfaces, we will be able to relate their symmetries to the isometries between the generating curves, therefore reducing the problem to the case of space rational curves, as we also did in the previous chapter. A remarkable subfamily of translational surfaces are *minimal surfaces*, better known for the characterizing property of having zero mean curvature and the fact that they can span a given closed space curve with minimal area. This optimal property makes them useful in Architecture, not only for aesthetic reasons but also to minimize material usage. Minimal surfaces are translational surfaces with conjugate complex generating curves, an impressive result already known to Karl Weierstrass and Sophus Lie, so the results for translational surfaces are also applicable to them.

DOI: 10.1201/9781003648710-9

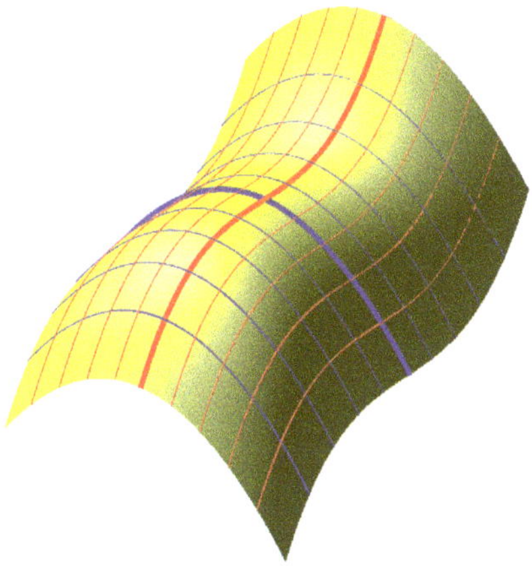

FIGURE 9.1 Translational surface parametrized by $\boldsymbol{x}(t,s) = (t, s, -t^2 + s^3)$

9.1 BACKGROUND ON TRANSLATIONAL AND MINIMAL SURFACES

A first description of a *translational surface* is the following: consider two rational curves, parametrized by $\boldsymbol{u}(t), \boldsymbol{v}(s)$; we will assume that both curves intersect at a point. Now, consider the rational surface S parametrized by

$$\boldsymbol{x}(t,s) = \boldsymbol{u}(t) + \boldsymbol{v}(s). \tag{9.1}$$

From this parametrization, we observe that the curves parametrized by $\boldsymbol{u}(t), \boldsymbol{v}(s)$ belong to S, and that in fact S carries two families of curves, one of them parallel to $\boldsymbol{u}(t)$ and the other one parallel to $\boldsymbol{v}(s)$, forming a net that covers S. Hence, S is constructed by letting $\boldsymbol{u}(t)$ slide along $\boldsymbol{v}(s)$, or conversely. The process is illustrated in Fig. 9.1: here we have taken $\boldsymbol{u}(t) = (t, 0, -t^2)$, $\boldsymbol{v}(s) = (0, s, s^3)$, so $\boldsymbol{x}(t,s) = (t, s, -t^2 + s^3)$. The curves $\boldsymbol{u}(t), \boldsymbol{v}(s)$ are plotted in thick blue and red line, respectively. Furthermore, we have plotted also in blue several curves parallel to $\boldsymbol{u}(t)$ contained in the surface; similarly for $\boldsymbol{v}(s)$, in red.

Although this description is really intuitive, we will consider a more formal description, though, that will be useful later when we introduce *minimal surfaces*. Thus, let $\mathbb{K}$ be a field, which in our case will be either $\mathbb{R}$ or $\mathbb{C}$, and let $\mathbb{K}^n$ be the corresponding n-dimensional affine space over $\mathbb{K}$. Following [131], we equip $\mathbb{K}^n$ with the binary operation $\oplus$ defined by taking the average, i.e.,

$$\oplus : \mathbb{K}^n \times \mathbb{K}^n \longrightarrow \mathbb{K}^n, \qquad p \oplus q := \frac{p+q}{2}. \tag{9.2}$$

By an abuse of notation, we can also consider $\oplus$ as a binary operation on the space of rational parametrizations,

$$\begin{aligned} \oplus : \mathrm{Hom}(\mathbb{K}, \mathbb{K}^n) \times \mathrm{Hom}(\mathbb{K}, \mathbb{K}^n) &\longrightarrow \mathrm{Hom}(\mathbb{K}^2, \mathbb{K}^n), \\ (\boldsymbol{p} \oplus \boldsymbol{q})(t, s) &:= \boldsymbol{p}(t) \oplus \boldsymbol{q}(s), \end{aligned} \tag{9.3}$$

where $\mathrm{Hom}(\mathbb{K}^m, \mathbb{K}^n)$ denotes the space of rational maps $\mathbb{K}^m \dashrightarrow \mathbb{K}^n$. Note that composition with an affine map $f(\boldsymbol{x}) = M\boldsymbol{x} + \boldsymbol{b}$ is distributive with respect to $\oplus$, i.e.,

$$f \circ (\mathbf{c}_1 \oplus \mathbf{c}_2) = (f \circ \mathbf{c}_1) \oplus (f \circ \mathbf{c}_2), \tag{9.4}$$

because, for any t, s,

$$\begin{aligned} f \circ (\mathbf{c}_1 \oplus \mathbf{c}_2)(t, s) &= \frac{M\mathbf{c}_1(t) + \boldsymbol{b} + M\mathbf{c}_2(s) + \boldsymbol{b}}{2} \\ &= (f \circ \mathbf{c}_1)(t) \oplus (f \circ \mathbf{c}_2)(s). \end{aligned} \tag{9.5}$$

Definition 170. *We say that $S \subset \mathbb{K}^3$ is a* translational surface *if S admits a parametrization $\boldsymbol{x}(t, s) = \mathbf{c}_1(t) \oplus \mathbf{c}_2(s)$, where $\mathbf{c}_1(t), \mathbf{c}_2(s)$ parametrize two space curves $\mathcal{C}_1, \mathcal{C}_2 \subset \mathbb{K}^3$. We say that $(\mathcal{C}_1, \mathcal{C}_2)$ is a* generator pair *of S.*

Here we will consider rational translational surfaces; in particular, we will assume that $\mathbf{c}_1(t), \mathbf{c}_2(s)$ are rational parametrizations, in which case $\boldsymbol{x}(t, s)$ is also rational. The converse is not necessarily true, see Section 2.2 in [131]. We will also assume that $\mathbf{c}_1(t), \mathbf{c}_2(s)$ intersect. Notice that comparing with Eq. (9.1) we have $\boldsymbol{u}(t) = \frac{1}{2}\mathbf{c}_1(t)$, $\boldsymbol{v}(s) = \frac{1}{2}\mathbf{c}_2(s)$; also, the intersection of $\mathbf{c}_1(t), \mathbf{c}_2(s)$ is a point of S. However, Definition 170 adds a more "arithmetical" flavor, since translational surfaces can be seen now as the result of "averaging" the points in $\mathcal{C}_1, \mathcal{C}_2$; these two curves, though, need not lie on S anymore. Definition 170 also allows for a simple interpretation of the case when both generating curves are equal: if $\mathbf{c}_1 = \mathbf{c}_2 = \mathbf{c}$, the surface $\boldsymbol{x}(t,s) = \mathbf{c}(t) \oplus \mathbf{c}(s)$ consists of the averaging of all the pairs of different points in $\mathbf{c}(t)$. Thus, if $\mathbf{c}$ defines a closed space curve, what we get is a surface that spans this curve, which in this case does belong to the surface.

Definition 170 and this idea of averaging is illustrated by Figs. 9.2 and 9.3. In Fig. 9.2, we have the translational surface S of Fig. 9.1, with the curves $\mathbf{c}_1, \mathbf{c}_2$ plotted in green and purple, respectively; one may see that both curves intersect at a point of S, but do not lie on S anymore. In Fig. 9.3, we have the same surface S and the curves $\mathbf{c}_1, \mathbf{c}_2$; the three points shown in black lie on the same line, one of them in $\mathbf{c}_1$, the other one in $\mathbf{c}_2$, and the middle point between these two points in S.

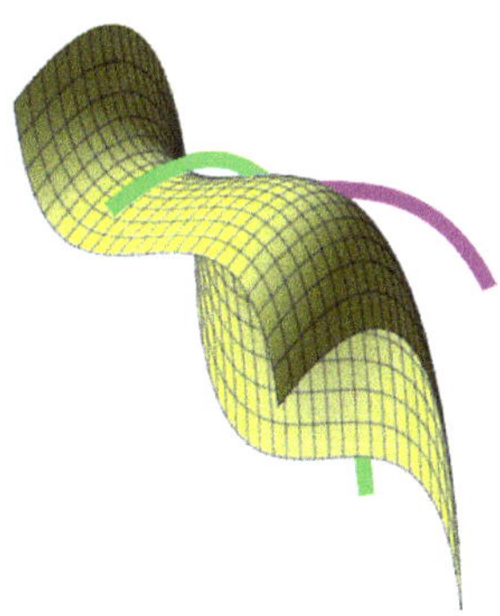

FIGURE 9.2 Illustrating Definition 170 (I)

Notice that the generator pair $(\mathcal{C}_1, \mathcal{C}_2)$ is not unique. Indeed, if we denote by $\tau_{\mathbf{v}}$ the translation by a vector $\mathbf{v}$, then replacing the curves $\boldsymbol{u}(t), \boldsymbol{v}(s)$ in Eq. (9.1) by $\tau_{\boldsymbol{v}(s_0)}(\boldsymbol{u}(t))$, $\tau_{\boldsymbol{u}(t_0)}(\boldsymbol{v}(s))$ for any t_0, s_0, yields the same surface S. Taking into account the relationship between $\boldsymbol{u}(t), \boldsymbol{v}(s)$ and $\mathbf{c}_1(t), \mathbf{c}_2(s)$, we observe that any other pair $(\tilde{\mathcal{C}}_1, \tilde{\mathcal{C}}_2)$, where $\tilde{\mathcal{C}}_1 = \tau_{\mathbf{c}_2(s_0)}(\mathcal{C}_1)$, $\tilde{\mathcal{C}}_2 = \tau_{\mathbf{c}_1(t_0)}(\mathcal{C}_2)$, is

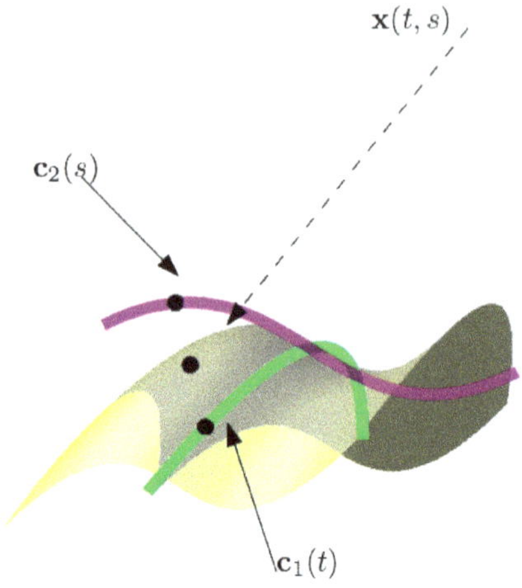

FIGURE 9.3 Illustrating Definition 170 (II)

also a generator pair of S. However, there are surfaces with other, exotic generator pairs that are not parallel curves to the original generators. These surfaces, called *multitranslational surfaces*, were already known to Sophus Lie, who in 1882 (c.f. [41, §1] or Theorem 2 in [131]) gave a necessary condition for them. This condition can be roughly put as follows: seeing the tangent vectors to the generator curves as projective planar points, all these points must belong to an algebraic curve of degree four.

Here we will assume that the surface S we are working with is not multitranslational. Additionally, we will assume that $\boldsymbol{x}(t, s)$ is a proper parametrization, so that $\mathbf{c}_1(t), \mathbf{c}_2(s)$ must be proper parametrizations as well. The converse is not necessarily true, though. Also, we will assume that $\mathcal{C}_1, \mathcal{C}_2$ are not planar curves contained in the same plane or in parallel planes, since in that case S would be a plane.

Finally, let us introduce a very special class of translational surfaces, namely *minimal* surfaces. Minimal surfaces are defined as surfaces with constant vanishing *mean curvature*. They are better known for the property that given a space Jordan curve Γ, there always exists a minimal surface having Γ in its boundary and spanning Γ with minimal area (see for instance [94], p. 67). However, the converse is not necessarily true, i.e., a minimal surface might have Γ in its boundary but might not be the surface of minimal area enclosing it (see for instance [107], p. 38). There are illustrative physical examples of minimal surfaces provided by soap films spanning a given wireframe; also, classical surfaces like the helicoid and the catenoid are minimal surfaces.

For us, however, the most relevant fact about minimal surfaces is that they are surfaces of translation with complex conjugate generating pairs (see [120, §5.4]). This follows from a classic representation of minimal surfaces, called the *Weierstrass form* of the surface. Weierstrass proved [52, §3.3] that any nonplanar minimal surface S defined over a simply-connected parameter domain can be parametrized as the real part

$$\mathbf{s}(t,s) = \Re\big(\Psi(z)\big) = \frac{\Psi(z) + \overline{\Psi(z)}}{2} = \Re\big(\Psi(t+is)\big), \qquad (t,s) \in \mathbb{R}^2, \tag{9.6}$$

of the complex curve $\mathcal{A} \subset \mathbb{C}^3$ with parametrization

$$\Psi = (\Psi_1, \Psi_2, \Psi_3) : U \subset \mathbb{C} \longrightarrow \mathbb{C}^3,$$

where, with $i^2 = -1$, $\Psi(z) = (\Psi_1(z), \Psi_2(z), \Psi_3(z))$,

$$\begin{aligned}
\Psi_1(z) &= \int_{z_0}^{z} f(\gamma)\frac{1-g(\gamma)^2}{2}\mathrm{d}\gamma, \\
\Psi_2(z) &= \mathbf{i}\int_{z_0}^{z} f(\gamma)\frac{1+g(\gamma)^2}{2}\mathrm{d}\gamma, \\
\Psi_3(z) &= \int_{z_0}^{z} f(\gamma)g(\gamma)\mathrm{d}\gamma.
\end{aligned} \tag{9.7}$$

Here f is holomorphic and g is meromorphic such that fg^2 is holomorphic in a simply-connected region $U \subset \mathbb{C}$ containing z_0. A straightforward calculation shows that Ψ is an *isotropic curve*, i.e.,

$$\Psi_1'(z)^2 + \Psi_2'(z)^2 + \Psi_3'(z)^2 = 0. \tag{9.8}$$

In the context of minimal surfaces, Ψ is sometimes called a *minimal curve*. We will say that Ψ *generates* the minimal surface. Thus, any minimal surface S can be parametrized as Eq. (9.6), where Ψ generates S. Notice that before performing the change $z \to t + is$, with $i^2 = -1$, we have a complex parametrization of S like $\boldsymbol{x}(z, \overline{z})$, where $z, \overline{z}$ are regarded as independent variables. However, after performing the transformation

$$\iota : (z, \overline{z}) \to (t + is, t - is), \tag{9.9}$$

with $i^2 = -1$, since we have the sum of a complex function and its conjugate we get a *real* parametrization of S, i.e., $\mathbf{s}(t, s)$ has real coefficients. In particular, we have $S = \mathcal{A} \oplus \overline{\mathcal{A}}$, so S is a surface of translation with complex generator pair $(\mathcal{A}, \overline{\mathcal{A}})$.

Example 25. *Let us illustrate the above ideas in the case of a well-known minimal surface, the* Enneper *surface. Let*

$$\Psi(z) = (\Psi_1(z), \Psi_2(z), \Psi_3(z)) = \left(z - \frac{z^3}{3}, i\left(z + \frac{z^3}{3}\right), z^2\right), \tag{9.10}$$

which comes from Eq. (9.7), taking $z_0 = 0$, $f(z) = 2$, $g(z) = z$. One can directly check that this $\Psi(z)$ satisfies Eq. (9.8), i.e., it is a minimal curve. Conjugating Eq. (9.10), we get

$$\overline{\Psi(z)} = \overline{\Psi}(\overline{z}) = (\overline{\Psi_1(z)}, \overline{\Psi_2(z)}, \overline{\Psi_3(z)}) = \left(\overline{z} - \frac{\overline{z}^3}{3}, -i\left(\overline{z} + \frac{\overline{z}^3}{3}\right), \overline{z}^2\right). \tag{9.11}$$

Notice the change of sign in the second component of Eq. (9.11), compared to Eq. (9.10), because of the conjugation. If we now take $\boldsymbol{x}(z, \overline{z}) = \frac{\Psi(z) + \overline{\Psi(z)}}{2}$, we get a complex parametrization of a surface S, where the parameters are $z, \overline{z}$, regarded as independent variables. The transformation ι in Eq. (9.9), which is nothing else but a reparametrization, leads to a real parametrization $\mathbf{s}(t, s)$ of the surface, namely

$$\mathbf{s}(t, s) = \left(s + \frac{s^3}{3} - t^2 s, t + \frac{t^3}{3} - ts^2, 2ts\right).$$

Fig. 9.4 shows two different views of the surface. From these pictures, one can guess that the surface has several symmetries, of different types. Let us mention that all minimal bicubic Bézier surfaces are affinely equivalent to this surface, a fact that makes it relevant in Computer Aided Geometric Design.

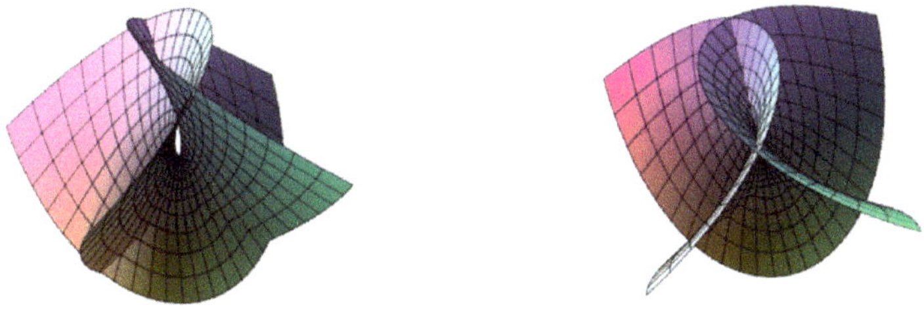

FIGURE 9.4 Two views of the Enneper surface, from different viewpoints

9.2 SYMMETRIES OF TRANSLATIONAL SURFACES

The ideas, results and examples in this section and the coming section are taken from [19]. Let S be a translational surface, not multitranslational, parametrized by $\boldsymbol{x}(t,s) = \mathbf{c}_1(t) \oplus \mathbf{c}_2(t)$, where $\mathbf{c}_1(t), \mathbf{c}_2(s)$ are proper, rational parametrizations defining two space curves $\mathcal{C}_1, \mathcal{C}_2$. We want to find the symmetries of S. Thus, let $f(\boldsymbol{x}) = A\boldsymbol{x} + \boldsymbol{b}$ be an isometry leaving S invariant. Therefore, $f \circ (\mathbf{c}_1(t) \oplus \mathbf{c}_2(s))$ is another parametrization of S. By the distributive property illustrated in Eq. (9.5), $f(\mathbf{c}_1(t) \oplus \mathbf{c}_2(s)) = (f \circ \mathbf{c}_1)(t) \oplus (f \circ \mathbf{c}_2)(s)$, so $(f \circ \mathbf{c}_1)(t), (f \circ \mathbf{c}_2)(s)$ is a generator pair of S. Thus, from the observations in the previous section we deduce that $f \circ \mathbf{c}_1, f \circ \mathbf{c}_2$ must coincide with $\tau_{\mathbf{c}_1(t_0)}(\mathbf{c}_2), \tau_{\mathbf{c}_2(s_0)}(\mathbf{c}_1)$ for some t_0, s_0, in whatever order. Then we have the following result.

Theorem 171. *Let $f(\boldsymbol{x}) = A\boldsymbol{x} + \boldsymbol{b}$ be an isometry. Then f is a symmetry of S if and only if at least one of the following situations occurs:*

1. *There exists t_0, s_0 such that $f(\mathcal{C}_1) = \tau_{\mathbf{c}_2(s_0)}(\mathcal{C}_1)$, $f(\mathcal{C}_2) = \tau_{\mathbf{c}_1(t_0)}(\mathcal{C}_2)$.*

2. *There exists t_0, s_0 such that $f(\mathcal{C}_1) = \tau_{\mathbf{c}_1(t_0)}(\mathcal{C}_2)$, $f(\mathcal{C}_2) = \tau_{\mathbf{c}_2(s_0)}(\mathcal{C}_1)$.*

Thus, if $f(\boldsymbol{x}) = A\boldsymbol{x}+\boldsymbol{b}$ satisfies case (1) in Theorem 171, taking Theorem 45 in Chapter 3 into account and the fact that $\mathbf{c}_1, \mathbf{c}_2$ are proper, we get that

$$\begin{aligned} A\mathbf{c}_1(t) + \boldsymbol{b} &= (\mathbf{c}_1 \circ \varphi)(t) + \mathbf{c}_2(s_0), \\ A\mathbf{c}_2(s) + \boldsymbol{b} &= (\mathbf{c}_2 \circ \psi)(t) + \mathbf{c}_1(t_0) \end{aligned}$$

for Möbius transformations φ, ψ, and therefore

$$\begin{aligned} A\mathbf{c}_1(t) + (\boldsymbol{b} - \mathbf{c}_2(s_0)) &= (\mathbf{c}_1 \circ \varphi)(t), \\ A\mathbf{c}_2(s) + (\boldsymbol{b} - \mathbf{c}_1(t_0)) &= (\mathbf{c}_2 \circ \psi)(t). \end{aligned}$$

Notice that the first (second) equality is saying that $f_1(\boldsymbol{x}) = A\boldsymbol{x} + \boldsymbol{b}_1$ (resp. $f_2(\boldsymbol{x}) = A\boldsymbol{x} + \boldsymbol{b}_2$), with $\boldsymbol{b}_1 = \boldsymbol{b} - \mathbf{c}_2(s_0)$ (resp. $\boldsymbol{b}_2 = \boldsymbol{b} - \mathbf{c}_1(t_0)$) is a symmetry of $\mathcal{C}_1$ (resp. $\mathcal{C}_2$). For case (2) in Theorem 171 we get something similar. Thus, we deduce the following corollary.

Corollary 172. *If S is symmetric, then at least one of the following situations occur:*

(1) $\mathcal{C}_1, \mathcal{C}_2$ are both symmetric curves, invariant by isometries $f_1(\boldsymbol{x}) = A\boldsymbol{x} + \boldsymbol{b}_1$, $f_2(\boldsymbol{x}) = A\boldsymbol{x} + \boldsymbol{b}_2$ (sharing the same matrix A), respectively, where $\boldsymbol{b}_1 - \boldsymbol{b}_2 = \mathbf{c}_1(t_0) - \mathbf{c}_2(s_0)$ for some t_0, s_0. In that case, $f(\boldsymbol{x}) = A\boldsymbol{x} + \boldsymbol{b}$, with $\boldsymbol{b} = \boldsymbol{b}_1 + \mathbf{c}_2(s_0) = \boldsymbol{b}_2 + \mathbf{c}_1(t_0)$ is a symmetry of S.

(2) $\mathcal{C}_1, \mathcal{C}_2$ are isometric curves, and there are two isometries $f_1(\boldsymbol{x}) = A\boldsymbol{x} + \boldsymbol{b}_1$, $f_2(\boldsymbol{x}) = A\boldsymbol{x} + \boldsymbol{b}_2$ (sharing the same matrix A), satisfying $f_1 = f_2^{-1}$, where $\boldsymbol{b}_1 - \boldsymbol{b}_2 = \mathbf{c}_2(s_0) - \mathbf{c}_1(t_0)$ for some t_0, s_0. In that case, $f(\boldsymbol{x}) = A\boldsymbol{x} + \boldsymbol{b}$, with $\boldsymbol{b} = \boldsymbol{b}_1 + \mathbf{c}_1(t_0) = \boldsymbol{b}_2 + \mathbf{c}_2(s_0)$ is a symmetry of S.

Furthermore, all the symmetries of S can be obtained this way.

Additionally, we also have the following corollary, which allows us to generate symmetric translational surfaces.

Corollary 173. *Let $f(\boldsymbol{x}) = A\boldsymbol{x} + \boldsymbol{b}$ be an isometry.*

(1) If f is a common symmetry of $\mathcal{C}_1, \mathcal{C}_2$, then f is also a symmetry of the translational surface $S = \mathcal{C}_1 \oplus \mathcal{C}_2$.

Algorithm 6 Symm-Trans

Require: A surface of translation $S = \mathcal{C}_1 \oplus \mathcal{C}_2$, properly and rationally parametrized by $\mathbf{s}(t,s) = \mathbf{c}_1(t) \oplus \mathbf{c}_2(s)$, where $\mathbf{c}_1, \mathbf{c}_2$ are non-coplanar or placed in parallel planes.

Ensure: The symmetries $f(\boldsymbol{x}) = A\boldsymbol{x} + \boldsymbol{b}$ of S, if any.

for $i = 1,2$ and $j = 1,2$ **do**
  Determine the isometries $f_{ij} : \mathcal{C}_i \longrightarrow \mathcal{C}_j$.
end for
for any pair $(f_{11}(\boldsymbol{x}) = A\boldsymbol{x}+\boldsymbol{b}_{11}, f_{22}(\boldsymbol{x}) = A\boldsymbol{x}+\boldsymbol{b}_{22})$ with equal matrix **do**
  check whether or not there exist t_0, s_0 such that $\boldsymbol{b}_{11} - \boldsymbol{b}_{22} = \mathbf{c}_1(t_0) - \mathbf{c}_2(s_0)$
  if the above condition holds **then**
    return $f(\boldsymbol{x}) = A\boldsymbol{x} + \boldsymbol{b}$, with $\boldsymbol{b} = \boldsymbol{b}_{11} + \mathbf{c}_2(s_0)$
  end if
end for
for any pair $(f_{12}(\boldsymbol{x}) = A\boldsymbol{x}+\boldsymbol{b}_{12}, f_{21}(\boldsymbol{x}) = A\boldsymbol{x}+\boldsymbol{b}_{21})$ with equal matrix **do**
  check whether or not there exist t_0, s_0 such that $\boldsymbol{b}_{12} - \boldsymbol{b}_{21} = \mathbf{c}_2(s_0) - \mathbf{c}_1(t_0)$
  if the above condition holds **then**
    return $f(\boldsymbol{x}) = A\boldsymbol{x} + \boldsymbol{b}$, with $\boldsymbol{b} = \boldsymbol{b}_{12} + \mathbf{c}_1(t_0)$
  end if
end for
if no symmetry has been found **then**
  return S `is not symmetric`
end if

(2) *If $f(\mathcal{C}_1) = \mathcal{C}_2$ and $f \circ f$ is the identity, then f is also a symmetry of the translational surface $S = \mathcal{C}_1 \oplus \mathcal{C}_2$.*

Proof. It follows from Corollary 172 taking t_0, s_0 so that $\mathbf{c}_1(t_0) = \mathbf{c}_2(s_0)$; recall that such t_0, s_0 exist because we are assuming that $\mathcal{C}_1, \mathcal{C}_2$ intersect. □

So we get an algorithm, `Symm-Trans`, for computing the symmetries of a rational translational surface.

Remark 18. *Because Möbius transformations are also the bi-analytic bijections on the complex projective line, results analogous to Theorem 171, Corollary 172 and Corollary 173 hold when we consider* $\mathbf{c}_1, \mathbf{c}_2$ *to be not birational, but bi-meromorphic.*

Example 26 (Crunode). *Let* $C_1 \subset \mathbb{R}^3$ *be the crunode curve parametrized by*

$$\mathbf{c}_1(t) = \big(x(t), y(t), z(t)\big) = \left(\frac{t}{t^4+1}, \frac{t^2}{t^4+1}, \frac{t^3}{t^4+1}\right). \tag{9.12}$$

Using the methods in Chapter 5, one can check that this curve is invariant under the half-turn about the y*-axis, as*

$$\begin{bmatrix} -1 & 0 & 0 \\ 0 & 1 & 0 \\ 0 & 0 & -1 \end{bmatrix} \begin{bmatrix} x(t) \\ y(t) \\ z(t) \end{bmatrix} = \begin{bmatrix} x(-t) \\ y(-t) \\ z(-t) \end{bmatrix}.$$

as well as reflections in the planes $z \pm x = 0$*, since*

$$\begin{bmatrix} 0 & 0 & \pm 1 \\ 0 & 1 & 0 \\ \pm 1 & 0 & 0 \end{bmatrix} \begin{bmatrix} x(t) \\ y(t) \\ z(t) \end{bmatrix} = \begin{bmatrix} \pm z(t) \\ y(t) \\ \pm x(t) \end{bmatrix} = \begin{bmatrix} x(\pm 1/t) \\ y(\pm 1/t) \\ z(\pm 1/t) \end{bmatrix}.$$

We consider now the curves $\mathbf{c}_2$ *parametrized as* $\mathbf{c}_2 = \mathbf{c}_1$, $\mathbf{c}_2 = \boldsymbol{M}_y \mathbf{c}_1$ *and* $\mathbf{c}_2 = \boldsymbol{M}_z \mathbf{c}_1$*, where the matrices* $\boldsymbol{M}_y$ *and* $\boldsymbol{M}_z$ *denote reflections in the planes* $y = 0$ *and* $z = 0$*. Corollary 173 implies that the translational surfaces* $\mathbf{c}_1 \oplus \mathbf{c}_2$ *are symmetric; these surfaces are shown in the first row of Table 9.1.*

Example 27 (Trefoil knot). *Let* $C_1 \subset \mathbb{R}^3$ *be the trefoil knot parametrized by*

$$\begin{aligned} \mathbf{c}_1(\theta) &= \big(x(\theta), y(\theta), z(\theta)\big) \\ &:= \big(\sin(\theta) + 2\sin(2\theta), \cos(\theta) - 2\cos(2\theta), -\sin(3\theta)\big). \end{aligned} \tag{9.13}$$

Since the parametrization is trigonometric (i.e., each component is polynomial in $\sin(\theta)$ *and* $\cos(\theta)$*), the reparametrization* $z = e^{i\theta}$ *provides a rational, complex parametrization* $\widehat{\mathbf{c}}_1 : S^1 \longrightarrow C_1$*, with* $S^1 := \{z \in \mathbb{C} : |z| = 1\}$ *the unit circle. Then, applying the*

TABLE 9.1 Translational surfaces inheriting symmetries shared by their generator pairs.

$\mathbf{c}_1$	$\mathbf{c}_1 \oplus \mathbf{c}_1$	$\mathbf{c}_1 \oplus \boldsymbol{M}_y\mathbf{c}_1$	$\mathbf{c}_1 \oplus \boldsymbol{M}_z\mathbf{c}_1$
Crunode (9.12)			
	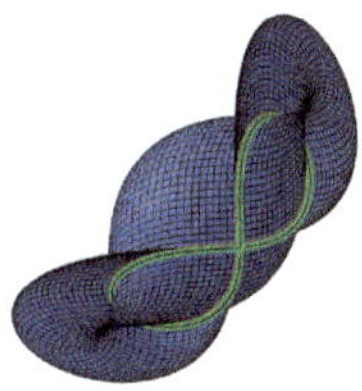		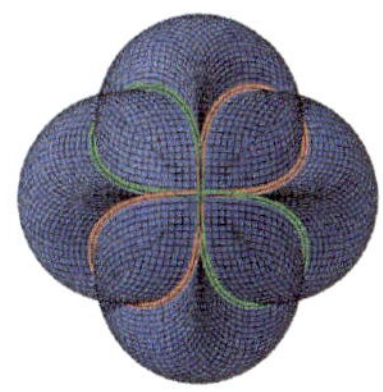
Trefoil knot (9.13)			
		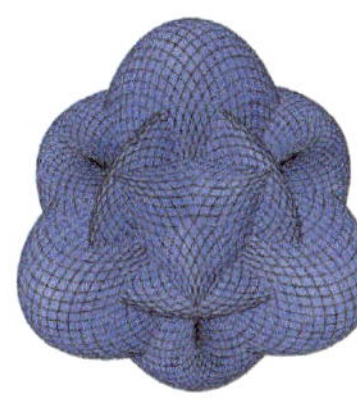	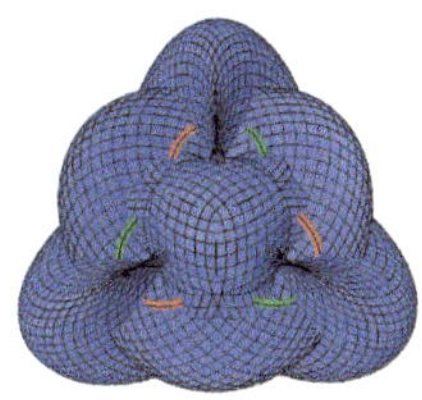

methods in Chapter 5 one obtains rotational symmetries about the z-axis by angles $\pm 2\pi/3$, as well as a half-turn about the y-axis. Alternatively, these symmetries are determined directly by applying a rotation matrix and standard trigonometric identities. In particular,

$$\begin{bmatrix} \cos\varphi & -\sin\varphi \\ \sin\varphi & \cos\varphi \end{bmatrix} \begin{bmatrix} x(\theta) \\ y(\theta) \end{bmatrix} = \begin{bmatrix} \sin(\theta-\varphi) + 2\sin(2\theta+\varphi) \\ \cos(\theta-\varphi) - 2\cos(2\theta+\varphi \end{bmatrix} = \begin{bmatrix} x(\theta-\varphi) \\ y(\theta-\varphi) \end{bmatrix}$$

holds identically if and only if $\varphi \equiv 0$ or $\varphi \equiv \pm 2\pi/3$ modulo 2π, while

$$\begin{bmatrix} -1 & 0 & 0 \\ 0 & 1 & 0 \\ 0 & 0 & -1 \end{bmatrix} \begin{bmatrix} x(\theta) \\ y(\theta) \\ z(\theta) \end{bmatrix} = \begin{bmatrix} x(-\theta) \\ y(-\theta) \\ z(-\theta) \end{bmatrix}.$$

Choosing $\mathbf{c}_2$ as in the previous example yields the symmetric translational surfaces $S = \mathbf{c}_1 \oplus \mathbf{c}_2$ shown in the second row of Table 9.1.

9.3 SYMMETRIES OF MINIMAL SURFACES

Here we consider minimal surfaces $S = \mathcal{A} \oplus \overline{\mathcal{A}}$ with rational parametrizations $\mathbf{s}(t,s) = (\mathbf{c} \circ \overline{\mathbf{c}}) \circ \iota$, where $\mathbf{c} = \mathbf{c}(z)$ is a minimal curve as in Eq. (9.7) parametrizing $\mathcal{A}$, and ι is the mapping in Eq. (9.9). Recall that the parametrization $\mathbf{s}(t,s)$ of S is, then, real. If $\mathbf{s}(t,s)$ is rational, then $\mathbf{c}$ must also be rational, see Corollary 22.25 in [65]; in this case, the functions f, g in Eq. (9.7) must be rational as well (although not every pair of rational functions f, g yields a rational parametrization of the type in Eq. (9.7), since the primitives might not be rational). We will make the additional assumptions that: (i) $\mathbf{c}$ is proper; (ii) $\mathbf{c}, \overline{\mathbf{c}}$ have a common (real) point, which by simplicity we will suppose to be the origin; (iii) S is not multitranslational.

We can certainly apply the results in the previous section to find the symmetries in this case. However, now we have extra benefits. Indeed, the first step in Algorithm `Symm-Trans` is to look for the isometries of $\mathcal{A}, \overline{\mathcal{A}}$, and the isometries mapping $\mathcal{A}$ to $\overline{\mathcal{A}}$. However, we are looking for real isometries $f(\boldsymbol{x}) = A\boldsymbol{x} + \boldsymbol{b}$, so that $A, \boldsymbol{b}$ are real. Thus, let f be a symmetry of $\mathcal{A}$,

$$A\mathbf{c}(z) + \boldsymbol{b} = (\mathbf{c} \circ \varphi)(z),$$

with φ a Möbius function. Conjugating this equality and since $A, \boldsymbol{b}$ are real, we get

$$A\overline{\mathbf{c}}(\omega) + \boldsymbol{b} = (\overline{\mathbf{c}} \circ \overline{\varphi})(\omega)$$

with $\omega = \overline{z}$, so f is also a symmetry of $\overline{\mathcal{A}}$, whose associated Möbius transformation is the conjugate of the Möbius transformation for $\mathbf{c}$. This means that we have fewer computations in this case. Furthermore, comparing with Corollary 172, we get that $\boldsymbol{b}_1 = \boldsymbol{b}_2$; and because we are assuming that $\mathcal{A}, \overline{\mathcal{A}}$ intersect at the origin, we can take t_0, s_0 so that $\boldsymbol{b}_1 = \boldsymbol{b}_2 = \boldsymbol{b}$. The situation is completely analogous for the isometries between $\mathcal{A}$ and $\overline{\mathcal{A}}$. Therefore, we reach the following result.

Proposition 174. *Let S be a rational minimal surface that is not multitranslational, rationally parametrized by $\mathbf{s} = (\mathbf{c} \oplus \overline{\mathbf{c}}) \circ \iota$, with $\mathbf{c}$ a proper parametrization of a minimal curve $\mathcal{C}$, where $\mathbf{c}, \overline{\mathbf{c}}$ intersect* *at a point. Then f is a (real) symmetry of S if and only if either f is a symmetry of* $\mathbf{c}$*, or f is an isometry between* $\mathbf{c}, \overline{\mathbf{c}}$.

We will now introduce an additional assumption, namely that the functions f, g in Eq. (9.7) are not only rational, but with real coefficients. The Enneper surface in Example 25 satisfies this condition, but in fact many of the classical algebraic minimal surfaces do as well; see for instance [91] and Chapter 5 of [65], where many examples of algebraic (and rational) minimal surfaces are given. If f, g have real coefficients, conjugating Eq. (9.7) shows that if $\mathbf{c}(z) = (c_1(z), c_2(z), c_3(z))$ parametrizes $\mathcal{A}$, then $\overline{\mathcal{A}}$ admits the parametrization

$$\overline{\mathbf{c}}(\omega) = (\overline{c_1}(\omega), \overline{c_2}(\omega), \overline{c_3}(\omega)) = (c_1(\omega), -c_2(\omega), c_3(\omega)).$$

This implies that for the Möbius transformation $\varphi(z) = z$ and the isometry $f(\boldsymbol{x}) = A\boldsymbol{x} + \boldsymbol{b}$, with

$$A = \begin{bmatrix} 1 & 0 & 0 \\ 0 & -1 & 0 \\ 0 & 0 & 1 \end{bmatrix}, \quad \boldsymbol{b} = \mathbf{0},$$

which defines a reflection in the plane $y = 0$, we have

$$f \circ \mathbf{c} = \overline{\mathbf{c}} \circ \varphi.$$

Hence Proposition 174 states that f is a symmetry of the surface S parametrized by $\mathbf{s} = (\mathbf{c} \oplus \overline{\mathbf{c}}) \circ \iota$. Thus we recover the following known result, which reveals that rational minimal surfaces generated by minimal curves $\mathbf{c}$ constructed from real rational functions f, g always have at least one mirror symmetry.

Corollary 175. *Every rational minimal surface parametrized by* $\mathbf{s} = (\mathbf{c} \oplus \overline{\mathbf{c}}) \circ \iota$ *with f, g rational functions with real coefficients is symmetric with respect to the plane $y = 0$.*

9.3.1 Higher-order Enneper surfaces

In this subsection we illustrate how Proposition 174 can be used to compute the symmetries of the *(higher-order) Enneper surfaces* S_k, for $k = 1, 2, \ldots$ (see [76]). These are the minimal surfaces obtained by taking constant $f = 2$ and monomial $g = z^k$ in Eq. (9.7); in particular, for $k = 1$ we get the Enneper surface in Example 25. The higher-order Enneper surfaces are classical examples of minimal surfaces with polynomial parametrizations.

We require an explicit parametrization of S_k, which we derive due to lack of a suitable reference. The proof involves the *Chebyshev polynomial* T_n of the first kind, defined recursively by

$$T_0(x) := 1, \qquad T_1(x) := x, \qquad T_n(x) := 2xT_{n-1}(x) - T_{n-2}(x),$$

with $n \geq 2$, or implicitly by

$$T_n\big(\cos(\theta)\big) = \cos(n\theta), \qquad n \geq 0. \tag{9.14}$$

Substituting $\theta = \frac{\pi}{2} - \theta'$ in (9.14), one obtains

$$\begin{aligned} T_n\big(\sin(\theta')\big) = T_n\big(\cos(\theta)\big) = \cos(n\theta) &= \cos\left(\frac{n\pi}{2} - n\theta'\right) \\ &= \cos\left(\frac{n\pi}{2}\right)\cos(n\theta') + \sin\left(\frac{n\pi}{2}\right)\sin(n\theta'), \end{aligned}$$

yielding the lesser-known identity

$$T_n\big(\sin(\theta)\big) = (-1)^k \sin(n\theta), \qquad n = 2k+1 \geq 1. \tag{9.15}$$

Proposition 176. *For $k \geq 1$, the higher-order Enneper surface S_k admits the parametrization, denoted $\mathbf{s}_k(t, s)$,*

$$\left(t - \frac{r^{2k+1}T_{2k+1}\left(\frac{t}{r}\right)}{2k+1}, -s - (-1)^k\frac{r^{2k+1}T_{2k+1}\left(\frac{s}{r}\right)}{2k+1}, 2\frac{r^{k+1}T_{k+1}\left(\frac{t}{r}\right)}{k+1}\right). \tag{9.16}$$

Proof. Write $z = re^{i\theta} = t + is$. With $f(z) = 2$ and $g(z) = z^k$, the expression (9.7) yields the minimal curve

$$\mathbf{c}_k(z) = \left(z - \frac{z^{2k+1}}{2k+1}, iz + i\frac{z^{2k+1}}{2k+1}, 2\frac{z^{k+1}}{k+1}\right). \tag{9.17}$$

From (9.14) it follows that

$$\frac{z^n + \overline{z}^n}{2} = r^n \cos(n\theta) = r^n T_n\big(\cos(\theta)\big) = r^n T_n\left(\frac{t}{r}\right)$$

for $n \geq 0$, while (9.15) implies

$$\frac{z^n - \overline{z}^n}{2i} = r^n \sin(n\theta) = (-1)^k r^n T_n\big(\sin(\theta)\big) = (-1)^k r^n T_n\left(\frac{s}{r}\right)$$

for $n = 2k+1 \geq 1$. Hence the statement follows from $\mathbf{s}_k(t,s) = \Re\big(\mathbf{c}_k(t+is)\big)$. □

Remark 19. *With $r = \sqrt{t^2+s^2}$ and $n = 2k+\varepsilon$ with $k \geq 0$ and $\varepsilon \in \{0,1\}$, one can show by induction that*

$$r^n T_n\left(\frac{t}{r}\right) = \sum_{m=0}^{k} (-1)^{k+m} \binom{n}{2m+\varepsilon} s^{n-2m-\varepsilon} t^{2m+\varepsilon},$$
$$r^n T_n\left(\frac{s}{r}\right) = \sum_{m=0}^{k} (-1)^{k+m} \binom{n}{2m+\varepsilon} t^{n-2m-\varepsilon} s^{2m+\varepsilon},$$

which yields a polynomial parametrization of S_k.

Recall the orthogonal group $O(3)$ of $\mathbb{R}^3$, i.e., the symmetry group of the sphere consisting of orthogonal 3×3 matrices, and let

$$\begin{aligned} D_{2k+2} &:= \big\langle \rho, \sigma \,:\, \rho^{2k+2} = \sigma^2 = e,\ \sigma\rho\sigma = \rho^{-1} \big\rangle \\ &= \{\sigma^n \rho^m \,:\, n = 0,1,\ m = 0,\dots,2k+1\} \end{aligned}$$

be the *dihedral group of order* $4k+4$ (here e denotes the neutral element). Let

$$\boldsymbol{S} := \begin{bmatrix} 1 & 0 & 0 \\ 0 & -1 & 0 \\ 0 & 0 & 1 \end{bmatrix}, \qquad \boldsymbol{R}_k := \begin{bmatrix} \cos\left(\frac{\pi}{k+1}\right) & \sin\left(\frac{\pi}{k+1}\right) & 0 \\ -\sin\left(\frac{\pi}{k+1}\right) & \cos\left(\frac{\pi}{k+1}\right) & 0 \\ 0 & 0 & -1 \end{bmatrix}. \tag{9.18}$$

Proposition 177. *The symmetry group* $\{f_{m,n}(\boldsymbol{x}) := A_{m,n}\boldsymbol{x}\}$ *of the higher-order Enneper surface* S_k *is parametrized by the group monomorphism*

$$D_{2k+2} \longrightarrow O(3), \qquad \sigma^n \rho^m \longmapsto A_{m,n} := \boldsymbol{S}^n \boldsymbol{R}_k^m. \tag{9.19}$$

Moreover, with $\mathbf{s}_k$ *as in* (9.16) *and Möbius transformations* $\varphi^m(z) := \zeta^m z$*, where* $\zeta = \zeta_{2k+2} := e^{2\pi i/(2k+2)}$ *is a* $(2k+2)$*-th root of unity,*

$$f_{m,n} \circ \mathbf{s} = \mathbf{s} \circ \varphi^m, \qquad n = 0, 1, \qquad m = 0, 1, \ldots, 2k+1.$$

Proof. Applying Proposition 174 to compute the symmetries of S_k, we first compute the symmetries of the complex space curve $\mathcal{A}_k$ parametrized by $\mathbf{c}_k$. Each symmetry of $\mathcal{A}_k$ corresponds to an isometry $f(\boldsymbol{x}) = A\boldsymbol{x} + \boldsymbol{b}$ and a Möbius transformation φ satisfying that $f \circ \mathbf{c}_k = \mathbf{c}_k \circ \varphi$. Since $\mathbf{c}_k$ is polynomial, $\varphi(z) = az + b$ is also polynomial, and we obtain the polynomial system

$$A\mathbf{c}_k(z) + \boldsymbol{b} = \mathbf{c}_k(az + b). \tag{9.20}$$

Writing $A = [m_{ij}]_{ij}$ and $\boldsymbol{b} = [b_i]_i$, the last equation of this system is

$$(m_{31} + m_{32}i)z + \frac{2}{k+1} m_{33} z^{k+1} + \frac{1}{2k+1}(-m_{31} + im_{32}) z^{2k+1} + b_3 = \frac{2}{k+1} \sum_{j=0}^{k+1} \binom{k+1}{j} a^j b^{k+1-j} z^j. \tag{9.21}$$

Setting equal coefficients of (highest) order $2k+1$ yields $m_{31} = m_{32} = 0$. Hence, since A is orthogonal, it follows that $m_{33} = \pm 1$. Setting equal coefficients of order $k+1$ yields $a^{k+1} = m_{33} = \pm 1$, so that $a = \zeta^m$ for some $m \in \{0, 1, \ldots, 2k+1\}$. Moreover, setting equal linear coefficients yields $2ab^k = m_{31} + m_{32}i = 0$, implying $b = 0$. Evaluating (9.20) at $z = 0$ yields $\boldsymbol{b} = \mathbf{c}_k(b) = \mathbf{0}$. Differentiating (9.20) l times and substituting $z = 0$ yields $A\mathbf{c}_k^{(l)}(0) = a^l \mathbf{c}_k^{(l)}(0)$, which provides the matrix equation

$$A = \boldsymbol{C}\boldsymbol{M}_m\boldsymbol{C}^{-1},$$

where, since $\zeta_{2k+2}^{(2k+1)} = \zeta_{2k+2}^{-1}$ and $\zeta_{2k+2}^{k+1} = -1$,

$$\boldsymbol{C} := \left[\mathbf{c}_k'(0), \mathbf{c}_k^{(k+1)}(0), \mathbf{c}_k^{(2k+1)}(0)\right], \quad \boldsymbol{M}_m := \begin{bmatrix} \zeta^m & 0 & 0 \\ 0 & (-1)^m & 0 \\ 0 & 0 & \zeta^{-m} \end{bmatrix}.$$

It follows that $A_{m,0}\mathbf{c}_k = \mathbf{c}_k \circ \varphi^m$ for $m = 0, 1, 2, 3$, where $\varphi^m(z) = \zeta^m z$ and

$$\begin{aligned} A_{m,0} &:= \begin{bmatrix} 1 & 0 & -(2k)! \\ i & 0 & i(2k)! \\ 0 & 2 & 0 \end{bmatrix} \begin{bmatrix} \zeta^m & 0 & 0 \\ 0 & (-1)^m & 0 \\ 0 & 0 & \zeta^{-m} \end{bmatrix} \frac{1}{2} \begin{bmatrix} 1 & -i & 0 \\ 0 & 0 & 1 \\ \frac{-1}{(2k)!} & \frac{-i}{(2k)!} & 0 \end{bmatrix} \\ &= \begin{bmatrix} \frac{\zeta^m + \zeta^{-m}}{2} & \frac{\zeta^m - \zeta^{-m}}{2i} & 0 \\ \frac{\zeta^{-m} - \zeta^m}{2i} & \frac{\zeta^m + \zeta^{-m}}{2} & 0 \\ 0 & 0 & (-1)^m \end{bmatrix} \\ &= \begin{bmatrix} \cos(\frac{\pi m}{k+1}) & \sin(\frac{\pi m}{k+1}) & 0 \\ -\sin(\frac{\pi m}{k+1}) & \cos(\frac{\pi m}{k+1}) & 0 \\ 0 & 0 & (-1)^m \end{bmatrix} \\ &= \boldsymbol{R}_k^m. \end{aligned}$$

Next we compute the isometries mapping $\mathcal{A}_k$ onto the complex curve $\overline{\mathcal{A}}_k$ parametrized by $\overline{\mathbf{c}}_k$. This time we have:

$$A\mathbf{c}_k(z) + \boldsymbol{b} = \overline{\mathbf{c}}_k(az + b). \tag{9.22}$$

Proceeding as before, one demonstrates that $\varphi(z) = \varphi^m(z) := \zeta^m z$ and $\boldsymbol{b} = \mathbf{0}$. It follows that $A_{m,1}\mathbf{c}_k = \overline{\mathbf{c}}_k \circ \varphi^m$ for $m = 0, 1, \ldots, 2k+1$, where

$$\begin{aligned} A_{m,1} &= \overline{\boldsymbol{C}}\boldsymbol{M}_m\boldsymbol{C}^{-1} = \boldsymbol{S}\boldsymbol{C}\boldsymbol{M}_m\boldsymbol{C}^{-1} = \\ &= \begin{bmatrix} 1 & 0 & 0 \\ 0 & -1 & 0 \\ 0 & 0 & 1 \end{bmatrix} \begin{bmatrix} \cos(\frac{\pi m}{k+1}) & \sin(\frac{\pi m}{k+1}) & 0 \\ -\sin(\frac{\pi m}{k+1}) & \cos(\frac{\pi m}{k+1}) & 0 \\ 0 & 0 & (-1)^m \end{bmatrix} = \boldsymbol{S}\boldsymbol{R}_k^m. \end{aligned} \tag{9.23}$$

One verifies that the map $\sigma^n \rho^m \longmapsto A_{m,n}$ is a monomorphism by comparing multiplication tables, or simply by verifying that its generators $\boldsymbol{R}_k$ and $\boldsymbol{S}$ satisfy $\boldsymbol{R}_k^{2k+2} = \boldsymbol{S}^2 = \boldsymbol{I}$ and $\boldsymbol{S}\boldsymbol{R}_k\boldsymbol{S} = \boldsymbol{R}_k^{-1}$. □

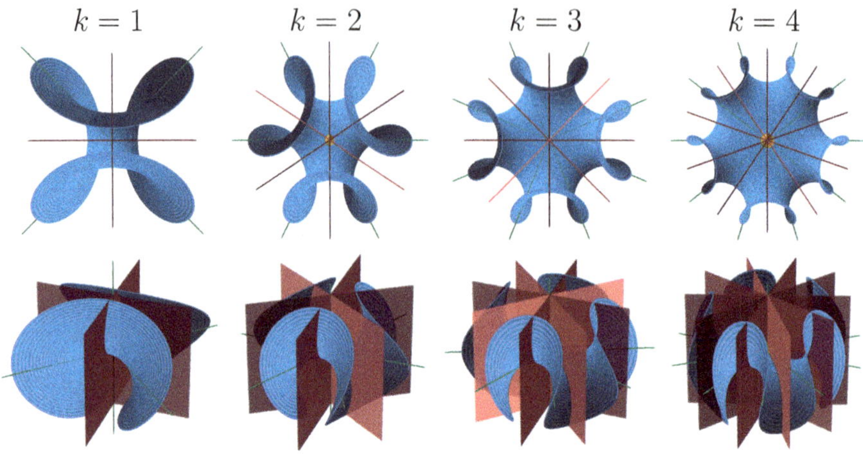

FIGURE 9.5 For $k = 1, 2, 3, 4$, top view (top) and side view (bottom) of higher-order Enneper surfaces S_k, together with symmetry planes, symmetry rotation axes, and symmetry point (the latter for $k = 2, 4$).

Remark 20. *Since* $\boldsymbol{S}\boldsymbol{R}_k^m = \boldsymbol{R}_k^{-m}\boldsymbol{S}$, *precomposing* (9.19) *with the group automorphism* $\sigma^n\rho^m \longmapsto \sigma^n\rho^{-m}$ *of* D_{2k+2} *yields an alternative group monomorphism*

$$D_{2k+2} \longrightarrow O(3), \qquad \sigma^n\rho^m \longmapsto A_{-m,n} := \boldsymbol{R}_k^m\boldsymbol{S}^n.$$

Note that $A_{m,n}$ is a rotation of angle $-\frac{\pi m}{k+1}$ about the z-axis, composed by a reflection in the plane $z = 0$ when $m \equiv 1 \pmod 2$, and in addition composed by a reflection in the plane $y = 0$ in the case $n = 1$. For $k = 1, 2, 3, 4$, Fig. 9.5 shows the higher-order Enneper surface S_k, together with its symmetry elements.

The particular form $\varphi^m(z) := \zeta^m z$ of the Möbius transformation was the inspiration for the results in the next subsection.

9.3.2 Prescribing symmetries

Inspired by Section 9.3.1, in this subsection we will see that imposing certain parity-like properties on the functions f, g in (9.7) will result in a minimal surface S with certain symmetries.

Let $\mathcal{F}$ be the space of meromorphic functions on a simply connected region $U \subset \mathbb{C}$, which certainly includes rational functions, and let $\varphi : U \longrightarrow U$ be meromorphic. Consider the composition operator

$$T_\varphi : \mathcal{F} \longrightarrow \mathcal{F}, \qquad T_\varphi(f) := f \circ \varphi. \tag{9.24}$$

The eigenvalue equation

$$T_\varphi(f) = \lambda f$$

is called *Schröder's equation*; it is known to have solutions under general conditions.

In our case, for any integer $K \geq 2$, we consider the Möbius transformation $\varphi(t) = \varphi_K(t) := \zeta_K \cdot t$, where ζ_K denotes a K-root of the unity. Notice that φ leaves invariant any complex disk centered at the origin; in particular, in this case we can take U to be any such disk, or even the entire complex plane $\mathbb{C}$. The corresponding composition operator $T = T_\varphi$ generalizes the parity operator. The K-fold composition satisfies $T^K(f) = f \circ \varphi^K = f$, which implies that the eigenvalues of T are the K-th roots of unity ζ_K^m, with $m = 0, \ldots, K-1$. These provide an eigendecomposition of the function space $\mathcal{F} = \mathcal{F}_0 \oplus \cdots \oplus \mathcal{F}_{K-1}$ into K parts.

For simplicity we restrict ourselves to $K = 4$, in which case $\zeta_K = i$; a similar analysis can be carried out for any $K \geq 2$. The following proposition states that choosing f, g in (9.7) as eigenfunctions of T (and hence of T^q, with $q \geq 1$) results in certain symmetries of the corresponding minimal surface. More precisely, we obtain a symmetry for every pair of eigenpairs $(i^r, f), (i^s, g)$ of T^q for which $q + r + s \equiv 0 \pmod 2$. The functions f, g and domain U satisfy the requirements for the Weierstrass representation (9.7), described in more detail in Section 9.1.

Proposition 178. *Let $\varphi(z) = iz$ and $T = T_\varphi$ be as above. Suppose that in a simply-connected region $U \subset \mathbb{C}$ invariant under φ, f is holomorphic, g is meromorphic with no pole at $z = 0$, and fg^2 is* *holomorphic. In addition, suppose f, g satisfy*

$$T^q(f)(z) = f(i^q z) = i^r \cdot f(z), \qquad T^q(g)(z) = g(i^q z) = i^s \cdot g(z) \tag{9.25}$$

for some $q, r, s \in \mathbb{Z}/4\mathbb{Z}$ satisfying $q + r + s \equiv 0 \pmod 2$. Let $\mathcal{A}$ be the corresponding curve parametrized by $\mathbf{c}$ as in (9.7), with complex conjugate $\overline{\mathcal{A}}$ parametrized by $\overline{\mathbf{c}}$. Let $S = \mathcal{A} \oplus \overline{\mathcal{A}}$ be the corresponding minimal surface parametrized by $\mathbf{s} = (\mathbf{c} \oplus \overline{\mathbf{c}}) \circ \iota$. Then S has the symmetry $f^{\pm}_{q+r,s}(\boldsymbol{x}) = A^{\pm}_{q+r,s}\boldsymbol{x}$ as in Table 9.2, for each choice of the sign $\pm$.

Proof. Suppose (9.25) holds for some $q, r, s \in \mathbb{Z}/4\mathbb{Z}$ satisfying $q + r + s \equiv 0 \pmod 2$. By Proposition 174 and Remark 18, the symmetries of S are the symmetries of $\mathcal{A}$ and the isometries mapping $\mathcal{A}$ onto $\overline{\mathcal{A}}$. Then, we examine when the reparametrization $\varphi^q(z) := i^q z$ of $\mathbf{c}$ can be expressed as the composition of such an isometry with either $\mathbf{c}$ or $\overline{\mathbf{c}}$. Applying the change of variable $\gamma = i^q \cdot \eta$ and using (9.25),

$$\begin{aligned}\Psi_1(i^q z) &= \int_0^{i^q \cdot z} f(\gamma)\frac{1 - g^2(\gamma)}{2}\mathrm{d}\gamma \\ &= i^{q+r}\int_0^z f(\eta)\frac{1 - (-1)^s g^2(\eta)}{2}\mathrm{d}\eta,\end{aligned} \tag{9.26}$$

$$\begin{aligned}\Psi_2(i^q z) &= i \cdot \int_0^{i^q \cdot z} f(\gamma)\frac{1 + g^2(\gamma)}{2}\mathrm{d}\gamma \\ &= i^{q+r+1}\int_0^z f(\eta)\frac{1 + (-1)^s g^2(\eta)}{2}\mathrm{d}\eta,\end{aligned} \tag{9.27}$$

$$\Psi_3(i^q z) = \int_0^{i^q \cdot z} f(\gamma)g(\gamma)\mathrm{d}\gamma = i^{q+r+s}\int_0^z f(\eta)g(\eta)\mathrm{d}\eta. \tag{9.28}$$

With $k := q + r$ and using that $\mathbf{c} = \boldsymbol{S}\overline{\mathbf{c}}$, it follows that

$$\mathbf{c} \circ \varphi^q = \begin{bmatrix} i^k & 0 & 0 \\ 0 & i^k & 0 \\ 0 & 0 & i^{k+s} \end{bmatrix} \mathbf{c} \qquad = \begin{bmatrix} i^k & 0 & 0 \\ 0 & i^{k-2} & 0 \\ 0 & 0 & i^{k+s} \end{bmatrix} \overline{\mathbf{c}},$$

if $s \equiv 0 \pmod 2$,

$$\mathbf{c} \circ \varphi^q = \begin{bmatrix} 0 & i^{k-1} & 0 \\ i^{k+1} & 0 & 0 \\ 0 & 0 & i^{k+s} \end{bmatrix} \mathbf{c} = \begin{bmatrix} 0 & i^{k+1} & 0 \\ i^{k+1} & 0 & 0 \\ 0 & 0 & i^{k+s} \end{bmatrix} \overline{\mathbf{c}},$$

if $s \equiv 1 \pmod 2$,

so that

$$\mathbf{c} \circ \varphi^q(z) = f^+_{q+r,s} \circ \mathbf{c}(z) = A^+_{q+r,s}\mathbf{c}(z), \tag{9.29}$$

$$\mathbf{c} \circ \varphi^q(z) = f^-_{q+r,s} \circ \overline{\mathbf{c}}(z) = A^-_{q+r,s}\overline{\mathbf{c}}(z), \tag{9.30}$$

where $A^+_{q+r,s} = A^-_{q+r,s}\boldsymbol{S}$ is as in Table 9.2. The real isometries $f^\pm_{q+r,s}$ are obtained by discarding the cases $q + r + s \equiv 1 \pmod 2$, shown in gray. Thus, we deduce that in the remaining cases $f^+_{q+r,s}$ is a symmetry of $\mathcal{A}$ and $f^-_{q+r,s}$ is an isometry mapping $\mathcal{A}$ onto $\overline{\mathcal{A}}$. In either case, f is a symmetry of the surface S. □

The Enneper surface S_1 originates from taking $f(z) = 2$ and $g(z) = z$, in which case

$$f(i^q z) = 2, \qquad g(i^q z) = i^q z, \qquad q \in \mathbb{Z}/4\mathbb{Z}.$$

Hence (9.25) holds whenever $r \equiv 0$ modulo 4 and $s \equiv q \equiv q + r$ modulo 4. Therefore the symmetries $\boldsymbol{R}_1^m \boldsymbol{S}^n$ of the Enneper surface are recovered as the diagonal cases $m \equiv s \equiv q + r$, with $n = 0$ for the top sign and $n = 1$ for the bottom sign.

Remark 21. *Consider the (external) direct product group*

$$D_4 \times \mathbb{Z}/2\mathbb{Z} \simeq$$
$$\langle \rho, \sigma, \tau \, : \, \rho^4 = \sigma^2 = \tau^2 = e, \ \sigma\rho\sigma = \rho^{-1}, \ \tau\rho = \rho\tau, \ \sigma\tau = \tau\sigma \rangle,$$

where e denotes the neutral element. With $\boldsymbol{R}_1, \boldsymbol{S}$ as in (9.18) and with $\boldsymbol{T} := \operatorname{diag}(1, 1, -1)$ the reflection in the plane $z = 0$, the map

$$D_4 \times \mathbb{Z}/2\mathbb{Z} \longrightarrow O(3), \qquad \rho^m \sigma^n \tau^p \longmapsto \boldsymbol{R}_1^m \boldsymbol{S}^n \boldsymbol{T}^p$$

is a group monomorphism establishing a group structure on the set of real matrices in Table 9.2.

Remark 22. *Alternatively, consider Möbius transformations $\varphi(z) = z + b$ and $T = T_\varphi$ as in (9.24) for a space $\mathcal{F}$ of periodic or doubly periodic meromorphic functions. For $K \geq 2$, choosing $b = \omega/K$ for one of the periods ω of $\mathcal{F}$, the operator T again has order K and eigenvalues ζ_K^m, for $m = 0, \dots, K - 1$. Analogous to Proposition 178, solutions f, g to Schröder's equation again lead to symmetric minimal surfaces (c.f. [76]).*

TABLE 9.2 Real orthogonal (black) and imaginary unitary (gray) matrices $A^{\pm}_{q+r,s}$ in (9.29)–(9.30) with symmetry types and symmetry elements for the various cases (q,r,s), where $\equiv$ denotes equivalence modulo 4.

$A^{+}_{q+r,s}$	$q+r\equiv 0$	$q+r\equiv 1$	$q+r\equiv 2$	$q+r\equiv 3$
$s\equiv 0$	$\begin{bmatrix}1&0&0\\0&1&0\\0&0&1\end{bmatrix}$	$\begin{bmatrix}i&0&0\\0&i&0\\0&0&i\end{bmatrix}$	$\begin{bmatrix}-1&0&0\\0&-1&0\\0&0&-1\end{bmatrix}$	$\begin{bmatrix}-i&0&0\\0&-i&0\\0&0&-i\end{bmatrix}$
$s\equiv 1$	$\begin{bmatrix}0&-i&0\\i&0&0\\0&0&i\end{bmatrix}$	$\begin{bmatrix}0&1&0\\-1&0&0\\0&0&-1\end{bmatrix}$	$\begin{bmatrix}0&i&0\\-i&0&0\\0&0&-i\end{bmatrix}$	$\begin{bmatrix}0&-1&0\\1&0&0\\0&0&1\end{bmatrix}$
$s\equiv 2$	$\begin{bmatrix}1&0&0\\0&1&0\\0&0&-1\end{bmatrix}$	$\begin{bmatrix}i&0&0\\0&i&0\\0&0&-i\end{bmatrix}$	$\begin{bmatrix}-1&0&0\\0&-1&0\\0&0&1\end{bmatrix}$	$\begin{bmatrix}-i&0&0\\0&-i&0\\0&0&i\end{bmatrix}$
$s\equiv 3$	$\begin{bmatrix}0&-i&0\\i&0&0\\0&0&-i\end{bmatrix}$	$\begin{bmatrix}0&1&0\\-1&0&0\\0&0&1\end{bmatrix}$	$\begin{bmatrix}0&i&0\\-i&0&0\\0&0&i\end{bmatrix}$	$\begin{bmatrix}0&-1&0\\1&0&0\\0&0&-1\end{bmatrix}$
$A^{-}_{q+r,s}$	$q+r\equiv 0$	$q+r\equiv 1$	$q+r\equiv 2$	$q+r\equiv 3$
$s\equiv 0$	$\begin{bmatrix}1&0&0\\0&-1&0\\0&0&1\end{bmatrix}$	$\begin{bmatrix}i&0&0\\0&-i&0\\0&0&i\end{bmatrix}$	$\begin{bmatrix}-1&0&0\\0&1&0\\0&0&-1\end{bmatrix}$	$\begin{bmatrix}-i&0&0\\0&i&0\\0&0&-i\end{bmatrix}$
$s\equiv 1$	$\begin{bmatrix}0&i&0\\i&0&0\\0&0&i\end{bmatrix}$	$\begin{bmatrix}0&-1&0\\-1&0&0\\0&0&-1\end{bmatrix}$	$\begin{bmatrix}0&-i&0\\-i&0&0\\0&0&-i\end{bmatrix}$	$\begin{bmatrix}0&1&0\\1&0&0\\0&0&1\end{bmatrix}$
$s\equiv 2$	$\begin{bmatrix}1&0&0\\0&-1&0\\0&0&-1\end{bmatrix}$	$\begin{bmatrix}i&0&0\\0&-i&0\\0&0&-i\end{bmatrix}$	$\begin{bmatrix}-1&0&0\\0&1&0\\0&0&1\end{bmatrix}$	$\begin{bmatrix}-i&0&0\\0&i&0\\0&0&i\end{bmatrix}$
$s\equiv 3$	$\begin{bmatrix}0&i&0\\i&0&0\\0&0&-i\end{bmatrix}$	$\begin{bmatrix}0&-1&0\\-1&0&0\\0&0&1\end{bmatrix}$	$\begin{bmatrix}0&-i&0\\-i&0&0\\0&0&i\end{bmatrix}$	$\begin{bmatrix}0&1&0\\1&0&0\\0&0&-1\end{bmatrix}$

9.4 FINAL REMARKS AND OPEN PROBLEMS

The algorithms that we have presented here for translational and minimal surfaces require the parametrization S of the surface to be given in a standard form. In the case of translational surfaces, we need to have a parametrization of the type $\boldsymbol{x}(t,s) = \boldsymbol{u}(t) + \boldsymbol{v}(s) = \mathbf{c}_1(t) \oplus \mathbf{c}_2(s)$, which is standard in the sense that it clearly shows the translational nature of the surface, and makes a generator pair explicit. However, if the surface undergoes a rational reparametrization, this standard form is lost, and it is not easy anymore to recognize the surface as being translational.

Although there are some necessary conditions for a surface to be translational (see Proposition 1 of [131]), these conditions are not sufficient. In fact, it is an open problem to recognize a translational surface given in non-standard form, and reparametrize it into standard representation. A first step in this direction is the paper [103], where ideas for recognizing rational surfaces of translation are provided. However, the algorithm in [103] requires computing the implicit equation of the surface and, which is harder, a tangent direction to one of the curves in the generator pair. Thus, a complete and efficient solution to the problem is still absent.

The same problem arises in the case of minimal surfaces. The algorithms here require the minimal surface to be parametrized as $\mathbf{s} = (\mathbf{c} \oplus \overline{\mathbf{c}}) \circ \iota$. In that case, after composing with ι^{-1} we can write the parametrization as a standard, complex, parametrization of a translational surface, and recover the minimal curve $\mathbf{c}$. If $\mathbf{s} = (\mathbf{c} \oplus \overline{\mathbf{c}}) \circ \iota$ is reparametrized, the unsolved recognition problem appears again. In the case of minimal surfaces, having a standard parametrization provides additional advantages [59, 131]: such a parametrization is *isothermal*, sometimes also called *conformal*, and *harmonic*. Isothermal means that its first fundamental form is diagonal, and therefore that the parametrization preserves the angles of curves in the parametric domain when they are mapped by the parametrization onto the surface; in particular, squares in the parametric domain are mapped onto curvilinear orthogonal squares on the surface, so it is very simple to create an orthogonal grid on the surface. Harmonic means that the components ot the parametrization are harmonic functions, i.e., functions with vanishing Laplacians.

Thus, we can point out the following open problems, related to the topics in this chapter.

Open Problem 25. *Find an algorithm to recognize a rational translational surface, and to reparametrize it into standard form.*

Open Problem 26. *Find an algorithm to reparametrize a rational minimal surface into standard, complex form, or equivalently to find a rational minimal curve of the surface.* Obviously, if the recognition and standard reparametrization question is solved for translational surfaces, then this problem is solved too. But there might be special strategies to solve the problem for minimal surfaces, taking advantage of their special properties. As a generalization, one can also consider the problem of recognizing whether or not a surface, not necessarily minimal, admits a rational isothermal parametrization, and to find an algorithm for reparametrizing it into standard form, i.e., so that the first fundamental form is diagonal.

CHAPTER 10

Finite sets of points

We close with a case that we have omitted so far, namely that of finite sets of points, $\mathcal{P} = \{\mathbf{p}_i\}_{i=1,\ldots,m}$, with $\mathbf{p}_i \in \mathbb{R}^n$, also called *point clouds* in $\mathbb{R}^n$. These are certainly algebraic sets as well: denoting $\mathbf{p}_i = (p_1^{(i)}, \ldots, p_n^{(i)})$, $i \in \{1, \ldots, m\}$, the set $\mathcal{P} = \{\mathbf{p}_i\}_{i=1,\ldots,m}$ is the set of common zeroes of the polynomials $f_j^{(i)}(x_j) = x_j - p_j^{(i)}$, where $i \in \{1, \ldots, m\}$, $j \in \{1, \ldots, n\}$. But there is no special benefit in thinking about the point cloud this way, which is the reason why we have postponed the analysis of these sets for the very end of the book; for its simplicity, we might also have started with this as one of our first chapters, but we wanted to present first the problems where the algebraic structure was key.

The symmetries of these sets are mostly interesting in applied fields, where quite often the input is just a point cloud, perhaps coming from a discrete or even noisy version of a continuous object, maybe an image or a 3D scan of a real object. In fact, the literature on (approximate) methods to address the symmetries of these sets is abundant, ranging from Statistics and Numerical Analysis to Approximation Theory: for a review of methods, we can point out the introduction to [10], as well as the introductions to the papers [32, 33]. Furthermore, methods for planar sets of points, with approaches close to the ones presented in this book, appear in [31, 32]; the extension of the method in [32] to the 3D case is given in [33].

DOI: 10.1201/9781003648710-10

In order to simplify the description, we will focus on the case where the $\mathbf{p}_i$ are elements of $\mathbb{R}^3$, so $\mathbf{p}_i = (x_i, y_i, z_i)$, although we will be making several observations on the planar case as well. We will present two very simple methods only using eigenvalues and tensors of high order, following the presentation of [22]. The first method uses the covariance matrix of the points, and the second one, which serves as a backup for pathological cases where the first method does not work, uses higher order tensors, which we will be reviewing.

10.1 USING THE COVARIANCE MATRIX

Let us recall a notion that has already appeared in some of the chapters of the book: the *barycenter*, also called *center of mass*, *centroid* or *center of gravity* of the points $\{\mathbf{p}_i\}_{i=1,\dots,m}$ is the point

$$\mathbf{p} = \frac{1}{m}\sum_{i=1}^{m}\mathbf{p}_i,$$

whose coordinates are the averages of each coordinate, i.e.,

$$\mathbf{p} = \left(\frac{\sum_{i=1}^{m} x_i}{m}, \frac{\sum_{i=1}^{m} y_i}{m}, \frac{\sum_{i=1}^{m} z_i}{m}\right) = (\bar{x}, \bar{y}, \bar{z}).$$

This notion was used when we addressed symmetries of implicit planar algebraic curves and also symmetries of algebraic vector fields. Let us see that the barycenter of the points is always a fixed point of any symmetry of $\mathcal{P}$.

Theorem 179. *Let $f(\boldsymbol{x}) = A\boldsymbol{x} + \boldsymbol{b}$ be a symmetry of the set $\mathcal{P} = \{\mathbf{p}_i\}_{i=1,\dots,m}$, and let $\mathbf{p}$ be the barycenter of the $\mathbf{p}_i$. Then $f(\mathbf{p}) = \mathbf{p}$.*

Proof. Since f is a symmetry of the $\mathbf{p}_i$, for each $\mathbf{p}_i$ there exists $\mathbf{p}_j$ such that $f(\mathbf{p}_i) = A\mathbf{p}_i + \boldsymbol{b} = \mathbf{p}_j$. Since f is a bijection in the set of the $\mathbf{p}_i$, summing over $i = 1, \dots, m$ yields

$$\sum_{i=1}^{m}(A\mathbf{p}_i + \boldsymbol{b}) = \sum_{j=1}^{m}\mathbf{p}_j.$$

By linearity, we get that

$$\sum_{i=1}^{m}(A\mathbf{p}_i + \boldsymbol{b}) = A\left(\sum_{i=1}^{m}\mathbf{p}_i\right) + m\boldsymbol{b}.$$

Furthermore, $\sum_{i=1}^{m}\mathbf{p}_i = \sum_{j=1}^{m}\mathbf{p}_j = m\mathbf{p}$, and therefore

$$mA\mathbf{p} + m\boldsymbol{b} = m\mathbf{p}.$$

Dividing by m, we get that $f(\mathbf{p}) = \mathbf{p}$. □

Corollary 180. *Let* $\mathbf{p}$ *be the barycenter of the set* $\mathcal{P} = \{\mathbf{p}_i\}_{i=1,\ldots,m}$.

(1) If $\mathcal{P}$ *exhibits central symmetry, then* $\mathbf{p}$ *is the center of symmetry.*

(2) If $\mathcal{P}$ *exhibits axial/rotational symmetry about a line* $\mathcal{L}$*, then* $\mathbf{p} \in \mathcal{L}$.

(3) If $\mathcal{P}$ *exhibits planar symmetry with respect to a plane* Π*, then* $\mathbf{p} \in \Pi$.

Remark 23. *For planar point clouds, the same argument proves that the center of rotation, if any, must be the barycenter, and that any axis of symmetry also contains the barycenter.*

To address the symmetries of $\mathcal{P}$, we first assume that the barycenter of the points has been moved, by applying if necessary a translation, to the origin of coordinates. Now if $\mathbf{p}$ is at the origin, then we can define the *covariance* matrix of the point set $\mathcal{P} = \{\mathbf{p}_i\}_{i=1,\ldots,m}$ as the following 3×3 symmetric matrix:

$$\mathbf{C} = \begin{bmatrix} c_{11} & c_{12} & c_{13} \\ c_{21} & c_{22} & c_{23} \\ c_{31} & c_{32} & c_{33} \end{bmatrix} = \frac{1}{m}\begin{bmatrix} \sum_{i=1}^{m} x_i^2 & \sum_{i=1}^{m} x_i y_i & \sum_{i=1}^{m} x_i z_i \\ \sum_{i=1}^{m} y_i x_i & \sum_{i=1}^{m} y_i^2 & \sum_{i=1}^{m} y_i z_i \\ \sum_{i=1}^{m} z_i x_i & \sum_{i=1}^{m} z_i y_i & \sum_{i=1}^{m} z_i^2 \end{bmatrix}. \tag{10.1}$$

If $\mathbf{p} = (\bar{x}, \bar{y}, \bar{z})$, not necessarily at the origin, the covariance matrix is defined as the result of replacing, in the above matrix, $x_i \to x_i - \bar{x}$, $y_i \to y_i - \bar{y}$, $z_i \to z_i - \bar{z}$. Furthermore, if we perform an orthogonal change of coordinates leaving the origin invariant, the covariance matrix $\mathbf{C}^\star$ with respect to the new coordinates satisfies that

$$\mathbf{C}^\star = P^T \mathbf{C} P, \tag{10.2}$$

with P orthogonal, i.e., $\mathbf{C}, \mathbf{C}^\star$ are similar matrices.

Theorem 181. *Let $\mathcal{P} = \{\mathbf{p}_i\}_{i=1,\ldots,m}$. Assume that the barycenter $\mathbf{p}$ is at the origin.*

(1) If $\mathcal{P}$ exhibits axial symmetry about a line $\mathcal{L}$, then $\mathcal{L}$ is parallel to one of the eigenvectors of $\mathbf{C}$.

(2) If $\mathcal{P}$ exhibits rotational symmetry about a line $\mathcal{L}$, then $\mathcal{L}$ is parallel to one of the eigenvectors of $\mathbf{C}$.

(3) If $\mathcal{P}$ exhibits planar symmetry with respect to a plane Π, then Π is perpendicular to one of the eigenvectors of $\mathbf{C}$.

Proof. Let us see (1). Assume that, after performing if necessary an orthogonal change of coordinates, $\mathcal{L}$ is the z-axis. Then, for each point $(x_i, y_i, z_i) \in \mathcal{P}$ there is another point of coordinates $(-x_i, -y_i, z_i) \in \mathcal{P}$. Thus, in the new reference system and for the corresponding covariance matrix $\mathbf{C}^\star$ we have

$$c^\star_{13} = c^\star_{31} = 0, \;\; c^\star_{23} = c^\star_{32} = 0. \tag{10.3}$$

Therefore, $(0, 0, 1)$, which is parallel to $\mathcal{L}$, is an eigenvector of $\mathbf{C}^\star$. Since the covariance matrices associated with $\mathcal{P}$ in different orthogonal systems of references are similar, the result follows. The proof of (3) is analogous: if we consider an orthogonal change of coordinates so that Π coincides with the xy-plane, for each point $(x_i, y_i, z_i) \in \mathcal{P}$ we must have $(x_i, y_i, -z_i) \in \mathcal{P}$, and therefore the entries of the covariance matrix $\mathbf{C}^\star$ satisfy Eq. (10.3) as well.

Finally, let us see (2). Assume that the rotation axis $\mathcal{L}$ is the z-axis, and let Π_i be the plane normal to $\mathcal{L}$ through $\mathbf{p}_i = (x_i, y_i, z_i)$. Let $\mathbf{C}^\star$ be the covariance matrix in the coordinate system we are using, where $\mathcal{L}$ is the z-axis. Now $\mathcal{P}_i = \mathcal{P} \cap \Pi_i$, which is a planar set of points, exhibits rotational symmetry about the point $\mathbf{q}_i$, intersection of Π_i with the z-axis. By Theorem 179, $\mathbf{q}_i = (0, 0, z_i)$ is the barycenter of the points in $\mathcal{P}_i$, which we can denote as $(x_{i,j}, y_{i,j}, z_i)$; notice also that since the rotation angle must be $\theta = \frac{2\pi}{n}$ for some $n \in \mathbb{N}$, j must be a divisor of n. Now, because the barycenter of the points in $\mathcal{P}_i$ is $\mathbf{q}_i = (0, 0, z_i)$, we deduce that

$$m_i = \sum_j x_{i,j} z_i = z_i \sum_j x_{i,j} = 0, \;\; n_i = \sum_j y_{i,j} z_i = z_i \sum_j y_{i,j} = 0.$$

Finally, let us denote by ℓ the number of different z-values appearing in the z-coordinates of the points in $\mathcal{P}$. Then

$$\begin{aligned} c^{\star}_{13} &= c^{\star}_{31} = \textstyle\sum_{i=1}^{\ell} m_i = 0, \\ c^{\star}_{23} &= c^{\star}_{32} = \textstyle\sum_{i=1}^{\ell} n_i = 0, \end{aligned}$$

so $(0, 0, 1)$ is an eigenvalue of $\boldsymbol{C}^{\star}$, and (2) follows. □

Remark 24. *In the plane, i.e., when $n = 2$, Theorem 181 implies that the directions of the axes of symmetry are eigenvectors of the (2×2) covariance matrix.*

Additionally, for rotational, non-axial, symmetries, we can go further.

Theorem 182. *Let $\mathcal{P} = \{\mathbf{p}_i\}_{i=1,\dots,m}$ be a 3D point cloud, whose barycenter $\mathbf{p}$ lies at the origin. If $\mathcal{P}$ exhibits nontrivial rotational symmetry about a line $\mathcal{L}$ and the rotation angle is $\theta \neq \pi$, then either $\mathbf{C}$ has a triple eigenvalue, or $\mathbf{C}$ has a simple eigenvalue associated with the axis of rotation and a double eigenvalue not associated with it.*

Proof. From Statement (2) of Theorem 181 we know that the direction of $\mathcal{L}$ is an eigenvector of the covariance matrix $\mathbf{C}$. Thus, let λ_1 be the eigenvalue of $\mathbf{C}$ associated with the axis of rotation, and let S be the coordinate system, centered at the origin, where the x-axis has the direction of the axis of rotation, and the y-axis and the z-axis are associated with the eigenvalues λ_2, λ_3 of $\mathbf{C}$. Then the covariance matrix of $\mathcal{P}$ in the coordinate system S is $\mathbf{C} = \operatorname{diag}(\lambda_1, \lambda_2, \lambda_3)$. Now if $\mathcal{P}$ has rotational symmetry about $\mathcal{L}$ by an angle θ, when we consider the coordinate system S' which results from rotating the system S around $\mathcal{L}$ by an angle θ, the covariance matrix $\mathbf{C}'$ in the new system S' must coincide with $\mathbf{C}$, i.e., $\mathbf{C}' = \mathbf{C}$ (because the point cloud defined by $\mathcal{P}$ looks exactly the same on both systems). On the other hand, $\mathbf{C}' = \mathbf{R}^T\mathbf{C}\mathbf{R}$ where $\mathbf{R}$ is the rotation matrix around $\mathcal{L}$ by an angle θ. Thus, we get that $\mathbf{C} = \mathbf{R}^T\mathbf{C}\mathbf{R}$. Therefore,

$$\underbrace{\begin{bmatrix} 1 & 0 & 0 \\ 0 & a & b \\ 0 & -b & a \end{bmatrix}}_{\mathbf{R}^T} \cdot \underbrace{\begin{bmatrix} \lambda_1 & 0 & 0 \\ 0 & \lambda_2 & 0 \\ 0 & 0 & \lambda_3 \end{bmatrix}}_{\mathbf{C}} \cdot \underbrace{\begin{bmatrix} 1 & 0 & 0 \\ 0 & a & -b \\ 0 & b & a \end{bmatrix}}_{\mathbf{R}} = \underbrace{\begin{bmatrix} \lambda_1 & 0 & 0 \\ 0 & \lambda_2 & 0 \\ 0 & 0 & \lambda_3 \end{bmatrix}}_{\mathbf{C}},$$

where $a^2 + b^2 = 1$. From the above matrix equality, and since $a^2 + b^2 = 1$, we deduce that $ab(\lambda_2 - \lambda_3) = 0$. Since by assumption the rotation angle $\theta \neq 0, \pi$, then $b \neq 0$, so either $a = 0$ or $\lambda_2 = \lambda_3$. However, if $a = 0$, then $b = \pm 1$ and we also conclude that $\lambda_2 = \lambda_3$. Thus, if $\lambda_1 = \lambda_2$, then λ_1 is a triple eigenvalue, and otherwise λ_1 is a simple eigenvalue, and there is just another eigenvalue, λ_2, with multiplicity two. □

Therefore, if the covariance matrix has simple eigenvalues we can check every eigenvector to find the symmetries of $\mathcal{P}$, although in that case $\mathcal{P}$ will not have rotational symmetries other than axial symmetries. If the covariance matrix has one simple eigenvalue and one double eigenvalue, then we can compute the rotational, non-axial symmetries of $\mathcal{P}$: in order to find the rotation angle in this case, we can project the points onto a plane normal to the direction of the axis of rotation, and apply some method for planar point clouds, e.g. [31, 32]. However, if $\mathbf{C}$ is a multiple of the identity, i.e., if $\mathbf{C}$ has a triple eigenvalue, then there is no hope of finding the symmetries by this method. In this case we can use the results in the next section. We finish this section with an example taken from [22].

Example 28. We demonstrate the symmetry detection method on a point cloud containing 3880 points. The points are taken as vertices of a triangular mesh, see Fig. 10.1(a). First, we translate the points so that their barycenter is at the origin. Next we construct the covariance matrix (10.1). It has one double eigenvalue

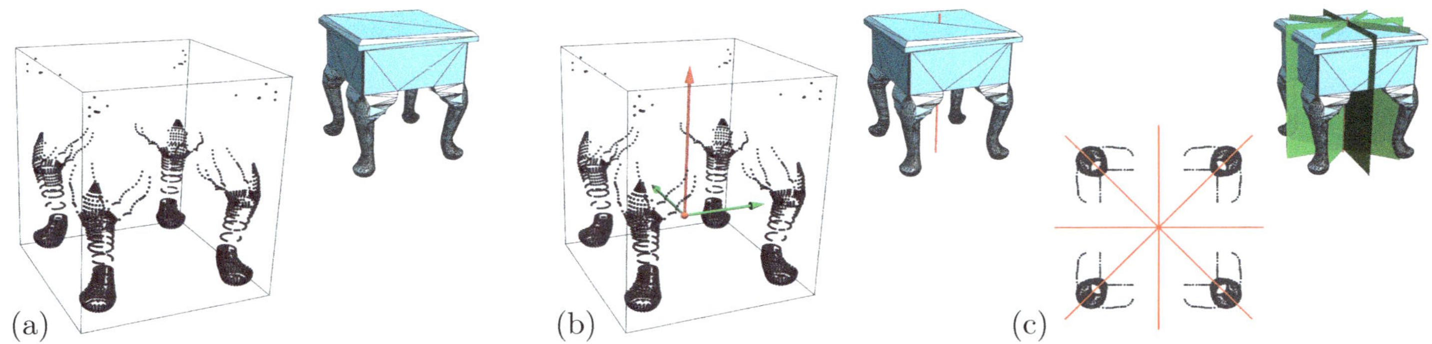

FIGURE 10.1 Symmetry detection of a point cloud. (a) The input point cloud is taken as vertices of a mesh (top). (b) The three eigenvectors of the corresponding covariance matrix. The red one corresponds to the single eigenvalue, while the two green ones correspond to the double eigenvalue. From Theorem 182 the red eigenvalue determines an axis of symmetry (top). (c) By projecting the given point cloud in the direction of the axis, we reduce the 3D problem to a 2D one. The group of symmetries of the planar point cloud is the dihedral group D_4. Therefore, the original object has rotational symmetry, and in fact 4 planes of symmetry (top).

and one single eigenvalue. Therefore, from Theorem 182 the eigenvector corresponding to the simple eigenvalue provides the tentative axis of symmetry, see Fig. 10.1(b). When the point cloud is projected onto the plane perpendicular to the axis and passing through the origin, we obtain a symmetric planar point cloud. Its symmetry group can be easily identified, e.g. by the method of [31]. In this case the symmetry group of the projected point cloud is the dihedral group D_4. From this we reconstruct the symmetry group of the 3D point cloud, consisting of four rotations and reflections on the planes given by the axis (direction of projection) and the axes from the planar configuration. Finally, we verify that the symmetries found are all the symmetries of the original point cloud.

10.2 USING HIGHER ORDER TENSORS

10.2.1 Brief review of tensors

The reason why the covariance matrix $\mathbf{C}$ satisfies that $\mathbf{C}^\star = P^T\mathbf{C}P$ is that $\mathbf{C}$ is, in fact, a *tensor*. Tensors are typically introduced in Physics to express *intrinsic* relationships, i.e., relationships which do not depend upon the system of coordinates we are using, which is what one should expect for something that can be called a "physical law". For instance, let us examine the rotational motion of a solid Ω about an axis of rotation; in Fig. 10.2, we have a solid sphere, upon which we exert the force $\mathbf{F}$. As a result, the sphere will rotate around the axis in red, according to the law

$$\mathbf{T} = \mathbb{I} \cdot \alpha. \tag{10.4}$$

Here, $\mathbf{T}$ is the *torque*, which intuitively is the turning effectiveness (about a certain axis of rotation) of a force, defined as

$$\mathbf{T} = \mathbf{r} \times \mathbf{F}.$$

Additionally, $\mathbf{r}$, called the *lever arm*, is the position vector of the point where we are exerting the force $\mathbf{F}$ in a system centered at the center of mass, and $\times$ denotes the cross product. Also, α is the *angular acceleration*, because the effect of applying a certain torque is a change in the angular velocity. Finally, $\mathbb{I}$ is called the *inertia tensor*, and soon we will see why it is called a tensor; $\mathbb{I}$ is related to the opposition that the solid exhibits to rotating, therefore playing with respect to rotations the part played by mass in Newton's second law. From a purely mathematical point of view, $\mathbf{T}$ and α are represented by vectors, while $\mathbb{I}$ is represented by a matrix,

$$\begin{bmatrix} I_{xx} & I_{xy} & I_{xz} \\ I_{xy} & I_{yy} & I_{yz} \\ I_{xz} & I_{yz} & I_{zz} \end{bmatrix}.$$

Assuming that the origin of the reference system lies at the center of mass of the solid,

$$\begin{array}{l} I_{xx} = \iiint_{\Omega}(y^2 + z^2)dm, \ I_{yy} = \iiint_{\Omega}(x^2 + z^2)dm, \\ I_{zz} = \iiint_{\Omega}(x^2 + y^2)dm, \end{array}$$

and

$$I_{xy} = -\iiint_{\Omega} xydm, \ I_{xz} = -\iiint_{\Omega} xzdm, \ I_{yz} = -\iiint_{\Omega} yzdm.$$

The quantities I_{xx}, I_{yy}, I_{zz} are called *moments of inertia* of the solid Ω with respect to the axes of coordinates, and I_{xy}, I_{xz}, I_{yz} are called *products of inerta*.

Now if Eq. (10.4) really corresponds to a law of Physics, then it should not depend on the system of reference we are using, because a system of reference is nothing else than a convenient artifact to write equations; in other words, it should be *intrinsic*, i.e., it should have exactly the *same form* in any other system of reference we might want to use. Next, if we change our system of reference, so that the change of basis matrix is the (orthogonal) matrix P, the torque and the angular acceleration, which are vectors, will change according to the usual rule in Linear Algebra,

$$\widehat{\mathbf{T}} = P^T\mathbf{T}, \ \widehat{\alpha} = P^T\alpha. \tag{10.5}$$

If Eq. (10.4) really corresponds to a physical law, in the new system of reference Eq. (10.4) should translate as

$$\widehat{\mathbf{T}} = \widehat{\mathbb{I}} \cdot \widehat{\alpha}, \tag{10.6}$$

where $\widehat{\mathbb{I}}$ is the inertia tensor in the new system. What should be the relationship between $\mathbb{I}, \widehat{\mathbb{I}}$ for this to happen? Since $\mathbf{T} = P\widehat{\mathbf{T}}$, $\alpha = P\widehat{\alpha}$, substituting in Eq. (10.4) we need that $\mathbb{I} = P \cdot \widehat{\mathbb{I}} \cdot P^T$, since in that case we get $P\widehat{\mathbf{T}} = P\widehat{\mathbb{I}}\widehat{\alpha}$, which implies that $P(\widehat{\mathbf{T}} - \widehat{\mathbb{I}}\widehat{\alpha}) = 0$, and in turn, Eq. (10.6). Therefore, when moving to a new system, the relationship between $\mathbb{I}$ and $\widehat{\mathbb{I}}$ must be $\widehat{\mathbb{I}} = P^T \cdot \mathbb{I} \cdot P$.

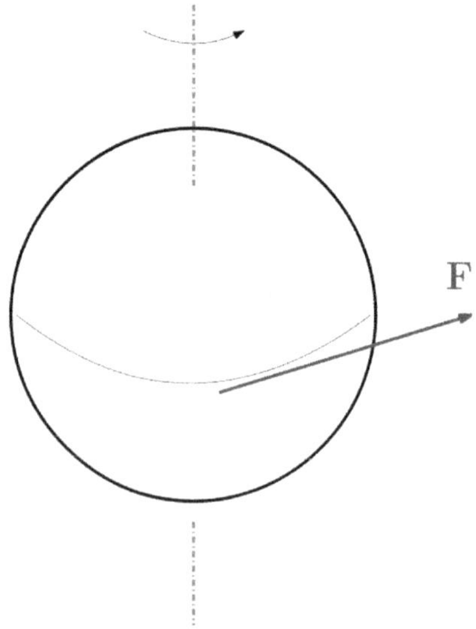

FIGURE 10.2 Rotational motion

Remark 25. *The matrix that defines* $\mathbb{I}$ *also defines the* ellipsoid of inertia *of the solid. The eigenvalues of* $\mathbb{I}$ *are the moments of inertia of the solid with respect to three directions, called* principal directions. *Intuitively, each of these eigenvalues represents the opposition of the solid to rotating about an axis through the center of mass, parallel to the corresponding principal direction. If we have not a solid, but a discrete distribution of equal masses, i.e., a point cloud* $\mathcal{P}$ *of points with unitary masses, the integrals in* $\mathbb{I}$ *are replaced by discrete sums, and we have something very close to the covariance matrix* $\mathbf{C}$*: the diagonal elements are the same, and the remaining elements are multiplied by a minus sign. In particular, when the coordinate axes are the principal directions, the* *covariance matrix and the matrix defining* $\mathbb{I}$ *coincide. This provides an intuitive explanation for the results in the previous section: the ellipsoid of inertia inherits the symmetries of* $\mathcal{P}$*, so that the reflection planes or the axes of symmetry of* $\mathcal{P}$ *are also reflection planes and axes of symmetry of the ellipsoid. Furthermore, according to Theorem 182, when* $\mathcal{P}$ *has rotational symmetry about an axis then the ellipsoid is an ellipsoid of revolution about that axis, which is perhaps not so intuitive. Also, the worst case in the previous section is the case when the ellipsoid of inertia becomes a sphere.*

Thus, in the case of Eq. (10.4), $\mathbb{I}$ changes in a very precise way when the basis, i.e., the reference system, is changed, and this guarantees that the form of Eq. (10.4) is preserved regardless of the system we use. Roughly put, a tensor, as it is usually introduced in Physics, is an entity, defined in a certain vector space, that changes in a very precise way when the basis is changed (and this property guarantees that, when being part of an equation defining a physical law, such equation preserves its form when we move to a different reference system). In fact, vectors and matrices are the simplest examples of tensors. Notice that vectors can be thought of as entities depending on one index (so that varying the index we get the different coordinates of the vector in a given basis), while matrices depend on two indexes (which allow us to move through the rows and columns). But we might have entities depending on more indexes, i.e., a *multidimensional array*, which is another intuitive way of introducing tensors.

Now it is definitely time to enter into the (slightly messy) details; part of the difficulty when presenting the notion in a more formal way is the fact that the indexes may have two different natures, *covariant* and *contravariant*, although we will see that in our case this will not be a problem. We start with the usual presentation in Physics; then we will provide the definition used in Mathematics. The interested readers can check, for instance, [118], for a more thorough introduction to tensors.

To get better acquainted, let us begin with the simplest examples of tensors, vectors and matrices. We consider an ambient vector space V, which is a Euclidean space of dimension M. A *contravariant vector* is an entity

$$T = \{t^i\},$$

where $i \in \{1, \ldots, M\}$, such that when a change if basis of the underlying vector space represented by a square matrix $P = (P_i^r)$ is applied, the components t^i of T transform by the rule

$$\hat{t}^i = \sum_{r=1,\ldots,M} P_r^i \cdot t^r.$$

For instance, the coordinates of a vector with respect to a basis provide an example of a contravariant vector; in fact, in the above expression one can recognize the usual formula for the new coordinates of a vector after a change of basis. Similarly, a *covariant vector* is an entity $T = \{t_j\}$, with $j \in \{1, \ldots, M\}$, whose components transform by the rule

$$\hat{t}_j = \sum_{r=1,\ldots,M} (P^{-1})_j^s \cdot t_s.$$

Notice that if P is orthogonal, then $P^{-1} = P^T$. An example of a covariant vector is the list of the covariant coordinates of a vector in a certain basis, which are the projections of the vector onto the directions of the basis vectors. Both in the case of contravariant and covariant vectors we have just one index; therefore, the vectors correspond to tensors of *order 1.*

Tensors of *order 2* can be represented by matrices. For instance, a *1-contravariant and 1-covariant tensor* is an entity $T = \{t_j^i\}$, where $i, j \in \{1, \ldots, M\}$, whose components t_j^i transform in the following way under a change of basis with change of basis matrix P:

$$\hat{t}_j^i = \sum_{r=1,\ldots,M} \sum_{s=1,\ldots,M} (P^{-1})_j^s \; P_r^i \; t_s^r. \tag{10.7}$$

In this formula one can recognize the matrix equality $\widehat{\mathbf{T}} = P^{-1} \cdot \mathbf{T} \cdot P$, where $\mathbf{T}$ is the matrix (t_j^i), and $\widehat{\mathbf{T}}$ is the matrix $(\hat{t}_j^i)$. Therefore, an example of a 1-contravariant and 1-covariant tensor is the set of entries of a matrix associated with a linear mapping. Since P orthogonal implies that $P^{-1} = P^T$, the inertia tensor, or the covariance matrix, can now be recognized as actual tensors.

In order to simplify the notation, the *Einstein summation notation* is commonly used. Here, whenever a subindex and superindex agree, it is understood that we must sum over the range of variation of that index. Using this notation, we can rewrite Eq. (10.7) as

$$\hat{t}^i_j = (P^{-1})^s_j \cdot P^i_r \cdot t^r_s.$$

In general, a mixed *n-contravariant* and *m-covariant tensor* is a multidimensional array

$$T = \{t^{i_1,\ldots,i_n}_{j_1,\ldots,j_m}\},$$

where $i_p, j_p \in \{1, \ldots, M\}$ for $p = 1, \ldots, n$, $q = 1, \ldots, m$, such that when a change of basis defined by a matrix $Q = (Q^r_i)$ is applied, the components $t^{i_1,\ldots,i_n}_{j_1,\ldots,j_m}$ transform by the rule

$$\hat{t}^{i_1,\ldots,i_n}_{j_1,\ldots,j_m} = (P^{-1})^{i_1}_{r_1} \cdots (P^{-1})^{i_n}_{r_n} \cdot P^{s_1}_{j_1} \cdots P^{s_m}_{j_m} \cdot t^{r_1,\ldots,r_n}_{s_1,\ldots,s_m}.$$

Such a tensor is said to be of *order* (n, m). If $n + m = 1$, then the tensor is in fact a vector, a *contravariant vector* if $n = 1, m = 0$ and a *covariant vector* if $n = 0, m = 1$. If $n+m = 2$, the tensor can be represented by a matrix. We say that a tensor is *symmetric* if the components of the tensor are invariant under any permutation of the indices. In our case, the covariance matrix, for instance, corresponds to a symmetric tensor.

There are several operations that can be carried out with tensors; for example, linear combinations of tensors of the same order (n, m) can be defined in a natural way, and produce tensors of the same order. However, for us the most useful operation is *contraction.* Contraction reduces a tensor of order (n, m) to a tensor of order $(n - 1, m - 1)$ by summing over one contravariant and one covariant index. For instance, summing over the last contravariant and covariant indices of T produces a new tensor S with components

$$s^{i_1,\ldots,i_{n-1}}_{j_1,\ldots,j_{m-1}} = t^{i_1,\ldots,i_{n-1},1}_{j_1,\ldots,j_{m-1},1} + \cdots + t^{i_1,\ldots,i_{n-1},M}_{j_1,\ldots,j_{m-1},M}.$$

In particular, a tensor $T = \{t^i_j\}$ of order $(1, 1)$ can be contracted to a tensor of order 0, the scalar equal to

$$t^1_1 + t^2_2 + t^3_3,$$

i.e., the trace of the matrix representing the tensor. And there is no problem is contracting repeatedly: in this sense, observe that when $n = m$, by repeatedly contracting we eventually arrive at a tensor of order $(1, 1)$, which corresponds to a matrix. If $n - m = 1$ or $m - n = 1$, by repeatedly contracting we eventually arrive at a contravariant or covariant vector. This implies that contracting always bring us to the more familiar ground of either vectors or matrices, although to the price of losing information.

It is also possible to *raise an index* or *lower an index* by using the *metric tensor* of the ambient space V, which is nothing else than the *Gram matrix*, that defines the dot product we are considering. When we use the Euclidean metric, as in our case, the metric tensor is the identity, so this operation does not change the components of the tensor. Thus in our case any index can be considered as either covariant or contravariant (see for instance Chapter 10 of [56]), which obviously simplifies things.

So far we have provided the notion of a tensor as it is usually introduced in Physics. In Mathematics, a tensor is usually defined as a *multilinear map*, i.e., a mapping linear in each component (a good example of a multilinear map is, precisely, a dot product). Given a vector space V with $\mathbb{R}$ as the field of scalars, a mixed *n-contravariant* and *m-covariant tensor* can also be defined as a multilinear map

$$T : V^\star \times \overset{n}{\cdots} \times V^\star \times V \times \overset{m}{\cdots} V \to \mathbb{R},$$

where $\times$ denotes the cartesian product, and $V^\star$ represents the *dual* of V: the dual $V^\star$ of a vector space V over a field $\mathbb{K}$ is the vector space (over the same field) consisting of the linear mappings $h : V \to \mathbb{K}$; when V has finite dimension, $V^\star$ has the same dimension as V. Furthermore, as we noted before, if V is a Euclidean vector space, then we do not need to distinguish between covariant and contravariant indexes, so we can just think of a tensor as a multilinear map $T : V \times \cdots \times V \to \mathbb{R}$. This definition is useful, sometimes, to prove the tensor nature of an expression.

10.2.2 Using higher order tensors to find symmetries

Denoting $x_1 := x$, $x_2 := y$, $x_3 := z$, and therefore writing the points in coordinates as $\mathbf{p}_n = \left(x_1^{(n)}, x_2^{(n)}, x_3^{(n)}\right)$, the tensor behind the covariance matrix corresponds to

$$c_{ij} = \frac{1}{m} \sum_{n=1}^{m} x_i^{(n)} x_j^{(n)}, \tag{10.8}$$

where m is the number of points in the point cloud $\mathcal{P}$. The tensor nature of Eq. (10.8) is easily derived from the fact that Eq. (10.8) is a sum, where each term can be recognized as a tensor by using the definition of a tensor as a multilinear mapping; in this case we have a tensor of order two. As we saw in Section 10.1, the covariance matrix $\mathbf{C}$ is useful when $\mathbf{C}$ has simple eigenvalues, although in this case we need to test three different directions to find axial symmetries or reflection planes, or when $\mathbf{C}$ has one simple eigenvalue and one double eigenvalue, because in this last case the direction of rotational symmetry, if any, corresponds to the simple eigenvalue. However, if we have multiple eigenvalues, then we do not have any information on reflections or axial symmetries (other than the case of rotational, non-axial symmetry, when there is a simple eigenvalue and a double eigenvalue); in particular, if $\mathbf{C}$ has a triple eigenvalue, the results in Section 10.1 are useless.

In those cases we can try higher order tensors. A potentially useful third-order tensor is

$$d_{ijk} = \sum_{n=1}^{m} x_i^{(n)} x_j^{(n)} x_k^{(n)}. \tag{10.9}$$

For instance, d_{111} is the sum of the cubic powers x^3 of the first coordinates of the points over $\mathcal{P}$; d_{233} is the sum of the products yz^2 over $\mathcal{P}$; d_{123} is the sum of the products xyz over $\mathcal{P}$. Notice that $d_{ijk} = d_{\alpha(i,j,k)}$ for any permutation α of the indices i, j, k, so d_{ijk}

(as it also happens with the covariance) is a symmetric tensor. A potentially, also symmetric, useful fourth-order tensor is

$$e_{ijk\ell} = \sum_{n=1}^{m} x_i^{(n)} x_j^{(n)} x_k^{(n)} x_\ell^{(n)}, \tag{10.10}$$

so that, for instance, e_{1233} is the sum, over all the points, of the products xyz^2. In fact, the covariance matrix is related to what in Statistics (and also in Physics) are called *moments of order 2* (with respect to the barycenter); the tensors in Eqs. (10.9) and (10.10) are related to *moments of order 3* and *4*, respectively. Of course, we might go higher.

Now let us use contraction. Contracting the tensor in Eq. (10.9) over the last two indexes yields a tensor of order 1, i.e., a vector

$$\tilde{d} = (\tilde{d}_1, \tilde{d}_2, \tilde{d}_3), \quad \text{where} \quad \tilde{d}_i = d_{i11} + d_{i22} + d_{i33}. \tag{10.11}$$

Observe that since d_{ijk} is symmetric, choosing any other two indexes produces the same result. Analogously, contracting the tensor in Eq. (10.10) yields a tensor $\tilde{e}_{ij}$ of order 2, where

$$\tilde{e}_{ij} = e_{ij11} + e_{ij22} + e_{ij33}. \tag{10.12}$$

The following results show how to use these tensors for our purposes. The first one involves non-trivial, rotational symmetries, including axial symmetries.

Theorem 183. *Let $\mathcal{P} = \{\mathbf{p}_i\}_{i=1,\ldots,m}$ be a 3D point cloud, whose barycenter $\mathbf{p}$ is at the origin, let $\tilde{d}_i$ be the result of contracting the tensor d_{ijk} in Eq. (10.9), and let $\tilde{e}_{ij}$ be the result of contracting the tensor $e_{ijk\ell}$ in Eq. (10.10).*

(1) If $\mathcal{P}$ exhibits a non-trivial rotational, possibly axial, symmetry about a line $\mathcal{L}$ parallel to a vector $\vec{u}$, then either $\tilde{d}_i = \vec{u}$ or $\tilde{d}_i$ is the zero vector.

(2) If $\mathcal{P}$ exhibits a non-trivial rotational symmetry, possibly axial, about a line $\mathcal{L}$ parallel to a vector $\vec{u}$, then $\vec{u}$ is an eigenvector of $\tilde{e}_{ij}$.

Proof. We begin with (1). Let f be a rotation, possibly an axial symmetry, such that $f(\mathcal{P}) = \mathcal{P}$. When f is applied, the tensor d_{ijk} in Eq. (10.9) stays invariant. Therefore, the result of contracting d_{ijk}, which is the vector $\tilde{d}_i$, also stays invariant under f, i.e., $f(\tilde{d}_i) = \tilde{d}_i$. Thus, writing $f(\boldsymbol{x}) = \mathbf{A}\boldsymbol{x}$, the vector $\tilde{d}_i$ is an eigenvector of $\mathbf{A}$, which is a rotation matrix, associated with the eigenvalue $\lambda = 1$. So if $\tilde{d}_i$ is nonzero, $\tilde{d}_i$ is a fixed direction and therefore corresponds to the axis of rotation.

Now let us see (2). Let f be a non-trivial rotation about a line $\mathcal{L}$ such that $f(\mathcal{P}) = \mathcal{P}$. In a system of coordinates where $\mathcal{L}$ is the x-axis, $f(\boldsymbol{x}) = \mathbf{A}\boldsymbol{x}$ with

$$\mathbf{A} = \begin{bmatrix} 1 & 0 & 0 \\ 0 & a & -b \\ 0 & b & a \end{bmatrix} \tag{10.13}$$

and $a^2 + b^2 = 1$, where $(a, b) \neq (1, 0)$ (since otherwise we have the identity). Again, $e_{ijk\ell}$ stays invariant when f is applied, and so does $\tilde{e}_{ij}$. Let $\mathbf{E}$ denote the matrix $(\tilde{e}_{ij})$. Since $\tilde{e}_{ij}$ stays invariant under f, we get that $\mathbf{E} = \mathbf{A}^T\mathbf{E}\mathbf{A}$, i.e., taking into account that $\tilde{e}_{ij}$ is symmetric,

$$\begin{bmatrix} \tilde{e}_{11} & \tilde{e}_{12} & \tilde{e}_{13} \\ \tilde{e}_{12} & \tilde{e}_{22} & \tilde{e}_{23} \\ \tilde{e}_{13} & \tilde{e}_{23} & \tilde{e}_{33} \end{bmatrix} = \begin{bmatrix} 1 & 0 & 0 \\ 0 & a & b \\ 0 & -b & a \end{bmatrix} \cdot \begin{bmatrix} \tilde{e}_{11} & \tilde{e}_{12} & \tilde{e}_{13} \\ \tilde{e}_{12} & \tilde{e}_{22} & \tilde{e}_{23} \\ \tilde{e}_{13} & \tilde{e}_{23} & \tilde{e}_{33} \end{bmatrix} \cdot \begin{bmatrix} 1 & 0 & 0 \\ 0 & a & -b \\ 0 & b & a \end{bmatrix}.$$

By multiplying the matrices at the right-hand side, the above equality yields

$$a\tilde{e}_{12} + b\tilde{e}_{13} = \tilde{e}_{12}, \quad -b\tilde{e}_{12} + a\tilde{e}_{13} = \tilde{e}_{13}.$$

Thus, we get the homogeneous linear system

$$\begin{cases} (a-1)\tilde{e}_{12} + b\tilde{e}_{13} & = & 0 \\ -b\tilde{e}_{12} + (a-1)\tilde{e}_{13} & = & 0 \end{cases}.$$

Since $(a, b) \neq (1, 0)$, the determinant of the linear system is $(a - 1)^2 + b^2 \neq 0$, and therefore we conclude that $\tilde{e}_{12} = \tilde{e}_{13} = 0$. Thus, $(1, 0, 0)$, which corresponds to the direction of $\mathcal{L}$, is an eigenvector of $\mathbf{E}$. □

Theorem 183 can be generalized to tensors of order $N \geq 5$ (statement (1) for N odd, statement (2) for N even). Notice that unless we get the zero vector, statement (1) of Theorem 183 directly provides the axis, also in the case of axial symmetry. Furthermore, statement (2) of Theorem 183 is useful whenever the matrix $\mathbf{E}$ of the contracted tensor $\tilde{e}_{ij}$ has simple eigenvalues. We illustrate this result with an example also taken from [22].

Example 29. Consider the following points, see Fig. 10.3(a).

$$\left(2,0,-\sqrt{5}\right),\left(-1,-\sqrt{3},-\sqrt{5}\right),\left(-1,\sqrt{3},-\sqrt{5}\right),\left(4,0,\sqrt{5}\right),$$
$$\left(-2,-2\sqrt{3},\sqrt{5}\right),\left(-2,2\sqrt{3},\sqrt{5}\right).$$

First we check that that the barycenter is at origin. Next we construct the covariance matrix, which turns out to be $\mathbf{C} = 5\mathbf{I}$, with $\mathbf{I}$ the 3×3 identity matrix. Since $\mathbf{C}$ is a multiple of the identity matrix, it cannot be used for symmetry detection. So we construct the third-order tensor (10.9). Its contraction yields the vector $\tilde{d} = \left(0,0,36\sqrt{5}\right)$, see Fig. 10.3(b). Therefore, the z-axis is a tentative axis of rotation. If we project the point cloud in the direction of the axis, we get a planar configuration whose symmetry group is the dihedral group D_3, see Fig. 10.3(c). From here, we can recognize the rotational symmetry of the cloud, which also has three symmetry planes.

Let us show that the fourth-order tensor also provides the axis of rotation. Its contraction is a matrix

$$\begin{pmatrix} 558 & 0 & 0 \\ 0 & 558 & 0 \\ 0 & 0 & 450 \end{pmatrix} \tag{10.14}$$

with one double and one single eigenvalue. The eigenvector corresponding to the single eigenvalue provides the direction of the axis.

Next let us consider reflections. In this case the tensor d_{ijk} in Eq. (10.9) is, unfortunately, useless. The reason is that if $f(\boldsymbol{x}) = \mathbf{A}\boldsymbol{x}$ represents a reflection on a plane, the normal direction of the plane corresponds to the eigenvalue $\lambda = -1$ of $\mathbf{A}$. However, the

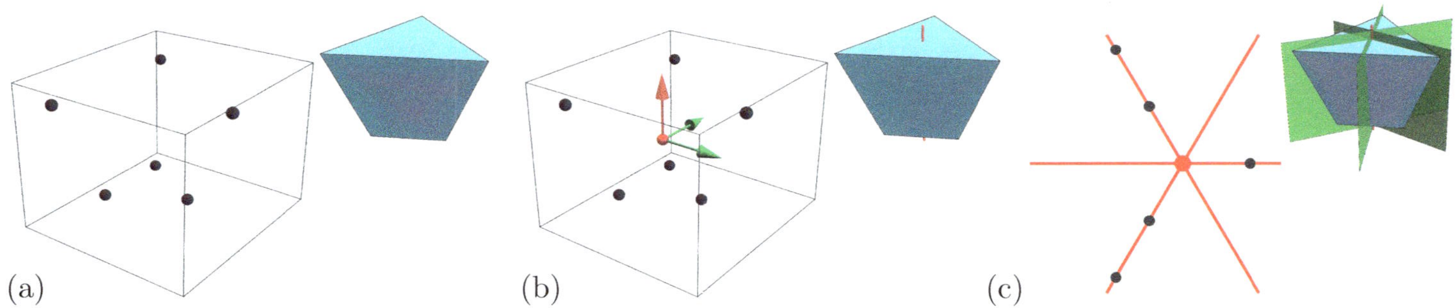

FIGURE 10.3 Symmetry detection of a point cloud in a pathological case.

argument in the proof of Statement (1) of Theorem 183, which also translates directly to this case, informs us that the result $\tilde{d}_i$ of contracting d_{ijk} satisfies that $f(\tilde{d}_i) = \tilde{d}_i$; thus, we conclude that $\tilde{d}_i$ is an eigenvector corresponding to $\lambda = 1$, but $\lambda = 1$ is not the eigenvalue we need. Nevertheless, the fourth-order tensor $e_{ijk\ell}$ in Eq. (10.10) can be useful.

Theorem 184. *Let $\mathcal{P} = \{\mathbf{p}_i\}_{i=1,\ldots,m}$ be a 3D point cloud, whose barycenter $\mathbf{p}$ is at the origin, and let $\tilde{e}_{ij}$ be the result of contracting the tensor $e_{ijk\ell}$ in Eq. (10.10). If $\mathcal{P}$ exhibits a non-trivial reflectional symmetry on a plane norma to a vector $\vec{u}$, then $\vec{u}$ is an eigenvector of $\tilde{e}_{ij}$.*

Proof. The proof is similar to that of Statement (2) of Theorem 183. Let f be a non-trivial reflection on a plane Π such that $f(\mathcal{P}) = \mathcal{P}$. In a system of coordinates where Π is the xy-plane, $f(\boldsymbol{x}) = \mathbf{A}\boldsymbol{x}$ with

$$\mathbf{A} = \begin{bmatrix} 1 & 0 & 0 \\ 0 & 1 & 0 \\ 0 & 0 & -1 \end{bmatrix}. \tag{10.15}$$

Next $e_{ijk\ell}$ stays invariant when f is applied, and so does $\tilde{e}_{ij}$. Let $\mathbf{E}$ denote the matrix $(\tilde{e}_{ij})$. Since $\tilde{e}_{ij}$ stays invariant under f, we get that $\mathbf{E} = \mathbf{A}^T\mathbf{E}\mathbf{A}$, i.e., taking into account that $\tilde{e}_{ij}$ is symmetric,

$$\begin{bmatrix} \tilde{e}_{11} & \tilde{e}_{12} & \tilde{e}_{13} \\ \tilde{e}_{12} & \tilde{e}_{22} & \tilde{e}_{23} \\ \tilde{e}_{13} & \tilde{e}_{23} & \tilde{e}_{33} \end{bmatrix} = \begin{bmatrix} 1 & 0 & 0 \\ 0 & 1 & 0 \\ 0 & 0 & -1 \end{bmatrix} \cdot \begin{bmatrix} \tilde{e}_{11} & \tilde{e}_{12} & \tilde{e}_{13} \\ \tilde{e}_{12} & \tilde{e}_{22} & \tilde{e}_{23} \\ \tilde{e}_{13} & \tilde{e}_{23} & \tilde{e}_{33} \end{bmatrix} \cdot \begin{bmatrix} 1 & 0 & 0 \\ 0 & 1 & 0 \\ 0 & 0 & -1 \end{bmatrix}.$$

After multiplying the matrices at the right-hand side, the above equality yields $-\tilde{e}_{13} = \tilde{e}_{13}$, $-\tilde{e}_{23} = \tilde{e}_{23}$, so $\tilde{e}_{13} = \tilde{e}_{23} = 0$. Therefore, $(0, 0, 1)$, which corresponds to the normal direction of Π, is an eigenvector of $\mathbf{E}$. □

Theorem 184 can be generalized to higher order tensors of even order. Now although Theorem 183 and Theorem 184 are surely useful in a generic situation, we must be aware that they have their own "bad cases": if by contraction we get the zero vector, or a matrix which is a multiple of the identity, we cannot say anything. We might fail, as well, if we get a matrix with double eigenvalues. One might wonder whether in that case the problem can be solved by moving to tensors of even higher order, but the answer is uncertain. If the point cloud $\mathcal{P}$ has many symmetries, then these methods are bound to fail: for example, if we have more than three axes of symmetry, or more than three reflection planes, we will not find them by contraction. So in super-symmetric cases, we must look for alternative methods. This will be, in fact, one of the last open problems we will suggest in the next, and final, section of this book.

Remark 26. *Results analogous to Theorem 183 and Theorem 184 for planar point clouds can also be given. The ideas are the same, so the task is left to the reader.*

10.3 FINAL REMARKS AND OPEN PROBLEMS

To finish this chapter, and before giving the usual battery of open problems, let us briefly consider a case which has actually appeared in several parts of this book, namely the case of finite sets of points which are the solutions of a polynomial system

$$F_1(x, y, z) = 0, \ldots, F_m(x, y, z) = 0. \qquad (10.16)$$

If the above system has just finitely many solutions, one might consider the problem of addressing the symmetries of its solutions. The problem is that even if one is able to compute the solutions, they might be difficult to manage, i.e., the expressions might involve complicated algebraic numbers, some real, some complex. However, at least the barycenter of these solutions can be computed exactly without having the solutions themselves. In order to do this, we must recall the correspondence between algebraic varieties and polynomial ideals that we recalled in Chapter 5 for space algebraic curves.

First, let us write the variety $\mathcal{V}$ defined by the set of common solutions of the polynomials in Eq. (10.16) as $\mathcal{V} = V(F_1, \ldots, F_m)$; if the number of solutions is finite, which is the case we are addressing here, we say that $\mathcal{V}$ is *0-dimensional.* The ideal $\mathbf{I}$ associated with $\mathcal{V}$ is the set of all polynomials in x, y, z that vanish at the solutions of Eq. (10.16). As we know, because of Hilbert Basis Theorem the ideal **I** is finitely generated, so we can find a basis for this ideal consisting of finitely many polynomials, and in fact infinitely many bases; in any case, if $\langle G_1, \ldots, G_\ell \rangle$ is a basis of **I**, the variety defined by the G_i coincides with the variety defined by the F_j, i.e., the solution set of $G_1(x, y, z) = \cdots = G_\ell(x, y, z) = 0$ coincides with the solution set of Eq. (10.16).

But there is a basis with remarkable properties, namely the *Gröbner* basis. Describing in full detail Gröbner bases is beyond the scope of this book; the interested reader can learn about this question, for instance, in [45]. However, Gröbner bases have an amazing property that we can exploit here. In order to describe it, let us observe first that to compute a Gröbner basis we need to specify a certain *ordering* in the set of monomials in the variables x, y, z, which we can identify with elements of $\mathbb{Z}^3$ with non-negative entries. This ordering is called the *lexicographic ordering*: given two monomials $m_1 = x^{\alpha_1} y^{\beta_1} z^{\gamma_1}$, $m_2 = x^{\alpha_2} y^{\beta_2} z^{\gamma_2}$, we say that $m_1 >_{\text{lex}} m_2$ if and only if the left-most nonzero entry of the vector $(\alpha_1 - \alpha_2, \beta_1 - \beta_2, \gamma_1 - \gamma_2)$ is positive. This way, for instance $x^3yz^2 >_{\text{lex}} xy^6z^8$ or $yz^2 >_{\text{lex}} z^6$. This is called the lexicographic order with $x > y > z$, and corresponds to a total ordering and well-ordering in the set $\mathbb{Z}^3$ with non-negative entries.

Now the amazing property is the following: if we compute a Gröbner basis of $\mathbf{I} = \langle F_1, \ldots, F_m \rangle$, where $\mathcal{V} = V(F_1, \ldots, F_m)$ is 0-dimensional (actually, the dimension of $\mathcal{V}$ can also be computed, see Chapter 9 of [45]), with the lexicographic ordering and $x > y > z$, then the Gröbner basis of **I** will contain a polynomial $r(z)$, i.e., only depending on z (see Chapter 3 of [45]), which is the last variable in the chain $x > y > z$. The property goes beyond, to ensure that the Gröbner basis is "triangular" in the sense that one has polynomials only depending on z, polynomials only depending on z, y, and finally polynomials depending on z, y, x. In particular, the z-coordinates of the solutions of Eq. (10.16) must be roots of $r(z)$. Conversely one needs to be a bit careful, because not every root z_i of $r(z)$ needs to extend to a solution of Eq. (10.16), which is something that came up when we introduced resultants. One can guarantee that a certain z_i is extended to a solution of Eq. (10.16) by using the *Extension Theorem*, see again [45]: the rough idea is to lift $z_i \to (z_i, y_i)$, and then $(z_i, y_i) \to (z_i, y_i, x_i)$. The first lift $z_i \to (z_i, y_i)$ is guaranteed whenever not all leading coefficients with respect to y of the polynomials in the Gröbner basis only depending on y, z vanish at $z = z_i$; the idea for the second lift is analogous.

If the lifting works, the z-coordinates of the solutions of Eq. (10.16) are exactly the roots of $r(z)$. We can do the same for the x-coordinates and the y-coordinates by considering the orderings $y > z > x$, $z > x > y$, to obtain polynomials $p(x), q(y)$. If the lifting is guaranteed in the three cases, the x, y and z coordinates of the solutions of Eq. (10.16) are stored as the roots of $p(x), q(y), r(z)$. And this means that the barycenter of the solutions of Eq. (10.16) can be computed exactly, even without actually computing the solutions of Eq. (10.16), using Cardano-Vieta relationships. For instance, for x we observe that

$$p(x) = p_n x^n + p_{n-1}x^{n-1} + \cdots + p_1 x + p_0.$$

Thus,

$$\bar{x} = \frac{1}{n}\sum_{k=1}^{n} x_i = -\frac{p_{n-1}}{np_n}.$$

Similarly for y, z. It is true that not all the solutions of Eq. (10.16) might be real, and we are including all the roots, real and complex, of, say, $p(x)$, in the above expression. However, a real orthogonal mapping f is a symmetry of the roots of Eq. (10.16) if and only if $\mathcal{V} = V(F_1, \ldots, F_m) = V(F_1 \circ f, \ldots, F_m \circ f)$. This means that f is both a symmetry of the real and complex solutions of Eq. (10.16), which in particular maps real points to real points. Thus, the barycenter of all, real and complex, solutions of Eq. (10.16) is the center of symmetry of the real points of Eq. (10.16), if any, and a point of any axis of symmetry, symmetry plane or rotational axis of such points, too. We employed a similar trick when we addressed the symmetries of implicit planar algebraic curves and also when computing the symmetries of planar algebraic vector fields; in this last case, though, resultant computations were enough.

Nevertheless, there is no clear method to exactly compute the elements of the covariance matrix of the solutions of Eq. (10.16) by a similar procedure. So here we have the first suggested open problem for this chapter.

Open Problem 27. *Symmetries of 0-dimensional solution sets of systems of polynomial equations.* Devise an algorithm for computing the symmetries of the solutions of a 0-dimensional system as Eq. (10.16) without actually computing the solutions themselves.

In addition to this question, we have the open problem mentioned at the end of Section 10.2.

Open Problem 28. *Symmetries of super-symmetric 3D point clouds.* Find an algorithm for computing the symmetries of a finite set of 3D points, super-symmetric in the sense that it has more than three axes of symmetry, or more than three reflection planes. For instance, any point cloud with the symmetries of one of the platonic solids falls in this case. A possibility might be to use tensors but *without* contracting (because contracting makes us lose information), which seems challenging.

Moreover, we can add two more open problems related to applied fields.

Open Problem 29. *Recognition of primitive forms.* A *primitive form*, or primitive, for short, is a very elementary shape possibly found when analyzing a scene: for instance, planes, spheres, cylinders, cones are all of them 3D primitives; in the planar case, lines, circles or ellipses are good examples. Then one can wonder how to efficiently recognize that a given point cloud corresponds to one of these shapes. This is certainly a problem that has been addressed in the literature, but notice that many of these primitives have symmetries, and that quickly finding their symmetries (and certainly the methods provided in this chapter are very fast) can help with the recognition. A harder problem is that of *extracting* these primitives from a more complex point cloud, i.e., without a previous segmentation.

Open Problem 30. *Projective equivalence between point clouds.* Checking whether or not two point clouds are affinely equivalent is a problem of interest in Computer Vision. Interestingly enough, the problem can be reduced to similarity checking, see for instance [43], using covariance matrices. However, checking not affine, but projective equivalence is harder. If the points stay on a line, it is just a matter of checking whether the *cross ratio* (the most popular projective invariant) of 4-tuples of points are preserved, but if the configuration of the points is more general, this approach is messy.

Bibliography

[1] Alcázar J.G., Sendra R. (2005), *Computation of the topology of real algebraic space curves*, Journal of Symbolic Computation Vol. 39, pp. 719–744.

[2] Alcázar J.G. (2014), *Efficient detection of symmetries of polynomially parametrized curves*, Journal of Computational and Applied Mathematics Vol. 255, pp. 715–724.

[3] Alcázar J.G., Dahl H., Muntingh G. (2018), *Symmetries of canal surfaces and Dupin cyclides*, Computer Aided Geometric Design Vol. 59, pp. 68–85.

[4] Alcázar J.G., Goldman R. (2017), *Detecting when an implicit equation or a rational parametrization defines a conical or cylindrical surface, or a surface of revolution*, IEEE Transactions on Visualization and Computer Graphics Vol. 23, Issue 12, pp. 2550–2559.

[5] Alcázar J.G., Goldman R. (2020), *Recognizing algebraic affine rotation surfaces*, Computer Aided Geometric Design Vol. 81, 101905.

[6] Alcázar J.G., Gözütok U., Çoban H.A., Hermoso C. (2022), *Detecting affine equivalences between implicit planar algebraic curves*, Acta Applicandae Mathematicae Vol. 182 (2).

[7] Alcázar J.G., Hermoso C. (2016), *Involutions of polynomially parametrized surfaces*, Journal of Computational and Applied Mathematics Vol. 294, pp. 23–38.

[8] Alcázar J.G., Hermoso C. (2021), *Computing projective equivalences of planar curves birationally equivalent to el-*

liptic and hyperelliptic curves, Computer Aided Geometric Design Vol. 91, 102048.

[9] Alcázar J.G., Hermoso C., Muntingh G. (2014), *Detecting similarity of rational plane curves*, Journal of Computational and Applied Mathematics Vol. 269, pp. 1–13.

[10] Alcázar J.G., Hermoso C., Muntingh G. (2014), *Detecting symmetries of rational plane and space curves*, Computer Aided Geometric Design Vol. 31, pp. 199–209.

[11] Alcázar J.G., Hermoso C., Muntingh G. (2015), *Symmetry detection of rational space curves from their curvature and torsion*, Computer Aided Geometric Design Vol. 33, pp. 51–65.

[12] Alcázar J.G., Hermoso C., Muntingh G. (2018), *Similarity detection of rational space curves*, Journal of Symbolic Computation Vol. 85, pp. 4–24.

[13] Alcázar J.G., Lávička M., Vršek J. (2019), *Symmetries and similarities of planar algebraic curves using harmonic polynomials*, Journal of Computational and Applied Mathematics Vol. 357, pp. 302–318.

[14] Alcázar J.G., Lávička M., Vršek J. (2024), *Symmetries of planar algebraic vector fields*, Computer Aided Geometric Design Vol. 111, 102290

[15] Alcázar J.G., Lávička M., Vršek J. (2023), *Computing symmetries of implicit algebraic surfaces*, Computer Aided Geometric Design Vol. 104, 102221.

[16] Alcázar J.G., Quintero E. (2020), *Affine equivalences of trigonometric curves*, Acta Applicandae Mathematicae Vol. 170, pp. 691–708.

[17] Alcázar J.G., Quintero E. (2020), *Affine equivalences, isometries and symmetries of ruled rational surfaces*, Journal of Computational and Applied Mathematics Vol. 364, 112339.

[18] Alcázar J.G., Quintero E. (2021), *Exact and approximate similarities of non-necessarily rational planar, parametrized curves, using centers of gravity and inertia tensors*, International Journal of Algebra and Computation Vol. 31, No. 4, pp. 581–603.

[19] Alcázar J.G., Muntingh G. (2022), *Affine equivalences of surfaces of translation and minimal surfaces, and applications to symmetry detection and design*, Journal of Computational and Applied Mathematics Vol. 411, 114206.

[20] Alcázar J.G., Gözütok U., Çoban H.A. (2024), *Detecting affine equivalences between certain types of parametric curves, in any dimension*, AIMS Mathematics Vol. 9(6), 13750–13769.

[21] Alcázar J.G., Hermoso C., Gözütok U., Çoban H.A. (2024), *Computation of symmetries of rational surfaces*, Electronic Research Archive (special issue on Applications of Symbolic Computation) Vol. 32, Issue 11, pp. 6087-6108.

[22] Alcázar J.G., Bizzarri M., Lávička M., Vršek J. (2025), *Rotational symmetries of 3D point clouds using the covariance matrix and higher order tensors*, Applied Mathematics Letters Vol. 162, 109381.

[23] Alonso C., Gutierrez J., Recio T. (1995), *A rational function decomposition algorithm by nearseparated polynomials*, Journal of Symbolic Computation Vol. 19, pp. 527–544.

[24] Axler S., Bourdon P., Ramey W. (2001), *Harmonic function theory*, Springer, Graduate Texts in Mathematics.

[25] Bajaj C.L., Holt R.J., Netravali A.R. (1998), *Rational parameterization of non-singular real cubic surfaces*, ACM Transactions on Graphics Vol. 17, pp. 1–31.

[26] Bastl, Jüttler B., Làvička M., Schulz T., Šír Z. (2014), *On the parameterization of rational ringed surfaces and rational*

canal surfaces, Mathematics in Computer Science Vol. 8, pp. 299–319.

[27] Bizzarri M., Làvička M (2015), *On modeling with rational ringed surfaces*, Computer-Aided Design Vol. 58, pp. 151–161.

[28] Bizzarri M., Làvička M., Vršek J. (2017), *Piecewise rational approximation of square-root parameterizable curves using the Weierstrass form*, Computer Aided Geometric Design Vol. 56, pp. 52–66.

[29] Bizzarri M., Làvička M., Vršek J. (2020), *Computing projective equivalences of special algebraic varieties*, Journal of Computational and Applied Mathematics Vol. 367, 112438.

[30] Bizzarri M., Lávička M., Vršek J. (2020), *Approximate symmetries of planar algebraic curves with inexact input*, Computer Aided Geometric Design Vol. 76, 101794.

[31] Bizzarri M., Làvička M., Vršek J. (2022), *Symmetries of discrete curves and point clouds via trigonometric interpolation*, Journal of Computational and Applied Mathematics Vol. 408, 114124.

[32] Bizzarri M., Làvička M., Vršek J. (2022), *Approximate symmetries of perturbed planar discrete curves*, Computer Aided Geometric Design Vol. 96, 102115.

[33] Bizzarri M., Hruda L., Làvička M., Vršek J. (2024), *Symmetry group detection of point clouds in 3D via a decomposition method*, Computer Aided Geometric Design Vol. 113, 102376.

[34] Brieskorn E., Knörrer H. (1986), *Plane algebraic curves*, Birkhauser, Basel.

[35] Brown J.W., Churchill R.V. (2008), *Complex variables and applications*, eighth edition, McGraw-Hill.

[36] Cederberg J. N. (1989), *A course on modern geometries*, Springer.

[37] Calabi E., Olver P.J., Shakiban C., Tannenbaum A., Haker S. (1998), *Differential and numerically invariant signature curves applied to object recognition*, International Journal of Computer Vision Vol. 26, pp. 107–135.

[38] Castelnuovo G. (1901), *Le trasformazioni generatrici del gruppo cremoniano nel piano*, Atti della R. Accad. delle Scienze di Torino Vol. 36, pp. 861–874.

[39] Castillo H., Marvá M. (2020), *The competition model with Holling type II competitive response to interfering time*, Journal of Biological Dynamics Vol. 14, No. 1, pp. 222–244.

[40] Chandru V., Dutta D., Hoffmann C. (1989), *On the geometry of Dupin cyclides*, Visual Computer Vol. 5, pp. 277–290.

[41] Chern S. (1982), *Web Geometry*, Bulletin of the American Mathematical Society Vol. 6, No. 1, pp. 1–8.

[42] Ching A., Baigent S. (2019), *Manifolds of balance in planar ecological systems*, Applied Mathematics and Computation Vol. 358, No. 1, pp. 204–215.

[43] Chi Y-T., Shahed S.M.N., Ho J., Yang M-H. (2008), *Higher dimensional affine registration and vision applications*, Proceedings of Computer Vision ECCV 2008, Part IV, pp. 256–269, Lecture Notes in Computer Science, Springer.

[44] Coffman A., Schwartz A.J., Stanton C. (1996), *The algebra and geometry of Steiner and other quadratically parametrizable surfaces*, Computer Aided Geometric Design Vol. 13, pp. 257–286.

[45] Cox D., Little J., O'Shea D. (2015), *Ideals, varieties and algorithms*, fourth edition, Springer.

[46] Coxeter H.S.M. (1969), *Introduction to geometry*, second edition, John Wiley & Sons, Inc., New York, London, Sydney.

[47] Dahl H. (2014), *Improved blends between primitive surfaces*, Ph. D thesis, http://urn.nb.no/URN:NBN:no-48202.

[48] Dahl H. (2014), *Piecewise rational parametrizations of canal surfaces*, Mathematical Methods for Curves and Surfaces, Lectur Notes in Computer Science, Springer, pp. 88–111.

[49] Daouda D.N., Mourrain B., Ruatta O. (2008), *On the computation of the topology of a non-reduced implicit space curve*, Proceedings ISSAC 2008, pp. 47–54.

[50] D'Arboux G. (1880), *Sur le contact des courbes et des surfaces*, Bulletin des Sciences Mathématiques et Astronomiques Vol. 4, No. 1, pp. 348–384.

[51] Degen W. (2002), *Cyclides*, Handbook of Computer Aided Geometric Design, pp. 575–601.

[52] Dierkes U., Hildebrandt S., Küster A., Wohlrab O. (1992). *Minimal surfaces.* vol. I. Springer.

[53] Do Carmo M.P. (2017), *Differential geometry of curves and surfaces*, second edition, Dover Books on Mathematics.

[54] Dolgachev I.V., Iskovskikh V.A. (2009), *Finite subgroups of the plane Cremona group*, in "Algebra, arithmetic, and geometry: in honor of Yu. I. Manin", Vol. I, pp. 443–548, Progr. Math., 269, Birkhäuser Boston, Inc., Boston, MA.

[55] Farin G. (1999), *Curves and surfaces for ACGD: a practical guide*, fifth edition, Academic Press.

[56] Farouki R.T. (2008), *Pythagorean-hodograph curves: algebra and geometry inseparable*, Springer.

[57] Fernández F., Domínguez J., González F., Martínez J.M., Palencia J., Salcedo A. (1988), *Curso de Geometría Métrica II* (in Spanish), Servicio de Publicaciones, Escuela Técnica Superior de Ingenieros de Caminos, Canales y Puertos de Madrid.

[58] Fischer G. (2000), *Plane algebraic curves*, American Mathematical Society, Student Mathematical Library Vol. 15

[59] Floater M., Hormann K. (2005), *Surface parameterization: a tutorial and a survey*, Advances in Multiresolution for Geometric Modelling, pp. 157–186, Springer.

[60] Goldman R., Wang W. (2004), *Using invariants to extract geometric characteristics of conic sections from rational quadratic parameterizations*, International Journal of Computational Geometry and Applications Vol. 14, No. 3, pp. 161–187.

[61] Goldman R. (2009), *Computer graphics and geometric modeling*, CRC Press.

[62] Goldman R. (2011), *Understanding quaternions*, Graphical Models Vol. 73 (2), pp. 21–49.

[63] Gözütok U., Çoban H.A., Sağiroğlu Y., Alcázar J.G. (2023), *A new method to detect projective equivalences and symmetries of 3D rational curves*, Journal of Computational and Applied Mathematics Vol. 419, 114782.

[64] Gözütok U., Çoban H.A. (2024), *Detecting isometries and symmetries of implicit algebraic surfaces*, AIMS Mathematics Vol. 9, No. 2, pp. 4294–4308.

[65] Gray A. (1999), *Modern differential geometry of curves and surfaces with mathematica*, CRC Press.

[66] Han M., Petek T., Romanovski G. (2020), *On some symmetries of quadratic systems*, Symmetry Vol. 12, No. 8.

[67] Han M., Petek T., Romanovski G. (2018), *Reversibility in polynomial systems of ODEs*, Applied Mathematics and Computation Vol. 338, pp. 55–71.

[68] Harris J. (1992), *Algebraic Geometry: a first course*, Springer-Verlag.

[69] Hauer M., Jüttler B. (2018), *Projective and affine symmetries and equivalences of rational curves in arbitrary dimension*, Journal of Symbolic Computation Vol. 87, pp. 68–86.

[70] Hauer M., Jüttler B., Schicho J. (2019), *Projective and affine symmetries and equivalences of rational and polynomial surfaces*, Journal of Computational and Applied Mathematics Vol. 349, pp. 424–437.

[71] Heard W.B. (2006), *Rigid body mechanics: mathematics, physics and applications*, Wiley-VCH Verlag.

[72] Hilbert D., Cohn-Vossen S. (1990), *Geometry and the imagination*, Chelsea Publishing Company, New York (USA).

[73] Hong H., Schicho J. (1998), *Algorithms for trigonometric curves (simplification, implicitization, parameterization)*, Journal of Symbolic Computation Vol. 26, pp. 279–300.

[74] de Jong T., Pfister G. (2000), *Local analytic geometry*, Springer.

[75] Jüttler B., Lubbes N., Shicho J. (2022), *Projective isomorphisms between rational surfaces*, Journal of Algebra Vol. 594, pages 571–596.

[76] Karcher H. (1989) *Construction of minimal surfaces.* Surveys in Geometry, 1–96, 1989. University of Tokyo, 1989, and Lecture Notes No. 12, SFB 256, Bonn.

[77] Kogan I., Ruddy M., Vinzant C. (2020), *Differential signatures of algebraic curves*, SIAM Journal on Applied Algebra and Geometry Vol. 4, Issue 1.

[78] Krasnov Y., Koulyshov K.K. (2020), *Symmetries in phase portrait*, Symmetry Vol. 12, No. 7.

[79] Krasauskas R. (2007), *Minimal rational parametrizations of canal surfaces*, Computing Vol. 79, No. 2-4, pp. 281–290.

[80] Kummer E. E. (1865), *Ueber die flächen vierten grades, auf welchen schaaren von kegelschnitten liegen*, Journal für die reine und angewandte Mathematik Vol. 64, pp. 66–76.

[81] Lamb J.S.W. (1992), *Reversing symmetries in dynamical systems*. Physica A, Vol. 25, pp. 925–937.

[82] Lamb J.S.W., Roberts M. (1999), *Reversible equivariant systems*, Journal of Differential Equations Vol. 159, pp. 239-279.

[83] Lebmair P., Richter-Gebert J. (2008), *Rotations, translations and symmetry detection for complexified curves*, Computer Aided Geometric Design Vol. 25, pp. 707–719.

[84] Lebmair P. (2009), *Feature detection for real plane algebraic curves*, Technische Universität München, Ph.D. Thesis.

[85] Lee J.M. (2012), *Introduction to smooth manifolds*, Springer.

[86] Lubbes N. (2021), *Surfaces that are covered by two pencils of circles*, Mathematische Zeitschrift Vol. 299, pp. 1445–1472.

[87] Lubbes N., Schicho J. (2018), *Kinematic generation of Darboux cyclides*, Computer Aided Geometric Design Vol. 64, pp. 11–14.

[88] Maxwell J.C. (1867), *On the cyclide*, The Quarterly Journal of Pure and Applied Mathematics Vol. 34.

[89] Mishra B. (1993), *Algorithmic algebra*, Springer.

[90] Monterde J. (2009), *Salkowski curves revisited: a family of curves with constant curvature and non-constant torsion*, Computer Aided Geometric Design Vol. 26, pp. 271–278.

[91] Ohdenal B. (2016), *On algebraic minimal surfaces*, KoG Vol. 20, pp. 61–78.

[92] Olver P.J. (1999), *Classical invariant theory*, London Mathematical Society, Student Texts 44, Cambridge University Press.

[93] O'Neill B. (1966), *Elementary differential geometry*, Academic Press.

[94] Oprea J. (2000), *The mathematics of soap films: explorations with maple*, American Mathematical Society.

[95] Patrikalakis N.M., Maekawa T. (2002), *Shape interrogation for computer aided geometric design and manufacturing*, Springer.

[96] Pellis D., Kilian M., Pottmann H., Pauly M. (2021), *Computational design of weingarten surfaces*, ACM Transactions on Graphics Vol. 40, No. 4, Article 114.

[97] Pérez-Díaz S., Schicho J., Sendra J.R. (2002), *Properness and inversion of rational parametrizations of surfaces*, Applicable Algebra in Engineering, Communication and Computing Vol. 13, pp. 29–51.

[98] Pérez-Díaz S., Sendra J.R. (2004), *Computation of the degree of rational surface parametrizations*, Journal of Pure and Applied Algebra Vol. 193/1-3, pp. 99–121.

[99] Pérez-Díaz S., Sendra J.R., Sendra J. (2004), *Parametrizations of approximate algebraic curves by lines*, Theoretical Computer Science Vol. 315, No. 2-3, pp. 627–650.

[100] Pérez-Díaz S. (2006), *On the problem of proper reparametrization for rational curves and surfaces*, Computer Aided Geometric Design Vol. 23, No. 4, pp. 307–323.

[101] Pérez-Díaz S., Sendra J.R., Sendra J., Rueda S.L. (2010), *Approximate parametrization of plane algebraic curves by linear systems of curves*, Computer Aided Geometric Design Vol. 27, pp. 212–231.

[102] Pérez-Díaz S., Sendra J.R., Villarino, C. (2010), *A first approach towards normal parametrizations of algebraic surfaces*, International Journal of Algebra and Computation Vol. 20, No. 8, pp. 977–990.

[103] Pérez-Díaz S., Sheng L-Y. (2014), *Characterization of rational ruled surfaces*, Journal of Symbolic Computation Vol. 63, pp. 21–45.

[104] Peternell M., Pottmann H. (1997), *Computing rational parametrizations of canal surfaces*, Journal of Symbolic Computation Vol. 23, No. 2-3, pp. 255–266.

[105] Pottmann H., Wallner J. (2001), *Computational line geometry*, Springer-Verlag.

[106] Pottmann H., Shi L., Skopenkov M. (2012), *Darboux cyclides and webs from circles*, Computer Aided Geometric Design Vol. 29 (1), pp. 77–97.

[107] Radó T. (1993), *On the problem of Plateau*, Springer.

[108] Romanovski V.G. (2008), *Time-reversibility in 2-dimensional systems*, Open Systems and Information Dynamics Vol. 15, No. 4, pp. 359–370.

[109] Rotman J.J. (1995), *An introduction to the theory of groups*, Graduate Texts in Mathematics Vol. 184, fourth edition, Springer-Verlag.

[110] Salmon G. (1914), *A treatise on the analytic geometry of three dimensions*, vol. I and II. Chelsea Publishing.

[111] Sánchez-Reyes J. (2011), *Simple determination via complex arithmetic of geometric characteristics of Bézier conics*, Computer Aided Geometric Design Vol. 28, pp. 345–348.

[112] Sánchez-Reyes J. (2015), *Detecting symmetries in polynomial Bézier curves*, Journal of Computational and Applied Mathematics Vol. 288, pp. 274– 283.

[113] Schläffi L. (1885), *An attempt to determine the twenty-seven lines upon a surface of the third order, and to divide such surfaces into species in reference to the reality of the lines upon the surface*, The Quarterly Journal of Pure and Applied Mathematics Vol. 2, pp. 110–120.

[114] Sendra J.R., Sendra J., Rueda S.L. (2008), *El reto de parametrizar curves y superficies de forma aproximada*, Gaceta de la Real Sociedad Matemática Española Vol. 11, pp. 521–541 (in Spanish).

[115] Sendra, J.R., Sevilla, D., Villarino, C. (2014), *Covering of surfaces parametrized without projective base points*, in Proceedings of the 39th International Symposium on Symbolic and Algebraic Computation, Kobe, Japan, 23–25 July 2014, ACM Press, New York, pp. 375–380.

[116] Sendra, J.R., Sevilla, D., Villarino, C. (2015), *Some results on the surjectivity of surface parametrizations*, Computer Algebra and Polynomials Vol. 8942 of the series LNCS, pp. 192–203.

[117] Sendra J.R., Winkler F., Pérez-Díaz S. (2008), *Rational algebraic curves*, Springer.

[118] Simmonds J.G. (1994), *A brief on tensor analysis*, Undergraduate Texts in Mathematics, Springer-Verlag.

[119] Strang, G. (2006), *Linear algebra and its applications*, fourth edition, Thomson Brooks/Cole.

[120] Struik, D.J. (1988), *Lectures on classical differential geometry*, Dover Publications.

[121] Sturmfels B. (2008), *Algorithms in invariant theory*, second edition, Texts and Monographs in Symbolic Computation, Springer.

[122] Sulanke R. (2020), *The fundamental theorem for curves in the n-dimensional Euclidean space*, http://www-irm.mathematik.hu-berlin.de/sulanke.

[123] Takeuchi N. (2000), *Cyclides*, Hokkaido Mathematical Journal Vol. 29 (1), pp. 119–148.

[124] Tarel J.P., Cooper D.B. (1998), *A new complex basis for implicit polynomial curves and its simple exploitation for pose estimation and invariant recognition*, Proceedings of the IEEE Computer Society Conference on Computer Vision and Pattern Recognition, CPVR 98. IEEE Computer Society, pp. 111–117.

[125] Tarel J.P., Cooper D.B. (2000), *The complex representation of algebraic curves and its simple exploitation for pose estimation and invariant recognition*, IEEE Transactions in Pattern Analysis and Machine Intelligence (7), pp. 663–674.

[126] Taubin G. (1991), *Estimation of planar curves, surfaces, and nonplanar space curves defined by implicit equations, with applications to edge and range image segmentation*, IEEE Transactions on Pattern Analysis and Machine Intelligence Vol. 13 (11), pp. 1115–1138.

[127] Unel M., Wolowich W.A. (1998), *Complex representations of algebraic curves*, Proceedings 1998 International Conference on Image Processing ICIP98, Vol. 2, pp. 272–276.

[128] Van der Waerden B.L. (1970), *Algebra I and II*, Springer-Verlag.

[129] Vršek J., Làvička M. (2015), *Determining surfaces of revolution from their implicit equations*, Journal of Computational and Applied Mathematics Vol. 290, pp. 125–135.

[130] Vršek J., Làvička M. (2016), *Recognizing implicitly given rational canal surfaces*, Journal of Symbolic Computation Vol. 74, pp. 367–377.

[131] Vršek J., Lávička, M. (2016), *Translation surfaces and isotropic transport nets on rational minimal surfaces*, Proceedings of the Congress Mathematical Methods for Curves and Surfaces (MMCS) 2016, Lecture Notes in Computer Science, vol 10521. Springer.

[132] Walker R.J. (1978), *Algebraic curves*, Springer.

[133] Wang W., Joe B., Goldman R. (2003), *Computing quadric surface intersections based on an analysis of plane cubic curves*, Graphical Models Vol. 64, pp. 335–367.

[134] Weiss I. (1993), *Noise-resistant invariants of curves*, IEEE Transactions on Pattern Analysis and Machine Intelligence Vol. 15 (9), pp. 943–948.

[135] Winkler F. (1996), *Polynomial algorithms in computer algebra*, Springer-Verlag.

[136] Wunderlich W. (1947), *Higher cycloid curves*, Osterr. Ing. Arch. 1, pp. 277–296.

[137] Zao M., Jia X., Tu C., Mourrain B., Wang W. (2019), *Enumerating the morphologies of non-degenerate Darboux cyclides*, Computer Aided Geometric Design Vol. 75, 101776.

Index

For Product Safety Concerns and Information please contact our EU representative GPSR@taylorandfrancis.com
Taylor & Francis Verlag GmbH, Kaufingerstraße 24, 80331 München, Germany

www.ingramcontent.com/pod-product-compliance
Ingram Content Group UK Ltd.
Pitfield, Milton Keynes, MK11 3LW, UK
UKHW022209190926
475702UK00003B/275
9781041088202